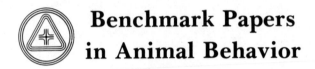 **Benchmark Papers
in Animal Behavior**

Series Editor: Martin W. Schein
West Virginia University

Published Volumes and Volumes in Preparation

HORMONES AND SEXUAL BEHAVIOR
  Carol Sue Carter
TERRITORY
  Allen W. Stokes
SOCIAL HIERARCHY AND DOMINANCE
  Martin W. Schein
CRITICAL PERIODS
  J. P. Scott
MIMICRY
  Joseph A. Marshall
IMPRINTING
  E. H. Hess
VERTEBRATE SOCIAL ORGANIZATION
  Edwin M. Bank

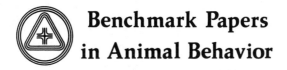

**Benchmark Papers
in Animal Behavior**

──────── A *BENCHMARK* TM Books Series ────────

HORMONES AND SEXUAL
BEHAVIOR

Edited by
CAROL SUE CARTER
University of Illinois

**Dowden, Hutchinson
& Ross, Inc.**

Stroudsburg, Pennsylvania

Library of Congress Cataloging in Publication Data

Carter, Carol Sue, 1944-   comp.
    Hormones and sexual behavior.

    (Benchmark papers in animal behavior, v. 1)
    Bibliography: p.
    1. Sexual behavior in animals--Addresses, essays,
lectures. 2. Hormones, Sex--Addresses, essays,
lectures. 3. Mammals--Behavior--Addresses, essays,
lectures. I. Title. [DNLM: 1. Hormones--Collected
works. 2. Sex behavior--Collected works. 3. Sex
hormones--Collected works. W1 BE514 v. 1 1973 /
WK102 C323h 1973]
QL761.C36   599'.01'66   73-16239
ISBN 0-87933-059-7

Manufactured in the United States of America.

Exclusive distributor outside the United States and
Canada: John Wiley & Sons, Inc.

Acknowledgments
and Permissions

ACKNOWLEDGMENTS

American Association for the Advancement of Science—*Science*
"Maternal and Sexual Behavior Induced by Intracranial Chemical Stimulation"
"The Theory of the Freemartin"
"The Experimental Production of Intersexuality in the Female Rat with Testosterone"

American Psychological Association
Journal of Comparative Psychology
"The Retention of Copulatory Ability in Male Rats Following Castration"
"Sexual Behavior and Sexual Receptivity in the Female Guinea Pig"
Psychological Bulletin
"Sexual Responsiveness in Female Monkeys After Castration and Subsequent Estrin Administration"
"Effect of Progesterone upon Sexual Excitability in the Female Monkey"

PERMISSIONS

The following papers have been reprinted with the permission of the authors and copyright owners.

American Association for the Advancement of Science—*Science*
"Male Sexual Behavior Induced by Intracranial Electrical Stimulation"
"Testosterone Regulation of Sexual Reflexes in Spinal Male Rats"
"Autoradiographic Localization of Radioactivity in Rat Brain After Injection of Tritiated Sex Hormone"

American Medical Association—*Journal of the American Medical Association*
"An Ovarian Hormone: Preliminary Report on Its Localization, Extraction and Partial Purification, and Action in Test Animals"

American Physiological Society
American Journal of Physiology
"Reduction of Sexual Behavior in Male Guinea Pigs by Hypothalamic Lesions"
"Rhinencephalon and Behavior"
"The Experimental Induction of Oestrus (Sexual Receptivity) in the Normal and Ovariectomized Guinea Pig"
Physiological Reviews
"Cerebral and Hormonal Control of Reflexive Mechanisms Involved in Copulatory Behavior"

American Psychological Association—*Journal of Comparative and Physiological Psychology*
"Effects of Different Concentrations of Androgen upon Sexual Behavior in Castrated Male Rats"

American Psychosomatic Society—*Psychosomatic Medicine*
"Effects of Injury to the Cerebral Cortex upon Sexually-receptive Behavior in the Female Rat"

Association for Research in Nervous and Mental Diseases—*Research Publication of the Association for Research in Nervous and Mental Diseases*
"The Hypothalamus and Sexual Behavior"

E. J. Brill, Publisher (Leiden, Holland)—*Behaviour*
  "Strain Differences in the Behavioral Responses of Female Guinea Pigs to Alpha-Estradiol Benzoate and Progesterone"

Journal of Endocrinology and H. H. Feder—*Journal of Endocrinology*
  "Progesterone Concentrations in the Arterial Plasma of Guinea-Pigs During the Oestrous Cycle"

J. B. Lippincott Co.
  Endocrinology
  "Differential Reactivity of Individuals and the Response of the Male Guinea Pig to Testosterone Propionate"
  "Development of Sexual Behavior in Male Guinea Pigs from Genetically Different Stocks Under Controlled Conditions of Androgen Treatment and Caging"
  "The Estrogen–Progesterone Induction of Mating Responses in the Spayed Female Rat"
  "The Abolition of Mating Behavior by Hypothalamic Lesions in Guinea Pigs"
  "Induction of Psychic Estrus in the Hamster with Progesterone Administered via the Lateral Brain Ventricle"
  "Neuroendocrine Correlates of Changes in Brain Activity Thresholds by Sex Steroids and Pituitary Hormones"
  "Organizing Action of Prenatally Administered Testosterone Propionate on the Tissues Mediating Mating Behavior in the Female Guinea Pig"
  "Evidence That the Hypothalamus is Responsible for Androgen-induced Sterility in the Female Rat"
  Journal of Clinical Endocrinology
  "Effect of Androgens upon Libido in Women"
  "The Role of Hormones in Human Behavior. I. Changes in Female Sexuality After Adrenalectomy"

Macmillan (Journals) Ltd.—*Nature*
  "Sexual Attractiveness and Receptivity in Rhesus Monkeys"
  "Distribution of Coitus in the Menstrual Cycle"

The Regents of the University of California—*Memoirs of the University of California Press*
  "The Oestrous Cycle in the Rat and Its Associated Phenomena"

The Williams & Wilkens Co.—*Journal of Nervous and Mental Disease*
  "Components of Eroticism in Man: The Hormones in Relation to Sexual Morphology and Sexual Desire"

Series Editor's Preface

It was not too many years ago that virtually all research publications dealing with animal behavior could be housed within the covers of a very few hard-bound volumes that were easily accessible to the few workers in the field. Times have changed! The present-day student of animal behavior has all he can do to keep abreast of developments within his own area of special interest, let alone within the field as a whole . . . and of course we have long since given up attempts to maintain more than a superficial awareness of what is happening "in biology," "in psychology," "in sociology," or in any of the broad fields touching upon or encompassing the behavioral sciences.

It was even fewer years ago that those who taught animal behavior courses could easily choose a suitable textbook from among the very few that were available; all "covered" the field, according to the bias of the author. Students working on a special project used *the* text and *the* journal as reference sources and, for the most part, successfully covered their assigned topics. Times have changed! The present-day teacher of animal behavior is confronted with a bewildering array of books from which to choose, some purporting to be all-encompassing, others confessing to strictly delimited coverage, and still others being simply collections of recent and profound writings.

In response to the problem of the steadily increasing and overwhelming volume of information in the area, the Benchmark Papers on Animal Behavior was launched as a series of single-topic volumes designed to be some things to some people. Each volume contains a collection of what an expert considers to be the significant research papers in a given topic area. Each volume, then, serves several purposes. For the teacher, a Benchmark volume serves as a supplement to other written materials assigned to students; it permits in-depth consideration of a particular topic while confronting the student (often for the first time) with original research papers of outstanding quality. For the researcher, a Benchmark volume saves countless hours of digging through the various journals to find the basic articles in his area of interest, journals that are often not easily available. For the student, a Benchmark volume provides a readily accessible set of original papers on the topic, a set that forms the core of the more extensive bibliography that he is likely to compile; it also permits him to see at first hand what an "expert" thinks is important in the area and to react accordingly. Finally, for the librarian, a Benchmark volume represents a collection of important papers from many diverse sources, thus making readily available materials that might otherwise be economically impossible to obtain or physically impossible to keep in stock.

This, the first volume in the Animal Behavior Series, is a good example of our approach to the problem of an overwhelming and scattered volume of information in a topic area. Few college libraries in the United States would have on their shelves the 1894 volume of *Pflüger's Archiv für die Gesamte Physiologie des Menschen und der Tiere* in which Steinach's original landmark paper was published; many students would have difficulty obtaining a copy of the *Memoirs of the Univeristy of California Press,* 1922, in which Long and Evans published their original study of oestrus cycles in rats. Indeed, the articles in the present volume are taken from a more diverse collection of journals than are usually available in many college libraries in America. But they are here, and now the student and researcher may have ready access to the basic writings of the "greats" in the area of hormones and sexual behavior.

Since the merit of the volume rests so heavily upon the judgment of the volume editor, it was important to ensure that volume editors are truly experts in their areas, as well as persons of sound judgment and scholarship. You will find in Dr. Carter's volume evidence of our success in these endeavors.

M. W. Schein

Contents

II. HORMONES AND THE BEHAVIOR OF THE FEMALE

III. HORMONES AND THE BEHAVIOR OF BOTH SEXES

IV. HORMONES AND SEXUAL DIFFERENTIATION

Contents by Author

Introduction

Reproduction is an essential aspect in the life cycle of organisms. In sexual reproduction gamete transfer and fertilization are fundamental to the process. A variety of species-specific behavioral patterns may accompany or precede the union of the egg and sperm and these are referred to, variously, as reproductive, courtship, mating, copulatory, or sexual behaviors.

In mammals, a growing body of research deals with relationships among hormones (biochemically active substances produced in one organ and acting on another) and the manifestation of sexual behavior. In particular, it has become increasingly evident that many mammalian behaviors depend in large measure upon the presence of ovarian, testicular, or other hormones. In the absence of appropriate gonadal hormones, many behavioral patterns characteristic of courtship and copulation may not occur. The concordance between the hormones of reproduction and mating behavior is not surprising if we consider the important role of endocrine secretions in other aspects of reproduction, such as ovulation, implantation, pregnancy, and lactation. However, the behavior of a given animal has frequently appeared to be more complex, or at least more labile, than the physiology of that organism. For this reason, the reliability of the relationship between gonadal secretions and sexual behavior appears the more remarkable to the student of behavior.

The historical development of research dealing with sexual behavior differs from the history of science in one important respect; sexual responses, perhaps more than other aspects of behavior or biology, have been considered unfit or unrespectable topics for scientific investigation.

European psychiatrists, such as Freud, have been credited with preparing the way for the modern study of sexual behavior, and many of the investigators whose works are reproduced in this volume were responsible for early research legitimizing the study of sexual behavior within the scientific community itself. Anyone aware of the difficulties encountered rather recently by Masters and Johnson (1966) in acquiring

1

initial scientific and public acceptance for their research on human sexual responses can scarcely believe that all cultural barriers to the scientific study of sexual behavior have been eliminated. Nonetheless, animal research describing and measuring sexual behavior is now widely accepted and insights into the psychology and physiology of sexuality are gradually emerging.

The current volume is designed to offer a sample of the kinds of basic research that have contributed to our present understanding of the relationship between hormones and sexual behavior. Time and subsequent discoveries have often rendered invalid the assumptions included in these papers, but the original papers have nonetheless molded the current state of the science and are of both theoretical and historical interest in this context.

Although endocrine secretions have been shown to be important to the display of reproductive behaviors in many vertebrate and invertebrate species, the selection of papers offered herein has been lmited to work dealing with mammals. This is an arbitrary decision imposed in part by the existing body of literature and in part by the desire of the editor to provide a degree of continuity among the papers selected. Many excellent and theoretically important papers have been excluded by this decision and the reader is referred to Young (1961) for a selection of review papers dealing with important developments arising from the study of nonmammalian species. In addition, following this introduction is a list of relevant journals, books, and review articles that offer background related to the current status of questions regarding hormones and behavior.

A Brief History of Research Relating Hormones and Sexual Behavior

Research dealing with hormones and sexual behavior has focused upon a series of interrelated questions. As a beginning point, one may simply ask if the gonads influence sexual behavior. This question was probably first directed toward the male and tested, in part, at various periods in antiquity by the relatively simple procedure of castration. Aristotle (ca. 350 B.C.) was aware of reductions in sexual behavior following castration in both men and domestic animals. Some information relating to the question of how the gonads might influence behavior was also available in medieval times. Medieval Chinese alchemists prepared testicular, placental, and urinary extracts containing concentrated sources of gonadal steroids, and these substances were used for therapeutic treatments much as they are today. The empirical knowledge and chemical techniques available in this period were not, however, transmitted to Western civilization.

In addition, misconceptions regarding the factors capable of influencing sexual impulses have gained favor at various times. For example, it was once believed that irritation or distention resulting from swelling in accessory sex glands, such as the seminal vesicles, was responsible for initiating male sexual behavior. The distention hypothesis received apparent support from experiments as late as the 1880s (see Steinach, Paper 1, this volume).

In the early 1820s, the "science" of phrenology, developed by Gall and Spurzheim, promoted the theory that the sexual impulses arose in the cerebellum. This theory apparently originated with the observation by Gall that warmth radiated from the skin over the cerebellum of a particularly passionate patient (a widow) of his acquaintance. A relationship between cerebellar size and presumed libido was also offered as evidence for this theory.

The modern history of theoretical research relating hormones and behavior was preceded by the development of a more precise knowledge of reproductive physiology. Although the role of the egg in the reproductive physiology of such vertebrates as birds and reptiles was obvious, it was not until 1672 that Regner de Graaf recognized a corresponding reproductive function for the mammalian ovary. Leeuwenhoek identified spermatozoa in semen in 1674, but it was nearly a century and a half before von Baer (1827) described the mammalian ovum.

The importance of the gonads in gamete production had been established by the midnineteenth century, but the discovery of other possible functions for these organs was first supported by the classic experiment of a German physician, A. A. Berthold. He demonstrated in 1849 that transplanted testicular tissue was capable of supporting both comb growth and male sexual behavior in chickens. Berthold's study not only established a function for the testis in the maintenance of male behavior but also is credited as the first example of experimental endocrinology.

Shortly thereafter, interest in the medical importance of endocrine secretions was stimulated by the reported success in self-rejuvenation described by the aging physician Brown-Séquard (Paper 2) following injections of an aqueous testicular extract. The renewed sexual interest and energy described by Brown-Séquard are generally attributed to placebo effects since androgens are not water soluble. Nonetheless, attention was drawn to the possibility of glandular influences on both the brain and body, and Brown-Séquard helped prepare the scientific community for the new science of endocrinology.

The class of chemicals now known as hormones (from the Greek *hormon,* to arouse or excite) was named by E. H. Starling in 1905 and endocrinology was formally born. The work of Bayliss and Starling on the digestive hormone secretin (1902–1905) offered clear evidence for biochemical regulators produced in one organ and capable of influencing the activity of a distant organ, and the concept of the hormone thus was defined. The methods instituted by Bayliss and Starling also established a paradigm for the study of endocrinology. The types of evidence necessary to demonstrate the humoral nature of a substance remain valid today and involve, basically, examining the effects of organ ablation and subsequent organ extract replacement.

Steinach, for example (Paper 1), in a series of papers in the late nineteenth and early twentieth century, applied the techniques of the new science of endocrinology to other questions of reproductive physiology and behavior. Steinach's efforts were not the earliest attempts to examine possible endocrine functions of the testis, but his behavioral experiments were more detailed than those of most of his predecessors. The interpretations of these early efforts are often not consistent with our current understanding, but the questions asked have changed very little.

Progress in the understanding of female reproductive behavior was hampered by the difficulties encountered in studying cyclic changes in the state of the animal. The early 1900s, however, brought the development of techniques which greatly facilitated this research. The first of these involved the astute observations of Stockard and Papanicolau (1917) relating cyclic changes in ovarian function to vaginal histology in the guinea pig, which permitted the making of inferences regarding the condition of the ovary in the intact female. The technique was extended to the laboratory rat, and basic features of the sexual behavior of the female and male were described in the classic monograph of Long and Evans (Paper 12). Allen and Doisy (Paper 13) subsequently identified an active ovarian hormone and established bioassay techniques for estrogenic hormones.

The next decade saw rapid progress in the area of steroid biochemistry. Ascheim and Zondek (1927) discovered (or rediscovered, in deference to the Chinese) large amounts of estrogen in the urine of pregnant women, and in 1930 Doisy and Butenandt simultaneously isolated the estrogen estrone. Shortly thereafter, Marrian (1930) isolated estriol; proof for the structure of estriol was provided in 1935 by Cook. Corner and Allen had obtained progesterone from ovarian extracts in 1929 and Butenandt (1934) isolated progesterone in a pure form.

Pézard (1911) had demonstrated an active extract of testicular tissue and in 1927 Koch concentrated testicular extracts. In 1931 Butenandt obtained from urine the first crystalline androgen, which he named androsterone. Finally, in 1934, David, Dingemanse, Freud, and Saqueur isolated the testicular hormone testosterone.

As a result of this remarkable period of biochemical discovery, active and relatively pure hormones could be isolated from gonadal extracts and other sources. In addition, the function and mechanism of action of such substances could be explored with much greater precision. Hormonal influences on sexual behavior could be approached more directly, and this is, in fact, the direction that the more recent research has assumed.

In the early studies, observations of mammalian sexual behavior were generally incidental to questions of reproductive physiology. During the 1920s, however, a number of experimental psychologists began to describe and question the importance of the sexual "drive." The earliest selection in this volume representative of the influence of psychology on the study of sexual behavior is a paper by Stone (Paper 3). Calvin Stone applied the techniques of behavioral observation to the question of the influence of the testes on male sexual behavior and, for the first time, quantified the duration of postcastrational copulatory behaviors. In subsequent papers, influenced by his association with the great neuropsychologist Karl Lashley, Stone examined directly the importance of various parts of the central nervous system in sexual behavior. The research strategies of Lashley and Stone have been extended to the present time by a series of prolific psychologists, including, particularly, Frank Beach and his students.

Following the isolation of the ovarian hormones estrogen and progesterone, reports that dealt with the importance of these substances in the regulation of female sexual behavior began to appear in the 1930s. Pioneering in these efforts were a group of young anatomists and physiologists associated with W. C. Young. Young and his associates (Paper 14) examined the correlation between the ovarian cycle and sexual behavior in the guinea pig and provided extensive parametric data on that species. Although Allen and Doisy (Paper 13) and others had observed some estrous behaviors in female rodents injected only with follicular extracts (presumably containing primarily estrogens), Dempsey, Hertz, and Young (Paper 15) discovered the further importance of the luteal hormone progesterone to the reliable induction of estrus in the guinea pig. Boling and Blandau (Paper 16) confirmed the ability of progesterone to facilitate the appearance of female receptivity in estrogen-primed rats. These papers are particularly important since they established the experimental paradigm for the study of female sexual behavior in laboratory rodents upon which many of our current theories are based.

Corresponding research with the Rhesus monkey by Ball (Paper 19) revealed the importance of estrogen to sexual responsivity in primates. However, Ball's later observation (Paper 20) of an inhibitory function for progesterone in the Rhesus brought into question the generality of a stimulatory action for progesterone in female sexuality. Further complexities regarding the action of progesterone on the nervous system were introduced by the report of both stimulatory and inhibitory actions of progesterone in the control of ovulation in the female rabbit (Makepeace, Weinstein, and Friedman, 1937). In addition, Kawakami and Sawyer (Paper 28) demonstrated changes in cortical stimulation thresholds as a function of hormone level in the rabbit, suggesting that progesterone may act to alter general arousal levels.

Beach and Holz-Tucker (Paper 4) applied the basic dose–response approach of the endocrinologist to examine the ability of different concentrations of testosterone to maintain male sexual behavior in the rat. Despite the behavioral response curves for various levels of testosterone in the Beach and Holz-Tucker study, it was clear that other factors could influence the ability of a male to respond to the female. Following the observation of individual differences in hormone responsivity by Grunt and Young (Paper 5), Riss, Valenstein, Sinks, and Young (Paper 6) further implicated genetic factors in determining responsivity to hormones in the male guinea pig. Corresponding findings were demonstrated in the female guinea pig by Goy and Young (Paper 17).

As the importance of hormones to the initiation and maintenance of sexual behavior in both males and females became apparent, research questioning the site of hormone action became increasingly important. Early attempts to investigate the importance of peripheral hormonal stimulation by observing the behavior of spinal animals (cited in Bard, Paper 29; and in Beach, Paper 30) suggested that several components of female sexual behavior were reflexive and, at least in cats and guinea pigs, could be elicited even in anestrus animals. Estrogen was proposed to act at a subcortical level at a site above the level of the lower mesencephalon (Bard, Paper 29). Brookhart and Dey (Paper 7) and Brookhart, Dey, and Ranson (Paper 25) used lesions to further implicate specific areas of the hypothalamus in the control of both male and female behavior in the guinea pig. In addition, Schreiner and Kling (Paper 8) reported that hormonally controlled centers in the limbic system might be involved in the inhibition of male sexual responses since lesions in that area released what was described as hypersexuality.

Attempts to further localize the site of hormone action have often adopted the approach first demonstrated by Kent and Liberman (Paper 27). A very small quantity of progesterone was implanted directly into the third ventricle of the brain and the subsequent short latency to lordosis in estrogen-primed female hamsters was offered as evidence of a local site of action for that hormone. A similar approach was used by

5

Fisher (Paper 9) in an attempt to identify the target area of the brain (in particular, of the hypothalamus) in which testosterone was presumed to function. Vaughan and Fisher (Paper 10) later attempted to elicit male behavior by electrical stimulation of the hypothalamus. Although these studies have proved difficult to replicate, their importance in stimulating research in this area is unquestionable. Additional information regarding the localization of hormone activity has been provided by new techniques for tracing the movement of radioactively labeled hormones. A sample of this approach is the use of autoradiography by Pfaff (Paper 31).

The development of sensitive and relatively simple methods of assessing circulating levels of hormones has also provided a new tool for the study of the hormone and behavior relationship. A sample of one of the first attempts to apply such assay techniques (in this case, gas chromatography) to questions regarding the hormonal basis of sexual behavior is a paper by Feder, Resko, and Goy (Paper 18) that relates progesterone levels and the estrous condition in the female guinea pig. Even more recently, a variety of investigators have applied radioimmuno-assay techniques to answer similar questions.

Although attempts to understand the physiology of sexual behavior have often ignored the genitalia and the peripheral nervous system, many questions remain concerning the action of hormones on these structures (see Beach, Paper 30). A recent awareness of these problems has been stimulated by the work of Hart (Paper 11) showing, for example, that testosterone may regulate sexual reflexes in spinal male rats.

A valuable source of information regarding the relationship between hormones and sexual behavior lies in an understanding of the development of sex differences and similarities. F. R. Lillie (Paper 33) must receive primary credit for drawing wide attention to a role for early hormones in sex determination, following his identification of the phenomenon of freemartinism in cattle. Green and Ivy (Paper 34) further demonstrated that the hormone testosterone was capable of dramatically masculinizing the genetic female rat. It was not, however, until 1959 that Phoenix, Goy, Gerall, and Young (Paper 35) clearly related the presence of early androgen to increases in masculine behavior and concurrent decreases in feminine sexual behavior and cyclicity in the guinea pig. Barraclough and Gorski (Paper 36) subsequently implicated changes in specific areas of the hypothalamus in the observed sterilization of genetic female rats following testosterone treatment. More recently, a large number of studies have examined the effects of perinatal hormones on adult sexual behavior in several species. Research revealing the behavioral lability of animals of both sexes has provided further insight into the complexity of the processes influencing sexual development.

The reader will note the rather restricted treatment of primate, and particularly human sexual behavior. With regard to male behavior this is in part because there has been almost no systematic research looking at postcastrational changes in behavior. The published information that is available, even for monkeys, is primarily antedotal (see Money's chapter in *Sex and Internal Secretions*, Young, 1961).

Based upon very recent research with Rhesus monkeys, Phoenix, Slob, and Goy (in press, *Journal of Comparative and Physiological Psychology*) have tended to agree with the common assumption that the maintenance of sexual performance in male primates is relatively more free of gonadal influence than is comparable behavior in nonprimates (see Beach, Paper 30). A possible role for nongonadal hormones in male primate behavior remains in doubt.

Correlations between the phase of the menstrual cycle and female responsivity, such as those shown in monkeys (Michael, Saayman, and Zumpe, 1967, Paper 21) and humans (Udry and Morris, Paper 22), suggest relationships between ovarian hormones and female sexual behavior. However, the adrenal glands have also been implicated in the well-known maintenance, even following ovariectomy, of female primate sexual behavior (Waxenberg, Drellich, and Sutherland, Paper 24). This work and earlier claims, such as those of Salmon and Geist (Paper 23), that testosterone injections were capable of stimulating sexual responsivity in women have been used in support of the hypothesis that female primate sexual behavior is influenced by adrenal androgens.

It also seems likely that experience interacts with the physiological substrate in the development of primate sexuality. A survey of several other hormone and behavior interactions in humans is found in the work of John Money (Paper 32).

In summary, research suggests that both the development and adult expression of masculine patterns of sexual behavior are facilitated by the presence of testicular hormones (androgens), while female behavior is generally facilitated by the presence of ovarian hormones (estrogens and in some cases progestins). Nongonadal hormones have been implicated in the regulation of sexual behavior in both sexes, but the exact nature of this relationship remains largely equivocal. Questions of the mechanisms of hormone action have also not been resolved, but the remarkable proliferation of research directed at these problems is encouraging. The reader is referred to the review articles, books, and journals listed here for information regarding current developments in this dynamic area of research.

Although this volume emphasizes the hormonal and physiological changes capable of influencing behavior, one should not lose sight of the relevance of the variety of environmental forces impinging on living organisms. External forces must ultimately be transduced into physiological processes if they are to influence behavior. The paucity of our understanding of this process reflects, at least in part, its complexity, but does not lessen its importance as a factor modifying sexual behavior.

Acknowledgments

I am grateful to both Arnold A. Gerall and Robert W. Goy for suggestions regarding the articles included in this volume. Their recommendations were invaluable, although I alone must accept responsibility for any weaknesses of the book. In addition, I wish to express appreciation to Lynwood G. Clemens, under whose guidance I became interested in the history of problems related to the study of hormones and behavior. For the opportunity to work on this volume and for excellent editorial help, I am indebted to Martin W. Schein, and for the emotional and intellectual support necessary to complete the task, I wish to thank Stephen W. Porges of the German Department at the University of Illinois.

Bibliography

Books

Beach, F. A. (Ed.). *Sex and Behavior.* Wiley, New York (1965).
Diamond, M. (Ed.). *Perspectives in Reproduction and Sexual Behavior.* Indiana University Press, Bloomington, Indiana (1968).
Levine, S. (Ed.). *Hormones and Behavior.* Academic Press, New York (1972).
Martini, L., and Ganong, W. F. (Eds.). *Neuroendocrinology.* Academic Press, New York (1967).
Martini, L., and Ganong, W. F. (Eds.). *Frontiers in Neuroendocrinology.* Oxford University Press, New York (1969, 1971, 1973).
Masters, W. H., and Johnson, V. E. *Human Sexual Response.* Little, Brown, Boston (1966).
Sawyer, C., and Gorski, R. (Eds.). *Steroid Hormones and Brain Function.* University of California Press, Berkeley, Calif. (1972).
Turner, C. D., and Bagnara, J. T. *General Endocrinology.* W. B. Saunders, Philadelphia (1971).
Young, W. C. (Ed.). *Sex and Internal Secretions.* Third Edition. Williams & Wilkins, Baltimore (1961).

Review Articles

Beach, F. A. Hormonal factors controlling the differentiation, development and display of copulatory behavior in the ramstergig and related species. In *The Biopsychology of Development,* Tobach, E., Aronson, L. R., and Shaw, E. (Eds.), Academic Press, New York (1971).
Davidson, J. M., and Levine, S. Endocrine regulation of behavior. *Annual Review of Physiology,* **34**, 375–408 (1972).
Goy, R. W., and Resko, J. A. Gonadal hormones and behavior of normal and pseudo-hermaphroditic nonhuman female primates. *Recent Progress in Hormone Research,* **28**, 707–733 (1972).
Michael, R. P., Zumpe, D., Keverne, E. B., and Bonsall, R. W. Neuroendocrine factors in the control of primate behavior. *Recent Progress in Hormone Research,* **28**, 665–706 (1972).

Journals

Animal Behaviour, Baillière Tindall, London
Archives of Sexual Behavior, Plenum Press, New York
Brain Research, Elsevier, Amsterdam
Endocrinology, J. B. Lippincott, Philadelphia
Hormones and Behavior, Academic Press, New York
Journal of Comparative and Physiological Psychology, American Psychological Association, Washington, D.C.
Neuroendocrinology, Karger, Basel, Switzerland
Physiology and Behavior, Brain Research Publications, Fayetteville, New York
Science, American Association for the Advancement of Science, Washington, D.C.

Hormones and the Behavior of the Male

I

Editor's Comments on Papers 1 and 2

1 **Steinach:** *Investigations into the Comparative Physiology of the Male Sexual Organs, with Particular Reference to the Accessory Sexual Glands*

2 **Brown-Séquard:** *The Effects Produced on Man by Subcutaneous Injections of a Liquid Obtained from the Testicles of Animals*

Evidence from the work of the German investigator E. Steinach suggested the importance of testicular secretions in the establishment and maintenance of male sexual behavior. The portion of one of the early reports of Steinach (Paper 1) that deals primarily with mammalian reproductive behavior has been translated and included here. Steinach and his contemporaries were aware of the interaction between sexual experience and castration in determining the rate at which postcastrational decrements in male performance would occur.

Of little scientific value, but of considerable historical interest, is the report of Brown-Séquard (Paper 2) of his early attempts to demonstrate the rejuvenating ability of the testis. Brown-Séquard had previously gained renown for work on the function of the spinal cord and adrenal glands. However, his reports of dramatic behavioral changes following self-injection of an aqueous testicular extract were ridiculed for a variety of reasons and were almost certainly placebo effects since it is now known that the androgens secreted by the testis are not water-soluble. Nonetheless the attempted therapeutic injection by Brown-Séquard of a glandular extract received wide attention and influenced and stimulated subsequent medical research.

1

Investigations into the Comparative Physiology of the Male Sexual Organs, with Particular Reference to the Accessory Sexual Glands

E. STEINACH

*Translated by Martyn Clarke,
University of Illinois, "Untersuchungen
zur vergleichenden Physiologie der
männlichen Gerschlechtsorgane
insbesondere der accessorischen
Geschlechtsdrüsen*, Pflügers Arch.
Ges. Physiol. des Menschen und der
Tiere, **56**, *304–306, 315–322,
333–338 (1894)*

Introduction

The following study is the result of a series of investigations that I have carried out in the course of the last two years (1892–1893) and have now brought to a conclusion.

The first incentive for these studies was given by information that Tarchanoff (1887, p. 330) published under the title "On the Physiology of the Sexual Apparatus of the Frog." Tarchanoff discussed the question "from where the centripetal impulses proceed, which arouse the sexual drive and maintain the neuromuscular mechanism of the embrace in a condition of heightened tonic excitement." He found, in common with earlier researchers, that severe irritation, injury, or mutilation of the male neither interrupted the sexual act nor reduced the sexual drive. Even the removal of various inner organs did not disturb the mating process; even after excision of both testicles, significantly enlarged at mating time, and from which a major influence upon the clasp reflex would be expected, the embrace persisted. Goltz (1869, p. 20) had already observed this behavior in castrated frogs. What was new, however, was the experimentation on filled seminal vesicles. Tarchanoff reports that opening these and completely emptying the seminal fluid, or removing the seminal vesicles, rapidly led, in the majority of cases, to a separation of the pairs and to a permanent loss of the sexual drive. From his final results he thought he should conclude that it is mainly from the seminal vesicles that the impulses spring which arouse the sexual drive and send the males in search of the females, and that at mating time these organs continuously produce stimuli which ultimately cause the clasp reflex and sustain it for periods of days or weeks. According to his notion, the nerve endings of the filled seminal vesicles were mechanically stimulated, on the one hand by the self-perpetuating tension, and on the other hand by the continual "collision" of the energetically moving spermatozoa against the walls, and the embrace reflex was triggered by the distribution of these stimuli to the uppermost segments of the spinal cord. There was previously no apparent reason to doubt the possibility of such a process, and the thought suggested itself of investigating whether the sexual excitability of mammals was similarly bound up with the integrity of the seminal vesicles. Certainly it was clear that expressions of the sexual drive in the frog are not automatically comparable with those of mammals, and on no account could one proceed, like Tarchanoff, from the supposition that the seminal vesicles of frogs and those of mammals represent homologous organs.

Microscopic investigations, to which I will return, have recently furnished detailed and definite confirmation of isolated earlier indications that the seminal vesicles of mammals are not to be considered as seminal containers at all. Through this discovery, the question of the physiological significance of the seminal vesicles of mammals gained additional interest, and I decided, when Tarchanoff produced no further information, to take up the matter from different points of view.

As an experimental subject, the white rat seemed to me to be ideally suited, first, because of its conspicuously well-developed seminal vesicles, and second, because of its unusually strong sexual drive.

The series of experiments on white rats that I carried out in the spring of 1892 produced, as its first result, the fact that by the removal of the seminal vesicles the intensity of the sexual drive suffers no loss; neither the frequency nor the vigor of the mating act is impaired. After verification of this finding, my investigation of rats followed a new course.

* * * * * * *

On the Physiological Importance
of the Accessory Glands in Mammals

The studies that are concerned with the accessory glands in mammals have referred almost exclusively to the seminal vesicles and have mainly discussed the argument, traceable back to Fallopia, of whether the seminal vesicles are containers for the semen, or whether they are secreting glands, or whether they are both simultaneously. The majority of later authors has no longer regarded the seminal vesicles as seminal containers, but as containers of their own secretion. The most recent observations on the above question are to be found in the dissertation of H. Kayser (1889). He prefaces his investigations with a comprehensive and penetrating historical survey of the relevant literature, to which I refer the reader in order to avoid going into detail here. The original work, in which the designation "seminal vesicles" is given, was done by E. H. Weber, and then, with particular success, by Leydig, who, through his detailed investigations, laid the foundation for our present knowledge of the accessory sexual glands.

The system of Leydig, to which, among others, Aeby and Broesike subscribed, was amplified by Sedgwick-Minot and more recently by Kayser. The work of the latter led to the following result: "The seminal vesicles are organs whose walls produce an egg-white-like secretion which collects in the vesicle itself. They are thus receptacles for a secretion of their own of which the physiological importance is not yet known." In those mammals whose seminal vesicles contain sometimes more, sometimes less spermatozoa, they still do not occur "so regularly and so numerously that one may, therefore, regard the seminal vesicles also as containers for the spermatozoa."

The function of the secretions of the seminal vesicles, as well as of the other accessory glands, is practically unexplored. The task generally ascribed to them is that of increasing the volume occupied by the semen and thereby diluting it. Leuckart (in Hensen, 1881) conjectured, in the particular case of the guinea pig, that the seminal vesicle secretion, which clots after flowing into the vagina, serves to prevent the sperm from running out. Landwehr (1880) has carefully examined this secretion and found that it easily congeals with a slight admixture of blood and that it contains 27 percent fibrinogenous substance.

My investigations are concerned with the seminal vesicles and prostates and are performed on white rats.

Since, in view of the above arguments, the designation "seminal vesicles" does not seem at all accurate for mammals, and since I am also convinced, in the case of my own experimental animals (and other rodents) of the complete absence of spermatozoa in the contents of the seminal vesicles, I will follow the suggestion made by Owen and repeated by Oudemans and use instead of "seminal vesicles" the term "vesicle glands" in my discussion. The richly illustrated publication of Oudemans (1892; see also Leydig, 1857) contains an extensive comparative-anatomical treatment of the accessory sexual glands which provides exhaustive information about the structure, form, size, and the topographical arrangement of these organs in a wide variety of mammals.

Extirpation of the Vesicle Glands

Series of Breeding Experiments

By their conspicuous size and shape, the vesicle glands, with their tips projecting far into the abdominal cavity, are easy to find; they are visible as a pair of yellowish pouches with lobed edges behind the urinary bladder. In mature one-year-old rats of

250–300 grams, these glands (measured from the tip to the root) are 35–42 milli-meters long and 9–30 millimeters wide at their widest point; in some particularly large examples, I have seen them even longer and wider. At a sexually mature age, especially at the times of greatest sexual excitability (February to October, inclusive) the vesicle glands are filled to the bursting point with secretion, which is squeezed out when they are punctured. The yellow fluid soon thickens on contact with the air and shows signs of congealing.

I chose for my experiments four strong, ten to eleven-month-old males which had previously mated in the normal way[1], but which had been isolated for several weeks and as a result, as tests with various females showed, displayed extraordinarily strong sexual appetites.

On February 13, 1892, I carried out the extirpation. The operation was done under careful observation of strict antiseptic precautions. The animals were carefully etherized. I endeavored to make the incision through the abdominal wall no bigger than absolutely necessary. It proved advantageous to proceed along the linea alba, as in this way the straight muscles remained uninjured, which became useful in the stitching. After cutting the peritoneum, the edges of the incision were held apart by means of eyelid holders, and one of the vesicle glands was gripped by its tip, raised out of the abdominal cavity, doubly bound with catgut at the deepest possible point of the neck, and cut off between the two ligatures. The second gland was similarly removed. If the urinary bladder was full, I made my work easier by causing it to empty by means of gentle mechanical stimulation and, thereby, made the two glands accessible. I sewed up the muscles with catgut and strewed the suture with iodine powder. The skin wounds I closed with sterilized silk and covered the suture with a thick layer of iodine collodion. I did not draw the stitching through the skin too tightly, since an animal with the stitching pulled too tightly had torn the wound open again with its teeth because of the pain. After the operation, the animals were placed in individual glass cages (aquaria) in which the floors were covered with clean wadding. Initially, this was changed daily. On the day of the operation, the animals received no food; from the next morning on they received boiled milk and dry black bread again as usual. The temperature of the room was kept even. During the initial period, the operated animals behaved very quietly and seemed to be fatigued; after the third day, they became lively again and began to behave like healthy animals. The healing progressed perfectly. On February 21, the wounds were already scarred over; on March 4, new hair had already grown over the places.

At first it interested me to investigate the sexual drive of the operated rats. I began my observations on February 20, that is, seven days after the extirpation. I introduced a female into the cage for a quarter of an hour. The male immediately turned to her, gave her his entire attention, continuously sniffed at the genitals, mounted her, and made all the preparations for mating. Typical coitus, however, did not take place. All four males behaved in the same way. On the following day, I repeated this short pairing experiment. The sexual manifestations became increasing-ly unrestrained. On February 24, the eleventh day after the operation, normal coitus was resumed. This was performed several times successively in the few minutes of being together. After I had observed the same thing on the following day, I left the males undisturbed until they were completely healed.

---

[1]Male white rats become capable of breeding after 3 to 3½ months.

On March 4, I transferred the animals from their glass cages to large wooden cages with metal floors; they appeared to be completely recovered and reinvigorated. In two days they had become accustomed to their new surroundings. I now gave them each a female permanently. What I then witnessed in terms of sexual potency borders on the unbelievable. I made a point on the first evening of noting the sexual acts. After the usual foreplay, the males began sooner or later to mount their females, who sometimes stoutly resisted. I counted, and I include this as a curiosity, 80 matings by one rat in an hour; by another, 64; and by the other two, 40 to 50. After two days, this sexual frenzy had partially abated; but it was clear even later, particularly in the evening hours, that coitus was performed not only daily, but very frequently. This sexual excitability of the operated males, moderated by permanent association, but still relatively high, remained unimpaired, and this is what I would now like to turn to, with reference to the following breeding experiments, which extended over several months.

Here I must insert some remarks of a methodological nature and some observations on the sexual life of white rats. If the strength of the sexual drive in the animals is to be correctly judged, certain conditions, of which my experiments only gradually made me aware, must be fulfilled.

Above all, the rule must be applied that the test should take place in the cage of the male. If this is done, he immediately and completely devotes himself to the female; he ceases to eat and to clean himself; he forgets his other activities and ignores all sounds; his impulses and efforts are centered on the female and are solely devoted to the frenzied satisfaction of his sexual urge. The male behaves differently if he meets the female in strange surroundings or if he is transferred directly from his cage to that of the female. Here he is initially timid and inhibited by the new environment and the new smells (which are known to play an important role in the orientation of these animals) to such an extent that instead of exploiting the opportunity, he becomes interested in every conceivable object and only occasionally turns his interest for a few moments to the female. In this situation, the female is usually the aggressive partner and tries by the most various importunities to attract the male. Nevertheless, hours can often elapse until finally one or two matings take place. Only after considerable time, when the male feels at home again, does the initially suppressed sexual urge reappear. If this feature is not taken into account, one can easily be deceived, especially when dealing, not with normal animals, but with ones in which one might expect a reduction of the sexual capacity as a result of operative interference, or in which one wishes to ascertain the survival of former potency, as, for example, after castration.

If it is necessary for certain experiments to select the most sexually excitable animals, or if it is desired to test and to increase the potency of males that live with females, it is self-evident that they must be isolated for several days before the test, or that the females must be removed from their cages. If they are together with other males, they must also be separated from them, since I have frequently observed cases of homosexuality. Male rats tolerate their own sex only if they are raised together. If strangers are introduced into the cage, violent fights break out that can lead to the death of one or the other animal. Particularly in such cases, if the animals are enclosed together for a long period, I have observed mutual sexual gratification. The weaker of the two, having been forced to submit, was repeatedly pursued and mounted, like a female. If I placed a female in the cage, she was immediately courted and mated by the males. Some particularly lustful males, however, in their paroxysms, made no great distinction, but alternately mounted the female and the other males, whichever they happened to encounter. This type of homosexuality is clearly explained by the intensity of the heightened sexual urge, which craves gratification at any price. It has never been my experience that males who have lived together with both other males and females favored their own sex.

A primary influence on the sexual drive is temperature. If one extends the experiments, as I did, into the cold seasons, the room in which the rats' cages are kept must be well heated. In unheated rooms the act of coitus is seldom observed, and the same applies on cool summer days. The animals lie crouching in their straw and only after warming of the room do they become lively again, and begin to play and to mate. The animals behave most vigorously in the evening and night hours. Indeed, rats are night creatures and this quality has survived in tame white rats. Thus, the evening proved to be the most favorable time for observation. The strong artificial lighting of the room did not disturb their activities.

Since the sexual act is so extraordinarily frequent, it is correspondingly short. The male pursues the female, continuously sniffing at the genitals, and mounts her with a firm clasp. The act of coitus itself only lasts for a few seconds and is characterized by the remarkable speed of the rhythmic muscular actions. The erection is normally so strong that immediately after completion of the act, the male remains sitting on his hind legs and tries to push the penis back into the foreskin with his snout. Immediately after this has been done, the aroused animal begins the entire cycle again.

The importance of the sense of smell in sexual life is well known (Althaus, 1882). In rats this feature is particularly conspicuous; with them, smell seems to be an important means of arousal. I inhibited the sexual frenzy of a blind male rat, who was unceasingly mating with an aroused female, by painting the female's sexual parts with a mixture of cod-liver oil and petroleum. Each time the male approached the genitals he recoiled from the smell and finally left the female in peace. After half an hour, I cleaned the female with soap and, after she was dry, put her back in the cage; she was immediately smelt and mounted with great lustfulness. Iodine did not have any particular deterrent effect. The behavior of the four operated male rats has proved that the sexual drive and the mating ability are in no way connected to the integrity of the vesicle glands.

* * * * * * *

On the Sexual Drive of Rats Castrated Before and After Puberty, and on the Fate of the Accessory Sexual Glands as a Result of Castration

I would like to treat these observations as briefly as possible. In order to avoid repetition, I here refer the reader to the remarks which I made about the manifestations and testing of the sexual drive in the previous section.

I carried out the castrations, on the one hand in order to investigate whether and what changes the accessory sexual glands undergo, and on the other hand because I supposed that the extraordinary sexual excitability of rats would be retained after castration longer than has normally been the case in previous experience with domestic animals, where the mating capacity disappeared very soon after castration, and during the short time of its survival was expressed only to a greatly reduced degree. The way in which the behavior of the human being differs from other animals after castration was explained by Pflüger (1877, p. 83) on the basis of psychic qualities, the vigor of his fantasy, and his powerful memory.

For the castration of the sexually mature animals, I naturally selected the most excitable ones and operated upon them in the usual way. After cutting the covering skin, I drew out the testicle and the "secondary testicle" on each side and made the binding and excision in such a way that part of the vas deferens was also removed; I then stitched up the scrotum with silk.

On the fifth day after castration, the tests began. They were carried out in the spacious cages of the individually separated males initially every day (in the evenings), later once a week, and never lasted longer than from fifteen to thirty minutes. One rat mated during the first test, and the other two in the days immediately following. I was able to make the observation that the mating drive, which was initially somewhat reduced, rapidly increased with the healing of the wounds, and after a few days, returned to normal. The mating act, especially when the test female was in heat, was repeated several times successively during the short period of cohabitation. The vigor of the act and the strength of the erection was exactly as in normal animals. This degree of potency remained unimpaired in one rat for six months and in the other two for approximately four months; I should point out again that I never allowed them to satisfy their lust to the point of satiation. After this period of unchanged potency, the mating capacity gradually decreased, but the sexual excitability and sexual inclination remained unaffected, and this was shown by unmistakable signs. Late autumn and winter were approaching, the period when the mating drive is, in general, somewhat reduced. The operated animals might have been much more subject to these effects than normal ones. I therefore postponed the regular, time-consuming tests until February of the following year (1893); a year had now elapsed since the castration, and the potency seemed, in fact, to be increasing again. Several times I observed erection and vigorous mating. But very soon their strength disappeared again and was replaced by a condition in which a vigorous mating drive was observed in various manifestations but which never resulted in erection or true mating. During this stage, which proved to be permanent, the mounting of the females and the accompanying actions were, nevertheless, so deceptively similar to the genuine act of coitus that only the experienced observer could note a distinction.

In another animal, I simultaneously removed the vesicle glands and the reproductive glands; the period of unchanged potency lasted approximately two to three months. In general, the result was the same as in the other castrated animals. At any rate, it was clear that mating capacity after castration is not dependent upon the existence of the vesicle glands. My observations on rats accord quite well with the information that Pelikan (1876) has given about the sexual behavior of castrated humans.

A measure of the suitability of rats as experimental subjects, apart from their high potency, is that even in captivity they enjoy a freedom of movement that can hardly be granted to domestic and working animals, on which most observations of this sort have hitherto been made.

One and a half years after castration, I killed the animals. The vesicle glands had shrunk considerably, a natural result, in part, of the very much reduced secretion content; by means of injection it was possible to expand them considerably, but not to a normal size. Even more noticeable was the diminution of the prostate bundles. In this respect also, similar effects have been found in humans (Hyrtl, 1881, p. 767).

I undertook a further series of experiments with young rats before puberty. I operated on four animals forty-five days of age. Since the testicles were still located in the abdominal cavity and could easily be moved under the abdominal skin by hand, I had an assistant push them toward the inguinal region, where I made a small opening, drew them out, cut them off, and stitched up the wounds. The organs were still undeveloped. In the microscopic examination, therefore, I found spermatozoa neither in the testicles nor in the "secondary testicles."

As a control, the animals were reared together with a normal male from the same litter. In the first month, none of them displayed any trace of sexual inclination. If I introduced a sexually mature female into the cage, they initially displayed anxiety and withdrew into a corner in fear; after a while they became more adjusted and tried to play with her as they would among themselves.

When they were ninety days old, I observed the first clear signs of the sexual drive in the normal male; on the 100th day I saw him perform coitus for the first time, and in a longer test, the act was frequently repeated.[2] Within a few days the mating desire had increased to its maximum. Meanwhile, a considerable degree of sexual inclination had also developed in the four castrated rats. They lost their timidity; they pursued, sniffed, and licked the female as excitably as the normal male. I did not allow myself to be deceived by mere expressions of curiosity, which are among the normal characteristics of growing animals, and from now on I carried out observations on the castrated rats in various ways. They were placed individually in large cages that were divided from each other by walls of coarse wire netting. In the middle cage between each two males, a female of similar size was placed. They now had this female before their eyes for days and weeks; they could smell her through the screen and familiarized themselves with her, so that their curiosity was completely satisfied; nevertheless, when the same female was placed in their cage, they behaved even more lustfully than before. The mounting of the female and the coitus-like movements were externally differentiated from those of normal males only by their less violent manner; erection and ejaculation did not take place. I often seized the animals at the moment of mounting and a proper erection was never visible; it was, at best, extremely inadequate and this happened only in rare cases. It was noticeable that they preferred a female in heat. If, instead of the female, I placed a normal, strange male of the same age and size in the cage, it was immediately recognized as such and a fight developed in which the weaker and less spirited castrated animal came off worse. The four castrated animals displayed these features for about a year; after this they

---

[2]After this observation, I removed the normal male from the cage.

became lazier, their sexual inclination decreased, and they gave the impression of premature old age. I killed them (at one and a half years old) in autumn 1893 in order to examine the accessory glands.

In view of these observations it is beyond doubt that in rats castrated long before puberty, that is, independent of the reproductive glands, a certain degree of sexuality developed that is comparable in its manifestations to the sexual drive that, in rats castrated at the age of puberty, continues after the stage of unchanged potency; it is also comparable with the sexual drive that has been described[3] in animals in which the nervous connection between reproductive organ and spinal cord has been broken. If I collate these observations with the findings made in frogs castrated before rutting time, the following conclusion seems to be suggested. In general—sometimes more, sometimes less recognizably—an incipient sexuality exists before puberty or rutting time, independently of the semen-producing organs, but its development necessary for reproduction is produced only by the impulses proceeding from the swelling reproductive glands, which determine the great increase in excitability of the centers governing sexuality.

Once this process has begun, psychic influences or the faintest impressions from one or more sensory sources suffice to set the reproductive apparatus in action.[4] In animals that are subject to a periodic rutting, the increase in excitability of the centers recurs once or several times each year. In the others it begins with puberty, remains constant with occasional fluctuations during the years of normal breeding capacity, and gradually disappears, under normal conditions, only after the secretory functions of the reproductive glands have ceased. The continued copulatory capacity of humans and, as we saw, also of rats for months after castration[5] is consequently explained by the fact that the heightened excitability of the relevant central organs outlasts the decrease in the impulses flowing from the reproductive glands and only disappears quite gradually.

As an additional sign of preexisting sexuality, the frequently described sexual manifestations of children may also be mentioned.

-----

[3]Rémy, *Journal de l'anatom. et de la physiol.*, 1886, stimulated in guinea pigs certain nerves proceeding from a ganglion (nerfs éjaculateurs) and observed movements of the vesicle glands, erection, and ejaculation. If I correctly understand the study of Loeb (in Eckhard's Beiträge, 1866), the ganglionic swelling in the lumbar region and the excitement of the fibers were already known to him and to Budge. It is of interest to us that after cutting the nerves proceeding from the ganglion, Rémy observed that although the sexual urge continued, erection and ejaculation were impossible.

[4]With reference to the arousal of the clasp center of frogs, see the remark of Goltz cited above; with reference to the arousal of the nervus dorsalis penis, see the remark of Hensen (1881, *Physiologie der Zeugung,* p. 108).

[5]With reference to this point, I intend shortly to perform an experiment with rats, cutting the nervi spermatici before and after puberty, in order to check whether, as one would suppose, the same conditions occur with regard to the sexual drive as after castration. In connection with this matter, I am aware that in the above discussion many points are alluded to and many questions are raised which call for more detailed investigation. As a consequence of the results presented, some new tasks are also suggested, and these remain to be treated in later investigations.

Bibliography

Althaus, J., *Beiträge zur Physiologie und Pathologie des nervus olfactorius*, Arch. f. Psychiatrie, XII, 1882.

Goltz, F., *Beiträge zur Lehre von den Functionen der Nervencentren des Frosches*, p. 20, Berlin, 1869.

Hensen, V., *Physiologie der Zeugung, Handb. d. Physiologie*, Vol. 6, part 2, 1881.

Hyrtl, *Lehrbuch der Anatomie*, 15, p. 767, Auflage, 1881.

Kayser, H., *Untersuchungen über die Bedeutung der Samenblasen*, Berlin, 1889.

Landweher, *Pflügers Archiv für die Gesamte Physiologie des Menschen und der Tiere*, Vol. 23 1880.

Leydig, F., *Zeitschr. f. wissenschaftl. Zoologie*, Vol. 2, and *Lehrbuch der Histologie*, 1857.

Oudemans, J. T., *Die accessorischen Geschlechtsdrüsen der Säugethiere*, Haarlem, 1892.

Pelikan, E., *Gerichtl. med. Untersuchsuger über das Skopzentum in Russland.*

Pflüger, E., *Die teleologische Mechanik der lebendigen Natur. Pflügers Archiv für die Gesamte Physiologie des Menschen und der Tiere*, Vol. 15, p. 83, 1877.

Tarchanoff, *Pflügers Archiv für die Gesamte Physiologie des Menschen und der Tiere*, Vol. 40, p. 330, 1887.

Reprinted from *Lancet*, 105–106 (1899)

2

NOTE ON

THE EFFECTS PRODUCED ON MAN BY SUB-
CUTANEOUS INJECTIONS OF A LIQUID
OBTAINED FROM THE TESTICLES
OF ANIMALS.

BY DR. BROWN-SÉQUARD, F.R.S. &c.

ON the 1st of June last I made at the Société de Biologie
of Paris a communication on the above subject, which was
published in the *Comptes Rendus* of that Society on June 21st
(No. 24). I will give here a summary of the facts and views
contained in that paper and in two subsequent ones, adding
to them some new points.

There is no need of describing at length the great effects
produced on the organisation of man by castration, when it
is made before the adult age. It is particularly well known
that eunuchs are characterised by their general debility and
their lack of intellectual and physical activity. There is no
medical man who does not know also how much the mind
and body of men (especially before the spermatic glands
have acquired their full power, or when that power is
declining in consequence of advanced age) are affected by
sexual abuse or by masturbation. Besides, it is well known
that seminal losses, arising from any cause, produce a
mental and physical debility which is in proportion to their
frequency. These facts and many others have led to
the generally admitted view that in the seminal fluid, as
secreted by the testicles, a substance or several substances
exist which, entering the blood by resorption, have a most
essential use in giving strength to the nervous system and
to other parts. But if what may be called spermatic
anæmia leads to that conclusion, the opposite state, which
can be named spermatic plethora, gives as strong a testimony
in favour of that conclusion. It is known that well-
organised men, especially from twenty to thirty-five years
of age, who remain absolutely free from sexual intercourse
or any other causes of expenditure of seminal fluid, are in a
state of excitement, giving them a great, although abnormal,
physical and mental activity. These two series of facts
contribute to show what great dynamogenic power is
possessed by some substance or substances which our blood
owes to the testicles.

For a great many years I have believed that the weakness of old men depended on two causes—a natural series
of organic changes and the gradually diminishing action of
the spermatic glands. In 1869, in a course of lectures
at the Paris Faculty of Medicine, discussing the influence
possessed by several glands upon the nervous centres, I put

forward the idea that if it were possible without danger
to inject semen into the blood of old men, we should
probably obtain manifestations of increased activity as
regards the mental and the various physical powers. Led
by this view, I made various experiments on animals at
Nahant, near Boston (United States), in 1875. In some of
those experiments, made on a dozen male dogs, I tried
vainly, except in one case, to engraft certain parts or
the whole body of young guinea-pigs. The success obtained in the exceptional case served to give me great
hopes that by a less difficult process I should some day
reach my aim. This I have now done. At the end of last
year I made on two old male rabbits experiments which
were repeated since on several others, with results leaving
no doubt as regards both the innocuity[1] of the process used
and the good effects produced in all those animals. This
having been ascertained, I resolved to make experiments on
myself, which I thought would be far more decisive on man
than on animals. The event has proved the correctness of
that idea.

Leaving aside and for future researches the questions relating to the substance or substances which, being formed
by the testicles, give power to the nervous centres and other
parts, I have made use, in subcutaneous injections, of a
liquid containing a small quantity of water mixed with the
three following parts : first, blood of the testicular veins ;[2]
secondly, semen ; and thirdly, juice extracted from a
testicle, crushed immediately after it has been taken from
a dog or a guinea-pig. Wishing in all the injections made
on myself to obtain the maximum of effects, I have employed
as little water as I could. To the three kinds of substances I
have just named, I added distilled water in a quantity which
never exceeded three or four times their volume. The crushing was always done after the addition of water. When
filtered through a paper filter, the liquid was of a reddish
hue and rather opaque, while it was almost perfectly clear
and transparent when Pasteur's filter was employed. For
each injection I have used nearly one cubic centimetre of the
filtered liquid. The animals employed were a strong and,
according to all appearances, perfectly healthy dog (from two
to three years old), and a number of very young or adult
guinea-pigs. The experiments, so far, do not allow of a positive conclusion as regards the relative power of the liquid
obtained from a dog and that drawn from guinea-pigs. All
I can assert is that the two kinds of animals have given a
liquid endowed with very great power. I have hitherto
made ten subcutaneous injections of such a liquid—two in
my left arm, all the others in my lower limbs—from
May 15th to June 4th last. The first five injections were
made on three succeeding days with a liquid obtained from
a dog. In all the subsequent injections, made on May 24th,
29th, and 30th, and June 4th, the liquid used came from
guinea-pigs. When I employed liquids having passed
through Pasteur's filter, the pains and other bad effects
were somewhat less than when a paper filter was used.

Coming now to the favourable effects of these injections,
I beg to be excused for speaking so much as I shall do of
my own person. I hope it will easily be understood that,
if my demonstration has any value—I will even say any
significance—it is owing to the details concerning the state
of my health, strength, and habits previously to my experiments, and to the effects they have produced.

I am seventy-two years old. My general strength, which
has been considerable, has notably and gradually diminished
during the last ten or twelve years. Before May 15th last,
I was so weak that I was always compelled to sit down
after half an hour's work in the laboratory. Even when I
remained seated all the time, or almost all the time, in the
laboratory, I used to come out of it quite exhausted after
three or four hours' experimental labour, and sometimes after
only two hours. For many years, on returning home in a
carriage by six o'clock after several hours passed in the
laboratory, I was so extremely tired that I invariably had to go
to bed after having hastily taken a very small amount of food.
Very frequently the exhaustion was so great that, although

[1] This innocuity was also proved on a very old dog by twenty subcutaneous injections of a fluid similar to that I intended to employ on
myself. No apparent harm resulted from these trials, which were
made by my assistant, Dr. D'Arsonval.
[2] For reasons I have given in many lectures in 1869 and since, I consider the spermatic as also the principal glands (kidneys, liver, &c.) as
endowed, besides their secretory power, with an influence over the composition of blood, such as is possessed by the spleen, the thyroid, &c.
Led by that view, I have already made some trials with the blood
returning from the testicles. But what I have seen is not sufficiently
decisive to be mentioned here.

extremely sleepy, I could not for hours go to sleep, and I only slept very little, waking up exceedingly tired.[3]

The day after the first subcutaneous injection, and still more after the two succeeding ones, a radical change took place in me, and I had ample reason to say and to write that I had regained at least all the strength I possessed a good many years ago. Considerable laboratory work hardly tired me. To the great astonishment of my two principal assistants, Drs. D'Arsonval and Hénocque, and other persons, I was able to make experiments for several hours while standing up, feeling no need whatever to sit down. Still more : one day (the 23rd of May), after three hours and a quarter of hard experimental labour in the standing attitude, I went home so little tired that after dinner I was able to go to work and to write for an hour and a half a part of a paper on a difficult subject. For more than twenty years I had never been able to do as much.[4] From a natural impetuosity, and also to avoid losing time, I had, till I was sixty years old, the habit of ascending and descending stairs so rapidly that my movements were rather those of running than of walking. This had gradually changed, and I had come to move slowly up and down stairs, having to hold the banister in difficult staircases. After the second injection I found that I had fully regained my old powers, and returned to my previous habits in that respect.

My limbs, tested with a dynamometer, for a week before my trial and during the month following the first injection, showed a decided gain of strength. The average number of kilogrammes moved by the flexors of the right forearm, before the first injection was about 34½ (from 32 to 37), and after that injection 41 (from 39 to 44), the gain being from 6 to 7 kilogrammes. In that respect the forearm flexors re-acquired, in a great measure, the strength they had when I was living in London (more than twenty-six years ago). The average number of kilogrammes moved by those muscles in London in 1863[5] was 43 (40 to 46 kilogrammes).

I have measured comparatively, before and after the first injection, the jet of urine in similar circumstances—i.e., after a meal in which I had taken food and drink of the same kind in similar quantity. The average length of the jet during the ten days that preceded the first injection was inferior by at least one quarter of what it came to be during the twenty following days. It is therefore quite evident that the power of the spinal cord over the bladder was considerably increased.

One of the most troublesome miseries of advanced life consists in the diminution of the power of defecation. To avoid repeating the details I have elsewhere given in that respect, I will simply say that after the first days of my experiments I have had a greater improvement with regard to the expulsion of fecal matters than in any other function. In fact a radical change took place, and even on days of great constipation the power I long ago possessed had returned.

With regard to the facility of intellectual labour, which had diminished within the last few years, a return to my previous ordinary condition became quite manifest during and after the first two or three days of my experiments.

It is evident from these facts and from some others that all the functions depending on the power of action of the nervous centres, and especially of the spinal cord, were notably and rapidly improved by the injections I have used. The last of these injections was made on June 4th, about five weeks and a half ago. I ceased making use of them for the purpose of ascertaining how long their good effects would last. For four weeks no marked change occurred, but gradually, although rapidly, from the 3rd of this month (July) I have witnessed almost a complete return of the state of weakness which existed before the first injection. This loss of strength is an excellent counter-proof as regards the demonstration of the influence exerted on me by the subcutaneous injections of a spermatic fluid.

My first communication to the Paris Biological Society was made with the wish that other medical men advanced

in life would make on themselves experiments similar to mine, so as to ascertain, as I then stated, if the effects I had observed depended or not on any special idiosyncrasy or on a kind of auto-suggestion without hypnotisation, due to the conviction which I had before experimenting that I should surely obtain a great part at least of these effects. This last supposition found some ground in many of the facts contained in the valuable and learned work of Dr. Hack Tuke on the " Influence of the Mind over the Body." Ready as I was to make on my own person experiments which, if they were not dangerous, were at least exceedingly painful, I refused absolutely to yield to the wishes of many people anxious to obtain the effects I had observed on myself. But, without asking my advice, Dr. Variot, a physician who believed that the subcutaneous injections of considerably diluted spermatic fluid[6] could do no harm, has made a trial of that method on three old men—one fifty-four, another fifty-six, and the third sixty-eight years old.[7] On each of them the effects have been found to be very nearly the same as those I have obtained on myself. Dr. Variot made use of the testicles of rabbits and guinea-pigs.

These facts clearly show that it was not to a peculiar idiosyncrasy of mine that the effects I have pointed out were due. As regards the explanation of those effects by an auto-suggestion, it is hardly possible to accept it in the case of the patients treated by Dr. Variot. They had no idea of what was being done ; they knew nothing of my experiments, and were only told that they were receiving fortifying injections. To find out if this qualification had anything to do with the effects produced, Dr. Variot, since the publication of his paper, has employed similar words of encouragement, whilst making subcutaneous injections of pure water on two other patients, who obtained thereby no strengthening effect whatever.[8]

I believe that, after the results of Dr. Variot's trials, it is hardly possible to explain the effects I have observed on myself otherwise than by admitting that the liquid injected possesses the power of increasing the strength of many parts of the human organism. I need hardly say that those effects cannot have been due to structural changes, and that they resulted only from nutritive modifications, perhaps in a very great measure from purely dynamical influences exerted by some of the principles contained in the injected fluid.

I have at present no fact to mention which might serve to solve the question whether it would be possible or not to change structurally muscles, nerves, and the nervous centres by making during a good many months frequent injections of the fluid I have used. As I stated at the Paris Biological Society, I have always feared, and I still fear, that the special nutritive actions which bring on certain changes in man and animals, from the primitive embryonal state till death by old age, are absolutely fatal and irreversible. But in the same way that we see muscles which have from disease undergone considerable structural alterations regain sometimes their normal organisation, we may, I believe, see also some structural changes not essentially allied with old age, although accompanying it, disappear to such a degree as to allow tissues to recover the power they possessed at a much less advanced age.

Whatever may be thought of these speculations, the results I have obtained by experiments on myself and those which have been observed by Dr. Variot on three old men show that this important subject should be further investigated experimentally.[9]

Brighton.

---

3 I ought to say that, notwithstanding that dark picture, my general health is and has been almost always good, and that I had very little to complain of, excepting merycism and muscular rheumatism.

4 My friends know that, owing to certain circumstances and certain habits, I have for thirty or forty years gone to bed very early and done my writing work in the morning, beginning it generally between three and four o'clock. For a great many years I had lost all power of doing any serious mental work after dinner. Since my first subcutaneous injections I have very frequently been able to do such work for two, three, and one evening for nearly four hours.

5 I have a record of the strength of my forearm, begun in March, 1860, when I first established myself in London. From that time to 1862 I occasionally moved as much as 50 kilogrammes. During the last three years the maximum moved was 38 kilogrammes. This year, previously to the first injection, the maximum was 37 kilogrammes. Since the injection it has been 44.

6 In my third communication at the Biological Society, I said that both the intense pain each injection has caused me and the inflammation it has produced would be notably diminished if the liquid employed were more diluted. The three cases of Dr. Variot have proved the exactitude of my statement. He made use of a much larger amount of water, and his patients had to suffer no very great pain and no inflammation.

7 The paper of Dr. Variot and my remarks upon it have appeared in the " Comptes Rendus de la Société de Biologie," No. 26, 5 Juillet, 1889; pp. 451 and 454.

8 Since writing the above I have received a letter from Dr. Variot announcing that, after injecting the liquid drawn from the testicles into these two individuals, he has obtained the same strengthening effects I have myself experienced

9 It may be well to add that there are good reasons to think that subcutaneous injections of a fluid obtained by crushing ovaries just extracted from young or adult animals, and mixed with a certain amount of water, would act on old women in a manner analogous to that of the solution extracted from the testicles injected into old men.

1 Brit. Med. Jour., Dec. 9th, 1882.

Editor's Comments on Papers 3 Through 6

The following papers are particularly representative of the increasing emphasis on careful behavioral analysis as a method of describing and understanding hormone action. Much credit for the development of this approach must go to Calvin Stone, Frank Beach, Robert Yerkes, William Young, and the many outstanding students and associates of these men.

Although a number of accurate observations regarding the importance of the testes to the activation of male behavior had been made prior to the work of Stone (Paper 3), his efforts represent some of the earliest attempts to evalute this relationship quantitatively. Following the isolation and identification of testosterone as both a physiologically and behaviorally active agent influencing male reproduction, it became possible to investigate in greater detail hormone and behavior interaction. For example, the report by Beach and Holz-Tucker (Paper 4) demonstrated a dose–response relationship between testosterone and male sexual behavior in rats. In addition, this paper indicated that various components of the mating pattern might be differentially sensitive to androgen stimulation. As shown by Grunt and Young (Paper 5), not all individual differences in male behavior can be explained by variations in hormone levels; and the findings of Riss, Valenstein, Sinks, and Young (Paper 6) further stressed the potential importance of genetic differences in the expression of male sexual behavior.

Reprinted from *J. Comp. Psychol.*, **7**, 369–387 (1927)

3

THE RETENTION OF COPULATORY ABILITY IN MALE RATS FOLLOWING CASTRATION[1]

CALVIN P. STONE

Department of Psychology, Stanford University

Male rats castrated after puberty may retain copulatory[2] ability for various periods of time thereafter. Steinach reported this fact in 1894, yet, in the meantime, so few investigations of similar character have been made (Lipschutz, 1924) that one cannot safely generalize as to the period of time sexual vigor will persist in castrated rats or other mammals. The subject is of special importance to investigators analyzing the dynamic factors underlying sexual activities or measuring the efficacy of testicular transplants, injections, extract, or other agencies employed therapeutically to augment or restore sexual vigor. For this reason it deserves further study. One can better evaluate the potency of foreign agencies when their action is clearly differentiated from equivalent influences normally persisting after castration.

I. ANIMALS, HOUSING, AND FEEDING

The major part of the study involved a group of 45 males reared in the Psychological Laboratory. Their immediate ancestors had been fed a well balanced diet hence very large and healthy young were available for the experiments. When weaned at the age of twenty days their weights averaged 33 grams; at ninety days they averaged 217 grams. To simplify control of the age-factor, prospective mothers were bred within

[1] This study and a preliminary survey of similar character conducted under the direction of Dr. K. S. Lashley at the University of Minnesota in 1921 was financed by a grant from The Committee for Sex Research, National Research Council.

[2] The term "copulatory" is used to denote the overt elements of the copulatory response without indication as to insemination.

369

a space of four or five days so that the young would be ready for study at about the same date.   After weaning, groups of 8 or 10 males were housed together until the age of eighty-five days when they were ready for the experiment proper.   No opportunity for sexual stimulation or copulation was permitted by this arrangement beyond that afforded by cage-mates which we believe to be of negligible amount.

Since thorough taming of animals is a prime requisite in experiments concerned with sexual activity we tried to keep all individuals thoroughly tame and gentle by intentional handling at irregular intervals or by unavoidable handling in connection with weighing, cage cleaning, and feeding.   Neither excitement nor fear was observed as they were transferred from one cage to another.

At the age of eighty-five days, groups of 3 or 4 animals were transferred to observation cages made of hardware cloth of $\frac{1}{2}$ inch mesh (dimensions, 8 by 18 by 18 inches).   The latter served as home and experimental cages throughout the remainder of the experiment.

Both parents and young were fed ad libitum on a diet consisting of the following mixture (McCollum diet):

Whole wheat, ground fine	67.5
Casein	15.0
Powdered whole milk	10.0
Calcium carbonate	1.5
Sodium chloride	1.0
Butter fat	5.0

II. PUBERTY AND AGE AT TIME OF CASTRATION

To insure sexual aggressiveness and copulatory ability in most if not all of our animals it seemed advisable to defer castration until they had reached the age of ninety days, which is the upper extreme at which well developed young rats of our colony first demonstrate copulatory ability (Stone, 1924).   Upon examining the castrated animals it was found that the external genitals were adequately developed to permit of intromission and that the testes of all contained an abundance of free-moving spermatozoa as revealed by smears microscopically examined immediately after the testes were removed. Hence there is no doubt that the animals were sexually mature when castrated.

III. COPULATORY EXPERIENCE PRIOR TO CASTRATION

Half of the animals of each group was allowed copulatory experience between the ages of eighty-five and ninety days. Two reasons prompted this procedure. In the first place, we had no way of exactly forecasting the per cent of castrated animals that would fail to copulate when tested for the first time at the age of ninety days, and secondly no way of estimating the immediate or permanent shock effects of castration in sexually inexperienced males. Hence it seemed desirable that half of the animals be tested for the first time before and half after castration to secure data by which these two types of impotency might be differentiated if present and their frequency numerically estimated.

Two groups comprised the total of 45 males used. For simplicity of discussion they will be designated as groups A and B. The former was born about two months before the latter and consequently began its experiment three months earlier. Their sexual experience prior to castration was identical or differed in the following respects. Odd numbered males of group A were first tested for copulation three days before castration whereas those of group B were tested five days before castration. After a male of either group had demonstrated copulatory ability, he with two or three others was put with a receptive female for the remainder of the night in order that he might have unrestricted opportunity for copulation while the female remained receptive.

To avoid prejudice in selecting animals for precastration indulgence, the odd numbered individuals were allowed to copulate before castration and the even numbered males denied this opportunity. This subdivision splits litters and thus, in a measure, tends to equalize genetic and early nutritional factors.

IV. TESTS FOR COPULATORY ABILITY BEFORE AND AFTER
CASTRATION

To test for copulation a sexually receptive female was introduced into a male's home cage where the two were confined for subsequent observation. The duration of each period was arbitrarily limited to nine to twelve minutes, except when a male not copulating during that time continued to pursue the female. In the latter case he was allowed to continue until copulating or becoming inattentive to the female. Previous experiments served as a guide in determining the length of observation periods. We have observed that normal, *cage adapted males* with copulatory experience which neither copulate nor show sustained interest in receptive females during the first six or eight minutes of an observation period rarely copulate during the next half hour or more of observation. Hence we may expect that only a slight degree of inaccuracy results from limiting the observation periods to from nine to twelve minutes providing the male becomes inattentive to the female in that time. Repetition of the tests at three or four day intervals also tended to minimize this unavoidable experimental error.

Copulatory experience was allowed both odd and even numbered animals of each group on the day of castration. Animals operated during the forenoon were tested between seven and eleven o'clock that evening. At the conclusion of the first evening's tests all even numbered males were allowed to cohabit with the receptive females during the remainder of the night. This afforded them an equal opportunity to exercise the copulatory mechanism just as opportunity for extensive copulation had been allowed the odd numbered males prior to castration (see page 371). Our object in this procedure was to equalize roughly the copulatory experience of the two groups at the outset of the long series of tests which followed.

As a general rule each male demonstrating copulatory ability was allowed to copulate four or five times before removal of the female. At the outset tests were given from three to five days apart, but toward the end of the experiment, when only a few

ANIMAL NO	1	2	3	4	5	6	7	8	9
3-8	● 1:55		● 0:52		● 2:15		● 1:45		● 1:06
*3-9		● 2:05		● 3:02		● 4:55		● 1:40	
3-10		● 0:20		● 1:20		● 0:50		● 8:45	
*3-11	● 1:37		● 1:23		● 0:30		● 1:21		● 0:14
3-12	● 0:15		● 1:30		● 1:10	● 1:07	● 1:20		● 0:07
3-15		● 0:42		13,9,16. ●	● 3:55			9,16. ●	
3-16	● 1:22		● 0:25		12:0 ●		● 0:45		● 0:11
3-19		● 1:30		12,9. ●	12:0 ●	● 7:35		9,9. ●	
3-20	● 0:40		● 2:20		12:0 ●		9:0 ●		● 2:33
3-23	● 9:9	● 11:05	● 0:35	11,9. ●	10:0 ●	11,9. ●	● 2:30	9,9. ●	● 2:25
3-26	● 0:50	● 1:35	● 0:15	10,10. ●	11:0 ●	● 1:10	● 1:55	9,10. ●	● 0:20
3-30	● 3:3	● 10:09	● 0:40	12,9. ●	15:0 ●	● 2:30	9:0 ●	9,9. ●	● 0:07
4-3	10:0 ●	● 7:40	9:0 ●	10:0 ●	15:0 ●	● 2:00	9:0 ●	10:0 ●	● 0:30
4-7	11:0 ●	12:0 ●	11:0 ●	12:0 ●	10:0 ●	12:0 ●	11:0 ●	12:0 ●	● 2:00
4-11	11:0 ●	9:0 ●	11:0 ●	15:0 ●	12:0 ●	● 0:45	10:0 ●	10:0 ●	● 0:25
4-15	● 10:0	12:0 ●	14:0 ●		15:0 ●	10:0 ●	● 8:00		13:0 ●
4-19	11:0 ●	16:0 ●	11:0 ●			● 1:15	● 0:15		● 3:03
4-22	12:0 ●	15:0 ●	12:0 ●			12:0 ●	● 2:40		10:0 ●
4-26	11:0 ●					11:0 ●	10:0 ●		● 0:50
4-29	12:0 ●					● 4:45	12:0 ●		● 0:35
5-2	● 0:5						10:0 ●		● 5:15
5-8	12:0 ●					12:0 ●	● 0:58		● 0:20
5-14	● 0:25					9:0 ●	11:0 ●		● 6:20
5-22	● 5:0					12:0 ●	12:0 ●		12:0 ●
5-30	● 4:55					9:0 ●	9:0 ●		9:0 ●
6-7	● 2:0					12:0 ●	15:0 ●		10:0 ●
6-11	● 4:45					12:0 ●	12:0 ●		10:0 ●
6-15	10:0 ●						Photo		10:0 ●
6-19	11:0 ●								12:0 ●
6-25	10:0 ●								15:0 ●
7-1	● 3:25								
7-7	● 9:0								
7-13	10:0 ●								
7-18	10:0 ●								
7-23	10:0 ●								
7-28	● 8:20								
8-3	10:0 ●								
8-8	9:0 ●								
8-14	● 5:20								
8-20	12:0 ●								
8-22	9:0 ●								
8-25	● 0:15								
8-30	10:0 ●								
9-4	11:0 ●								
9-10	16:0 ●								
9-19	10:0 ●								
9-23	10:0 ●								
10-10	10:0 ●								
11-21	● 5:10								
12-2	9:0 ●								
1-12	15:0 ●								
1-23	15:0 ●								
2-10	25:0 ●								
2-14	25:0 ●								
4-3	25:0 ●								

FIG. 1. GRAPHICAL RECORD OF POST-CASTRATION COPULATION BY MALES WHOSE NUMBERS APPEAR AT THE TOPS OF THE COLUMNS

Odd numbered males had copulatory experience prior to castration; even numbered males were deprived of this experience. Dates of castration are indicated by an asterisk. In the columns dots at the left and right respectively signify that positive and negative results were gotten in tests for copulation. The number of minutes and seconds elapsing before the first copulatory act after putting the receptive female and castrated male together is indicated by the arabic numbers at the right of dots. Similarly, numbers at the left of dots indicate time elapsing before a test was terminated if the male failed to copulate. Whole numbers separated by commas signify repetition of the test one-half hour later. The foregoing description applies to figures 2 to 5, as well.

animals remained potent, the test intervals were somewhat lengthened (see figs. 1 to 5). The tests were arbitrarily terminated when negative results had been obtained on consecutive tests for a period of approximately one month. In some cases, however, the period of observation was extended beyond a month

ANIMAL NO	10	11	12	13	14	15	16	17	18
3-8		• 1:13		• 1:34		• 0:20		• 0:08	
3-9	9,9 •		3,9 •		• 11:00		• 0:50		• 3:00
3-10	• 5:10		• 1:15		• 1:58		• 0:14		• 5:00
3-11		• 0:05		• 0:20		• 1:45		• 0:10	
3-12		• 0:15		• 0:07		• 0:05		• 0:05	
3-15	9,16 •		• 0:40		• 3:10		• 0:22		• 0:40
3-16									
3-19	9,9 •	• 0:04	• 0:35	• 0:07	• 2:30	• 0:28		• 0:55	• 1:30
3-20		• 0:35			• 1:06	• 0:15		• 0:05	
3-23	10,9 •	• 0:11	• 1:30	• 2:10	10,9 •	• 3:50	5:15	• 2:30	10,9 •
3-26	8,10 •	• 2:30	• 0:05	• 0:25	• 2:15	• 1:45	5:20	• 2:15	• 4:45
3-30	• 8:10	• 0:20	• 1:15	9:0 •	• 0:20	• 1:10	10:15	12:0 •	• 2:20
4-3	9:0 •	11:0 •	• 0:45	11:0 •	• 1:15	• 0:25	9:0 •	10:0 •	• 1:00
4-7	12:0 •	• 1:00	• 8:00	• 7:00	• 2:10	• 2:00	12:0 •	11:0 •	12:0 •
4-11	13:0 •	• 2:25	9:0 •	11:0 •	•	• 0:45	10:0 •	10:0 •	11:0 •
4-15	13:0 •	• 0:35	• 3:15	• 0:55	• 3:50	• 4:10	12:0 •	• 8:10	• 1:00
4-19	13:0, •	• 0:30	• 1:10	• 3:00	• 3:20	• 2:45	12:0 •	• 0:10	• 2:00
4-22	13:0 •	11:0 •	• 1:02	• 0:50	• 4:10	• 1:15	12:0 •	13:0 •	• 3:02
4-26		• 1:05	• 1:45	• 0:10	• 5:00	• 0:25		11:0 •	• 0:20
4-29		• 0:45	12:0 •	• 1:45	• 5:25	12:0 •		12:0 •	18:0 •
5-2		• 0:20	15:0 •	• 0:15	• 0:45	• 0:30		12:0 •	• 1:10
5-8		12:0 •	12:0 •	• 0:50	• 0:30	• 0:40		• 0:50	12:0 •
5-14		9:0 •	• 5:04	9:0 •	• 1:00	9:0 •		9:0 •	9:0 •
5-22		• 0:10	• 0:15	• 9:00	• 2:00	• 2:00		10:0 •	12:0 •
5-30		• 0:30	• 0:30	9:0 •	9:0 •	• 0:30		9:0 •	9:0 •
6-7		• 0:10	• 0:04	10:0 •	11:0 •	• 1:10		12:0 •	• 1:30
6-11		10:0 •	12:0 •	• 1:30	• 9:00	• 0:15		10:0 •	• 2:14
6-15		10:0 •	• 8:10	10:0 •	12:0 •	• 1:00		10:0 •	• 0:10
6-19		10:0 •	11:0 •	• 9:00	10:0 •	• 1:50			10:0 •
6-25		• 0:35	10:0 •	• 1:15	• 3:35	10:0 •			10:0 •
7-1		• 0:40	11:0 •	• 0:20	• 1:25	• 2:50			10:0 •
7-7		9:0 •	• 8:10	9:0 •	• 2:20	• 2:15			• 3:30
7-13		9:0 •	• 0:13	9:0 •	• 8:00	• 2:00			• 1:35
7-18		10:0 •	• 4:20	• 2:15	• 2:55	• 7:45			10:0 •
7-23		12:0 •	9:0 •	10:0 •	• 2:55	• 8:20			• 0:30
7-28		12:0 •	10:0 •	• 5:20	11:0 •	10:0 •			10:0 •
8-3		12:0 •	10:0 •	• 5:00	10:0 •	10:0 •			10:0 •
8-8			12:0 •	10:0 •	• 8:00	12:0 •			10:0 •
8-14			12:0 •	9:0 •	9:0 •	9:0 •			• 2:15
8-20				11:0 •	• 8:00	18:0 •			10:0 •
8-22				10:0 •	9:0 •	10:0 •			11:0 •
8-25				12:0 •	• 1:55				• 0:17
8-30				12:0 •	30:0 •				10:0 •
9-4					20:0 •				10:0 •
9-10					10:0 •				12:0 •
9-19					12:0. •				10:0 •
9-23					Died				10:0 •
10-10									10:0 •
11-21									10:0 •
12-2									9:0 •
1-12									15:0 •
1-23									15:0 •

Fig. 2

for the purpose of ascertaining whether castrated males, after a long series of failures on repeated tests, would copulate sporadically in a series of tests, as many of them do after a relatively short series of failures. The results of this variation from the usual technique will be noted later.

V. DATA FROM THE EXPERIMENT

The results of this experiment are presented graphically in figures 1 to 5 inclusive. Each column embodies a complete record of post-castration copulation by the animal whose number

ANIMAL NO	19	20	21	22	23	24	25	26	26A
3-8	• 0.27		• 1:37	• 0:35			• 0:52		
•3-9		9,9, •		• 0.45		9,9. •			
3-10		• 1.00		• 0.10		• 12:00			
•3-11	• 0.04		• 0.15		• 0.05		9,9 •	16:0 •	• 2:00
3-12	• 0.05		• 0.05		• 0.10		• 0.45	• 11:15	• 0:12
3-15		• 0.37		• 0.07		• 5:15			
3-16	• 0.04		• 0.21		• 0.15		• 8:30	• 0.40	• 0:45
3-19		• 1.00		• 2.55		9,9 •			
3-20	• 2.30		• 0.25		• 0.09		9,12. •	• 1:10	18:0 •
3-23	• 5.23	• 0:30	• 3.20	• 4.55	• 1.10	9,9 •	• 1:25	• 0:10	• 0:30
3-26	• 1:20	• 3:05	• 0.05	• 0.10	• 0.15	• 1:35	• 0:40	• 1:20	• 0.20
3-30	• 0.40	• 0.25	• 0:07	• 0.20	• 3:00	9,9. •	• 0:11	12:0 •	• 1:00
4-3	• 0.20	9:0 •	• 1:30	• 0.22	9:0 •	• 1:30	• 0:06	• 2:40	• 0:15
4-7	• 0:15	12:0 •	11:0 •	• 6.00	11:0 •	12:0 •	13:0 •	11:0 •	• 0:40
4-11	11:0 •	10:0 •	10:0 •	• 6.05	11:0 •	10:0 •	12:0 •	• 4:55	11:0 •
4-15	• 1:14	11:0 •	12:0 •	• 0.50	14:0 •	11:0 •	12:0 •	• 6:50	12:0 •
4-19	19:0 •	• 1:30	• 0.09	• 0.30	• 0.45	12:0 •	12:0 •	• 0:35	18:0 •
4-22	• 1:30	• 0:37	10:0 •	• 9.00	• 0.50	16:0 •	12:0 •	• 2:30	11:0 •
4-26	• 0:20	• 0:30	• 0.35	• 1.20	11:0 •	14:0 •	12:0 •	15:0 •	12:0 •
4-29	• 0:30	• 3:00	• 0.15	• 0.15	• 0:25	16:0 •	12:0 •	12:0 •	12:0 •
5-2	• 3.25	• 0:30	• 2.20	• 1.20	• 1.10			• 3:00	12:0 •
5-8	• 0.30	• 0.40	• 1:20	• 1.00	• 0:15			11:0 •	
5-14	9:0 •	• 0:20	• 0:44	• 3:43	9:0 •			• 8:00	
5-22	• 0:45	• 0:15	• 0:15	• 4:50	• 0:30			• 9:00	
5-30	9:0 •	9:0 •	• 0:15	9:0 •	9:0 •			• 8:00	
6-7	• 4:55	• 1:30	• 0:20	9:0 •	10:0 •			• 0:20	
6-11	11:0 •	• 1:00	Photo	• 1:15	11:0 •			• 5:20	
6-15	10:0 •	10:0 •		• 0:50	12:0 •			• 0:40	
6-19	10:0 •	• 0:35		• 0:50	12:0 •			• 3:00	
6-25	• 2:30	• 0:50		• 6:30	12:0 •			10:0 •	
7-1	• 0:20	• 1:40		• 4:25				• 6:20	
7-7	(——)	9:0 •		• 0:15				• 7:40	
7-13	9: •	• 0:50		• 1:15				9:0 •	
7-18	9: •	• 0:50		• 8:20				• 3:10	
7-23	10: •	9:0 •		• 1:35				9:0 •	
7-26	10: •	11:0 •		10:0 •				11:0 •	
8-3	10: •	10:0 •		11:0 •				11:0 •	
8-8	10: •	10:0 •		• 3:00				• 2:30	
8-14	10: •	10:0 •		9:0 •				9:0 •	
8-20	10: •	10:0 •		• 6:40				14:0 •	
8-22	10: •	10:0 •		• 4:22				11:0 •	
8-25				• 2:45				10:0 •	
8-30				10:0 •				10:0 •	
9-4				20:0 •				10:0 •	
9-10				12:0 •				10:0 •	
9-19				10:0 •					
9-23				10:0 •					
10-10				10:0 •					
11-21				12:0 •					
12-2				(——)					
1-12				15:0 •					
1-23				15:0 •					

FIG. 3

appears at its top. Dots at the left and right of each column respectively signify that positive and negative results were gotten in tests for copulation. The number of minutes and seconds elapsing before the first copulatory act after placing the receptive female and castrated male together is indicated by the arabic numbers at the right of dots. Similarly, numbers at the left of dots indicate time elapsing before a test was terminated if the male failed to copulate. Whole numbers separated by commas signify repetition of the test one-half hour later.

To read the charts proceed as in the following example, for figure 1, column 1. Animal 1, given a pre-castration test for copulation on March 8, copulated within 1 minute 55 seconds

ANIMAL NO	28	29	30	31	32	33	34	35	36
4-22		● 0:07		● 0:25		● 1:40		● 0:35	
5-2	● 1:20	● 0:30	● 4:25	● 0:55	● 0:15	● 1:20	● 1:15	● 0:40	● 0:15
5-5	● 0:45	● 0:15	● 0:15	● 0:45	● 1:00	● 1:15	● 1:10	● 0:15	● 0:15
5-11	● 0:55	● 0:10	● 0:15	● 2:00	11:0 ●	● 0:10	● 11:00	● 0:10	● 0:15
5-16	● 0:50	● 0:20	● 3:20	● 1:40	10:0 ●	● 0:05	● 0:55	15:0 ●	● 2:00
5-22	11:0 ●	● 0:10	11:0 ●	9:0 ●	12:0 ●	● 0:10	● 1:20	● 0:10	● 0:25
5-30	10:0 ●	10:0 ●	9:0 ●	10:0 ●	10:0 ●	● 1:30	● 1:50	● 0:50	● 0:20
6-7	● 5:20	● 1:10	12:0 ●	12:0 ●	● 1:40	● 0:40	● 1:40	10:0 ●	● 0:30
6-11	● 9:10	12:0 ●	12:0 ●	12:0 ●	10:0 ●	● 1:00	12:0 ●	12:0 ●	● 0:15
6-15	● 7:50	● 2:50	12:0 ●	12:0 ●	12:0 ●	● 0:10	● 8:10	● 0:05	● 1:20
6-19	● 2:30	11:0 ●		10:0 ●	10:0 ●	● 1:10	● 9:50	10:0 ●	● 0:10
6-25	● 1:30	● 3:00		10:0 ●	● 8:10	● 0:10	● 0:20	● 4:00	● 2:10
7-11	● 0:20	10:0 ●		10:0 ●	10:0 ●	● 0:30	● 1:20	● 1:50	● 0:35
7-6	● 3:40	● 1:00		10:0 ●	10:0 ●	● 0:25	● 0:20	● 3:20	● 0:55
7-13	10:0 ●	● 2:35		10:0 ●	9:0 ●	● 2:30	● 1:15	9:0 ●	
7-18	● 1:15	● 0:10			10:0 ●	● 0:10	● 0:40	10:0 ●	● 0:20
7-23	● 1:45	● 1:50			10:0 ●	● 0:30	● 1:30	10:0 ●	● 0:05
7-28	● 2:30	● 5:00			10:0 ●	● 0:20	● 4:20	10:0 ●	● 0:15
8-3	● 5:00	12:0 ●				11:0 ●	● 5:45	10:0 ●	● 0:40
8-8	● 2:30	9:0 ●				● (——)	9:0 ●	10:0 ●	● 0:45
8-14	● 2:15	12:0 ●				9:0 ●	9:0 ●		● 0:45
8-20	10:0 ●	9:0 ●				10:0 ●	● 0:20		● 0:20
8-22	9:0 ●	10:0 ●				12:0 ●	● 0:40		● 2:20
8-25	10:0 ●	10:0 ●				12:0 ●	● 0:45		● 0:36
8-30	● 2:35	10:0 ●				12:0 ●	10:0 ●		● 2:30
9-4	10:0 ●						15:0 ●		11:0 ●
9-10	15:0 ●						● 1:30		● 0:15
9-19	10:0 ●						● 3:30		10:0 ●
9-23	10:0 ●						10:0 ●		10:0 ●
10-10	10:0 ●						● 2:45		● 8:20
10-22	10:0 ●						● 1:25		● 1:20
11-12	9:0 ●						● (——)		● (——)
11-23	9:0 ●						● 2:05		9:0 ●
12-2							10:0 ●		9:0 ●
1-12							● 0:30		15:0 ●
1-23							9:0 ●		9:0 ●
2-10							15:0 ●		15:0 ●
2-14							15:0 ●		15:0 ●
4-3							25:0 ●		15:0 ●

FIG 4

after the female was introduced into his cage (indicated by dot on left and arabic numerals on right of column.) Three days later, March 11, he was castrated (asterisk before this date) and on that same night tested again for copulation; he copulated 1 minute 37 seconds after the female was put into his cage. His next test fell on March 12 and positive results are indicated again. Continuing down the column it is apparent from circles at the left of the column that positive results were obtained regularly until April 3 when during a 10 minute period he failed to copulate. On April 7 and 11 negative results again followed in test periods of eleven minutes but positive results were gotten again on April 15. This, in turn, was followed by negative results thereafter until May 2 when positive results were obtained. After another failure on May 8, a series of positive results follow for the next five periods, etc. The last positive result for animal 1 was

ANIMAL NO.	37	38	39	40	41	42	43	44	45
4-22	● 1:47		● 6:04		● 1:10		● 1 12		● 2:10
●5-2	● 0:30	9,9 ●	● 8:00	● 2:50	● 0:15	● 1:30	● 0:15	● 2.0	● 0:15
5-5	● 0:15	● 2:00	9,9 ●	● 0:45	● 0:33	● 1:25	● 0:20	10:0 ●	● 0:32
5-11	● 0:30	● 0:45	● 0:45	● 0:46	● 1:30	● 0:15	● 0:10	9:0 ●	● 0:10
5-16	● 0:10	● 0:20	● 7:00	● 0:55	● 1:30	● 0:17	● 0:07	9:0 ●	● 0:10
5-22	● 1 45	● 5:30	● 2:00	● 8:00	● 1:40	9:0 ●	● 0:05	10:0 ●	● 0:46
5-30	10:0 ●	10:0 ●	● 4:45	● 1 30	10:0 ●	10:0 ●	● 0:20	15:0 ●	● 2:10
6-7	10:0 ●	● 3:45	● 5:00	● 1 50	10:0 ●	● 6:00	● 0:20	10:0 ●	● 1:30
6-11	11:0 ●	● 0:10	12:0 ●	Photo	12:0 ●	10:0 ●	● 1:11	10:0 ●	● 0:53
6-15	12:0 ●	12:0 ●	● 8:00		10:0 ●	10:0 ●	10:0 ●	Photo	
6-19	12:0 ●	● 0:45	● 1:20		● 0:10	10:0 ●	10:0 ●		10:0 ●
6-25	12:0 ●	11:0 ●	● 1:15		● 0:20	10:0 ●	● 0:50		● 0:35
7-1	12:0 ●	● 1:50	● 0:40		● 3:00	12:0 ●	10:0 ●		● 6:20
7-6		● 0:30	11:0 ●		9:0 ●	12:0 ●	● 2:30		● 1:35
7-13		Photo	● 3:00		9:0 ●		● 1:30		● 0:55
7-18			9:0 ●		● 5:00		● 3:05		● 1:48
7-23			● 1:25		10:0 ●		10:0 ●		● 0:30
7-28			● 0:30		11:0 ●		10:0 ●		10:0 ●
8-3			10:0 ●		10:0 ●		10:0 ●		10:0 ●
8-8			10:0 ●		● 0:35		● 0:45		● 1:30
8-14			12:0 ●		9:0 ●		9:0 ●		9:0 ●
8-20			9:0 ●		10:0 ●		10:0 ●		12:0 ●
8-22			● 2:50		11:0 ●		9:0 ●		11:0 ●
8-25			11:0 ●		10:0 ●		10:0 ●		10:0 ●
8-30			12:0 ●		10:0 ●		● 9:40		12:0 ●
9-4			10:0 ●		10:0 ●		11:0 ●		10:0 ●
9-10			12:0 ●				12:0 ●		10:0 ●
9-19			10:0 ●				10:0 ●		
9-23							10:0 ●		
10-10							10:0 ●		
10-22							10:0 ●		
11-12							9:0 ●		
11-23							9:0 ●		
12-2							15:0 ●		
1-12							12:0 ●		
1-29							15:0 ●		

Fig. 5

obtained on November 21, *somewhat over* eight months from the date of castration. He was last tested on April 3, or about thirteen months after castration; results were negative, as on all previous tests after December 2.

To indicate numerically the rate at which animals ceased to copulate after castration the data of figures 1 to 5 were condensed to tabular form. Table 1 gives the number of cases whose last copulatory act fell within the first, the second, the third, etc., month after castration. For purposes of comparison we have kept separate the records of animals with and those without pre-copulatory sexual experience, yet little stress is laid upon their apparent differences because relatively small numbers of cases are involved. This makes small differences highly unreliable. Taking the data as they stand, however, it is clear that a larger per

TABLE 1

Number and per cent of male rats last copulating during the first, second, third, etc., month after castration

| | MONTH AFTER CASTRATION WHEN ANIMAL LAST COPULATED | | | | | | | | |
	First	Second	Third	Fourth	Fifth	Sixth	Seventh	Eighth	TOTAL
With sex experience prior to castration	5	2	5	6	2	0	0	1	21
Without sex experience prior to castration	9	3	0	1	2	5	0	1	21
All animals	14	5	5	7	4	5	0	2	42
Per cent of total	33.3	11.9	11.9	16.7	4.8	11.9	0	2.4	100

TABLE 2

Reduction in per cent of positive results from tests for copulation given during successive months after castration

| | MONTHS AFTER CASTRATION | | | | | |
	First	Second	Third	Fourth	Fifth	Sixth
Per cent of animals copulating when tested in successive months after castration:						
Animals with sex experience prior to castration	78.2	60.0	50.0	20.1	11.7	14.0
Animals without sex experience prior to castration	65.8	61.9	54.5	68.2	37.9	16.5
All animals	72.0	60.9	51.9	45.9	24.1	15.8

cent of cases without pre-castration indulgence lost copulatory ability during the first month than was the case with those having pre-castration experience. After the first month, the animals of both groups are about equal although a slight superiority rests with that deprived of pre-castration experience.

Reviewing the combined data of the two groups it will be seen that one third of the original number of animals dropped out during the first month after castration; by the end of the second month, 45 per cent had dropped; 57 per cent by the end of the third; 79 per cent by the end of the fifth; and 90 per cent by the end of the sixth month. No cases were left at the end of the eighth month.

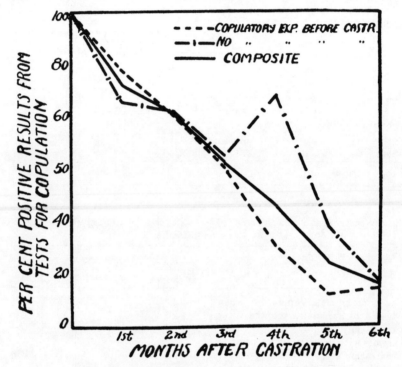

FIG. 6. PER CENT OF TESTS FOR COPULATION YIELDING POSITIVE RESULTS DURING SUCCESSIVE MONTHS AFTER CASTRATION

Considering individual performances it becomes apparent that from the first day after castration to the time all test results become negative there is, as a rule, an increasing number of negative test results, yet, as we shall see this is not the invariable course of events. Note for example the cases of animal 1, (fig. 1). In less than one month his record shows periods of sexual inactivity interspersed with periods of activity. For several months this irregular condition prevailed and eventually terminated with total sexual inactivity. Our laboratory notes indicate, furthermore, that after the first month the rate of copulatory performances per unit of time diminished rather markedly.

By way of contrast consider the record of either animal 2, 3, or 5, the littermates of no. 1. All of these animals copulated regularly for the first and second weeks, then, during the third week, dropped out abruptly. No. 4 and 8 copulated on only two test-nights after castration. Animals 12, 13, 14, 15 present a picture which I believe may be taken to be fairly typical of the composite data from the group as a whole. For a short time after castration the animals copulated readily when tested. following this period is one in which positive and negative results are interspersed somewhat irregularly. Finally, there is a period in which only sporadic instances of copulation are found and these in turn lead to total cessation of copulation. Summarily stated, the negative results of the third stage practically reverse the positive results of the first period. This tendency is clearly indicated by the data of table 2 and the graphs of figure 6. The data of this table are based upon the records of all animals remaining in the experiment during the whole of the first month after castration, the whole of the second month, the third month, etc. Compilations end with the sixth month since only two cases are active thereafter.

Immediately after castration the per cent of positive test results was close to 100 per cent but during the first month negative results increase to such an extent that for the month as a whole the per cent drops to 78.2 and 65.8 respectively for groups that had and did not have pre-castration indulgence. During the succeeding months positive results became fewer and fewer.

It is of interest to note from figure 6 that the group without pre-castration experience has a lower per cent of positive results for the first and second months but a higher per cent for the third, fourth, fifth, and sixth months. The most plausable explanation of this suggested by the data at hand is as follows. In the group without pre-castration indulgence were several cases, probably chance grouping, with weak potentia coeundi. Most of their tests yielded negative results yet they were carried through the first month and into the first of the second month before satisfying the requirements (see page 374) for being discarded. Likewise in this same group were several cases with strong potentia coeundi and they give a very high per cent of positive returns. Whereas the large number of weak cases pulled the average down at the outset these latter, after the weak ones were dropped, hold it up to a relatively high level. In the group with pre-castration experience there was not a great number of exceptionally weak or exceptionally strong cases, hence their curve drops more uniformly from the first to the sixth month. This interpretation may seem quite arbitrary since we have not definitely demonstrated that the apparent difference is not one that would persist even if larger groups of animals had been studied, in which case it might be attributed, other factors being eliminated, to the factor of pre-castration indulgence. Against the latter interpretation are the facts that both groups have individuals dropping out early because of weak copulatory ability and animals that copulate for a long time after castration; that both present great individual variability which frequently gives rise to apparent differences in group averages based upon relatively small numbers of cases; and that there was no characteristic difference between the sexual activity of the two groups observable at the time they underwent their periodic tests for copulation.

VI. ATROPHIC CHANGES IN THE ACCESSORY SEXUAL APPARATUS

We raised the questions as to whether frequent repetition of the copulatory act would serve to perpetuate the usual sexual aggressiveness and ability to copulate seen in recently castrated males and prevent the atrophic changes normally appearing in the accessory sexual apparatus soon after castration. Negative answers to these questions are indicated by the data at hand. Sexual aggressiveness dwindles and copulatory ability wanes and disappears under conditions adequate to insure them in normal animals. In this respect the copulatory function appears to differ radically from the so-called "habits" or acts of skill and precision

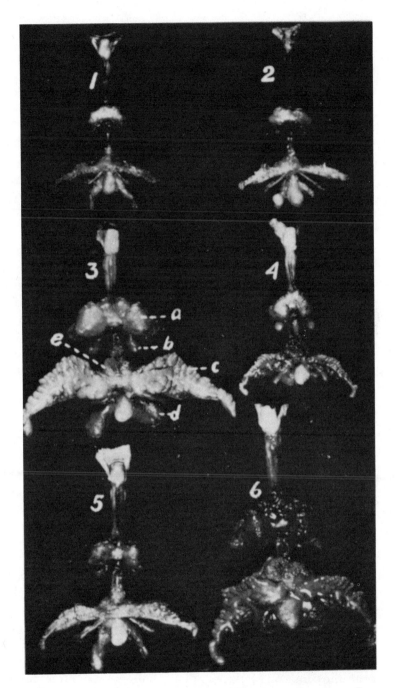

FIG. 7. ACCESSORY SEXUAL APPARATUS OF NORMAL AND CASTRATED RATS
(*a*) M. ischio cavernosum penis; (*b*) Cowper's gland; (*c*) seminal vesicle; (*d*) coagulatory gland; (*e*) prostate.

382

37

which appear to wax in strength, other things being equal, under conditions of repetition.

To answer the second question the reproductive organs of normal males were contrasted with those of potent and impotent castrated males. Figure 7 gives life-sized photographs of the reproductive tracts of three animals of the same age and approximately the same weight. Nos. 1 and 2 were castrated at the age of ninety-five days and had copulated continuously when tested for a period of seventy-five days, except for the sporadic negative results found so commonly with rats of otherwise good sexual vigor. No. 3, a normal control of the same age and weight as nos. 1 and 2 was in full possession of sexual vigor at the time of sacrifice. Not only is his reproductive tract normal in appearance but its size is similar to that of other healthy and potent young males. From it one can see that the tracts of 1 and 2 are far below the development of a normal male. This is due both to arrest of development and post-castration atrophy. Animal 4 is a normal male ninety-five days of age and of the approximate weight of animals 1 and 2 when they were castrated. We may infer that at the time the latter were castrated their accessory reproductive organs were approximately the size of that of no. 4 and in a similarly healthy condition. Since they are now somewhat smaller than his and much smaller than those of no. 3 the inference of arrest of development and atrophy would seem to be legitimate. Although these animals copulated with more than average vigor, their reproductive tracts underwent marked atrophy. Therefore, *copulatory activity alone is not sufficient to keep the accessory reproductive organs in a normal state of development.*

The same fact is illustrated again by the tract of no. 5, figure 7. This male, weighing 380 grams and over two years of age had copulated on alternate evenings for a period of seventy-seven days after castration. Note especially the delapidated state of the seminal vesicles and prostate and the relatively small size of the muscle ischio-cavernosum penis of this reproductive tract. Contrast with it the tract of no. 6, a male two and one-half years old, weighing approximately 360 grams, and sexually vigorous and fertile at the time of sacrifice. Since both reproductive tracts probably had reached their full development and the animals were apparently comparable in every way until no. 5 was castrated, we may infer that the apparent differences are primarily if not wholly due to atrophic changes caused by the castration of no. 5.

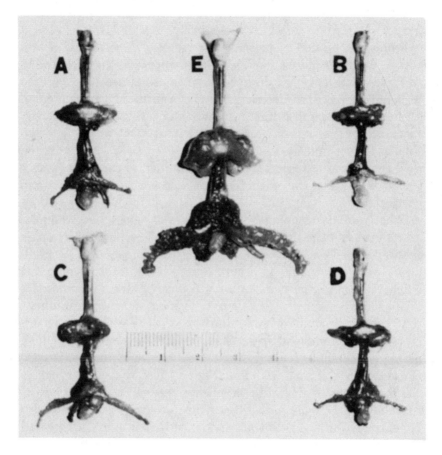

FIG. 8. ACCESSORY SEXUAL APPARATUS OF NORMAL AND CASTRATED RATS

If this inference is correct we may again conclude that *atrophy occurs despite the fact that ample opportunity for exercise of these organs was given after the time of castration.* Figure 8 brings out points essentially like the foregoing. Animal E is a control of the same age and approximately the same size as animals A and

B which belong to this series of experiments (no. 7, fig. 1 and no. 21, fig. 3, respectively). Animals C and D also belong to this series (nos. 44 and 40 fig. 5). Prior to sacrifice A and B had been castrated eighty-six days and C and D, thirty-one days. Referring to the charts containing their records it will be seen that B (no. 21) had, for a period of about twelve weeks, given an almost unbroken succession of positive test results. The record of animal A, although fairly good, was less impressive than that of B. In fact A had just been dropped from the tests because of failure to copulate for five successive tests. Animal D had copulated at the time of each test during the thirty-one days after castration. In marked contrast with his record is that of C which failed to copulate except on the night following castration. As compared with the reproductive tract of normal animal E the reproductive tracts of all give ample evidence of arrest of development and post-castration atrophy.

Other animals dropping from the group of sexually active males near the end of the first month after castration showed atrophic changes already advanced as much as those of animal C (fig. 8), hence we were unable to secure a series of protocols depicting different stages of post-castration atrophy. With a series of this kind at hand one might be able to determine whether the animals that cease to copulate early have poorly developed accessory reproductive organs whereas those remaining active for a long period of time have relatively better developed organs at the time of castration. If such were found to be the case it would suggest the probability of a causal relationship between the loss of sexual aggressiveness on the part of the male and atrophy of the organs of reproduction. As contrary evidence it is important to recall that animals B and D were copulating vigorously at the time of sacrifice despite the fact that their reproductive organs were no less atrophic than those of A and C which had ceased to copulate. Furthermore there is no evidence contrary to the accepted belief that post-castration atrophy becomes manifest in all rats soon after castration. Further intensive research along this line would lead to experimental evaluation of the theory expressed by Havelock Ellis (1900) and others that the source of one component of the sexual "drive" is afferent stimuli arising in distended glands of the reproductive tracts. If this theory has a factual basis one might infer that collapsing of these glands after castration reduces the amount of stimulation resulting from their turgidity and thus weakens the sexual drive by eliminating one of its components.

We were unable to determine the time at which reflexes for erection of the penis are lost after castration. The nature of the copulatory act of the rat is such and so rapid is its performances that one cannot, as a rule, be sure that intromission takes place except by subsequent inspection of the female reproductive tract for traces of the vaginal plug or spermatozoa. The latter we naturally would not look for in castrated males and the former were not looked for, as it was believed that they would be of relatively rare occurence. If accurate records of the history of the erection reflexes could be made for castrated rats eventually ceasing to copulate, one might determine experimentally the importance of contretactive impulses (Ellis, 1900) arising from afferent stimulation of the erect penis for the persistence of copulatory ability.

VII. PRACTICAL SIGNIFICANCE OF STUDIES OF THIS CHARACTER

It would seem desirable that the rate and variability with which copulatory ability normally wanes in castrated males (whether man or the lower animals) be rather accurately determined in a group sufficiently large to indicate central tendencies and amounts of variability for the species in question. These norms would facilitate more accurate measurement of the efficiency of testicular transplants, injections, extracts, and other agencies employed in gonadal therapy to restore or enhance sexual vigor because a background of experimental data is needed to differentiate between sexual vigor persisting after castration (and probably independent of the immediate influence of the gonadal hormone) and that induced by the therapeutic measures employed.

CONCLUSIONS

1. Forty-five male rats castrated at the age of ninety days demonstrated the ability to copulate after castration.

2. Much variability in the periods of time sexual aggressiveness and copulatory ability persisted was found. Some ceased to copulate during the first week after castration; others copulated during the eighth month thereafter. Considering the groups as a whole, 33 per cent of the castrates had ceased to copulate before the end of the first month; 45 per cent, at the end of the second month; 57 per cent, the third; 74 per cent, the fourth; 79 per cent, the fifth; and 91 per cent, the sixth.

3. No substantial and reliable difference in the retention of copulatory ability in males allowed a limited amount of copulation prior to castration versus males not allowed pre-castration indulgence was found.

4. Exercise of the copulatory function alone will not insure the persistence of sexual aggressiveness and the copulatory act in castrated males. In this respect sexual activity differs fundamentally from so-called "habits" or acts of skill and precision. Likewise exercise of the accessory sexual apparatus does not prevent onset of the usual post-castration atrophy.

REFERENCES

ELLIS, HAVELOCK: Studies in the Psychology of Sex. Vols. 3, 4, 1900.

LIPSCHÜTZ, ALEXANDER: The Internal Secretions of the Sex Glands. 1924.

STEINACH, E.: Untersuchungen zur Vergleichenden Physiologie der Männlichen Geschlechtsorgane. Arch. f. die ges. Physiol., 1894, lvi, 304–338.

STONE, CALVIN P.: The awakening of copulatory ability in the male albino rat. Amer. Jour. Physiol., 1924, lxviii, 407–424.

Reprinted from *J. Comp. Physiol. Psychol.*, **42**, 433–453 (1949)

EFFECTS OF DIFFERENT CONCENTRATIONS OF ANDROGEN UPON SEXUAL BEHAVIOR IN CASTRATED MALE RATS[1]

FRANK A. BEACH, *Yale University*

AND

A. MARIE HOLZ-TUCKER, *American Museum of Natural History*

4

Received October 11, 1948

Several investigators have employed testosterone propionate to evoke copulatory behavior in castrated male animals (Shapiro, 1937; Moore and Price, 1938; Stone, 1939). The majority of studies have dealt with the rat, and the dosages have varied from 100 to 1250 micrograms of hormone per day (1 microgram = .001 mg). It is well established that large amounts of androgen will maintain or restore normal sexual behavior in the adult castrate but we still do not know just how much hormone is required to achieve this result.

The present experiment was conducted to measure the sexual behavior of castrated male rats receiving various amounts of testosterone propionate by daily injection. We hoped to define the "adequate maintenance dose" which would keep behavior exactly at preoperative levels, and in addition to observe the behavioral effects of holding the hormone level below or above the minimal concentration needed for maintenance.

PROCEDURE

Subjects for the investigation were male rats, 90 to 100 days of age at the beginning of the tests. The animals were drawn from a colony which for ten years has been inbred to the extent of avoiding the introduction of any new stock, although no systematic plan of sibling crosses or back crosses has been followed. The strain was derived from a cross between wild *Rattus norvegicus* and tame albino rats from the Wistar Institute. For five or six years preceding this experiment males used as sires were individuals chosen for their willingness to mate promptly and vigorously. If one assumes that such characteristics are determined in part by heredity it follows that our practice may have resulted in a gradual increase in the sexual excitability of the strain.

Selection of the experimental animals was preceded by a series of preliminary sex tests in which males were placed singly with a female in heat and observed for the execution of mating responses. Individuals which failed to show any copulatory activity after two or three tests were discarded. This procedure eliminated sexually-sluggish animals and perhaps some others which were emotionally disturbed by the general testing situation.

The regular sex tests began after 52 suitably active males had been selected. These tests were conducted once each week in a quiet room with adequate ventilation and lighting. The circular observation cage was 34 inches in diameter, 30 inches high, and had no cover.

[1] This investigation was supported by a grant from the Committee for Research in Problems of Sex, National Research Council. The experiment was conducted while the senior author held the Chairmanship of the Department of Animal Behavior at the American Museum of Natural History. The junior author conducted most of the tests, performed the operations, and made the hormone injections. The senior author's responsibility included planning the experiment, interpretation of the data and preparation of the manuscript.

433

43

Stimulus animals were spayed females which had been brought into heat by the administration of ovarian hormones (Beach, 1942). Each day the injected females first were tested with non-experimental males of known sexual vigor. Only those individuals that displayed normal heat behavior in response to the mating attempts of the "indicator male" were employed in the experimental tests.

A male was placed in the observation cage and allowed a three-minute adaptation period and then a sexually-receptive female was quietly deposited in the center of the cage. Each test lasted for 10 minutes from the time of the first complete or incomplete copulation by the experimental male. In a complete copulation intromission is achieved. Incomplete copulations involve mounting the female and executing pelvic thrusts but insertion is lacking. Ejaculation is not involved in either type of response. If mounting responses did not occur within 10 minutes after the introduction of the female, the test was terminated and scored as negative. The behavioral items noted and the various measures employed will be described in the presentation of experimental results.

At the conclusion of the sixth preoperative test all males were castrated, and hormone injections were begun 48 hours later.

TABLE 1

Schedule of androgen treatment after castration

GROUP	N	MICROGRAMS OF TESTOSTERONE PROPIONATE PER DAY	
		First post operative period (9 weeks)	Second post operative period (10 weeks)
I	11	0	1
II	10	25	75
III	10	50	0
IV	10	100	omitted
V	11	500	0

The operated animals were divided into five experimental groups which were equated as closely as possible in terms of the average frequency of copulations per preoperative test. Each male was injected subcutaneously once every 24 hours with .2 cc. of sesame oil. The concentration of testosterone propionate contained in this amount of oil ranged from 25 to 500 micrograms for various groups and a control group received plain oil with no hormone.[2]

Tests were continued for a period of nine weeks, at which time the hormone dosages were changed. A final ten-week period concluded the experimental tests. The amounts of androgen administered daily to males in each group are shown in table 1.

Animals in Group IV received 100 micrograms per day during the first nine weeks after operation and 5 micrograms daily for the remainder of the experiment. These rats showed no response to the smaller dose. Their behavior was quite similar to that of castrates in Groups III and V who were given no hormone in the second post operative period. However, Group I exhibited definite increases in sexual activity in response to daily injections of 1 microgram of testosterone propionate. This suggests that treatment at a high dose level may render animals insensitive to much lower concentrations but our data are insufficient to establish the point. We do not here report the scores of Group IV rats during the second post operative period because they were in no way different from those of Groups III and V and because inclusion of their records would have obscured an otherwise clear

[2] The hormone preparations used in this study were generously supplied by Dr. Edward Henderson of Schering Corporation, Bloomfield, N. J.

relationship between hormone dosage and frequency of sexual responses. It is felt that exclusion of these data is justified by the fact that Group IV was the only group in which androgen concentration was reduced but not brought to zero,—a procedure which evidently yields results quite different from (1) increasing the dosage or (2) withdrawing the hormone altogether.

All animals were sacrificed within 24 hours after the final test. Completeness of testis removal was checked by macroscopic inspection and samples of seminal vesicle tissue were removed for histological study.

RESULTS

The experimental results will be discussed in terms of the various behavioral items studied.

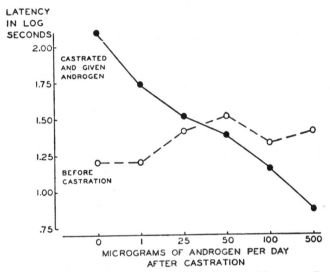

FIG. 1. AVERAGE DELAY PRECEDING THE FIRST SEXUAL MOUNT IN PRE- AND
POST-OPERATIVE TESTS

Changes in Latency

Latency is defined as the number of seconds elapsing between the time that the receptive female was placed in the observation cage and the time that the male executed his first sexual mount (complete or incomplete copulation). This measure proved to be a very sensitive indicator of the amount of hormone administered to castrated males.

In order to normalize the distributions, raw scores have been translated into logarithms. Figure 1 is a graphic representation of the average log of latencies for each group before and after castration. Some differences existed between the averages for the various groups in preoperative tests, and after castration the average latency was directly proportional to the size of the hormone dosage. During the first preoperative test many animals showed long latencies which seemed to reflect emotional disturbance due to the strangeness of the environment. Therefore the scores in this test were not used in computing group averages. Immediately after operation the latency scores tended to be quite variable, but they were stabilized after the castrates had been receiving hormone for a week or ten days. Accordingly the records of the first postoperative test have been omitted from these particular calculations. Preoperative averages are based upon the results of tests 2 to 6 and the values representing postoperative performances of all groups are based upon nine tests conducted from 2 to 10 weeks after castration. In addition we have inserted the average score for Group I during the second post operative period at which time these rats were receiving 1 microgram of testosterone propionate per twenty-four hours.

In order to determine the significance of changes in latency we have compared the average log latencies of each group before and after castration. Preoperative means are based upon scores made in the last three tests before castration, and post operative means are based upon the records of six successive tests conducted

TABLE 2

Mean log latency

GROUP	N	NORMAL (TESTS 4–6)	CASTRATE (TESTS 10–15)	DIFFERENCE	PROBABILITY	MICROGAMS OF ANDROGEN PER DAY
I	7	1.14	2.26	+1.12	< .01	0
I	8	1.22	1.66*	+0.44	.03	1
II	10	1.33	1.53	+0.20	.04	25
III	10	1.47	1.34	−0.13	.11	50
IV	10	1.28	1.15	−0.13	.06	100
V	11	1.30	0.78	−0.52	< .01	500

* Based on tests 19 to 25.

after the animals had been receiving androgen for one month. In addition there are included the records of Group I during the last six tests in the second postoperative period when 1 microgram per day was being injected. The post operative change in mean values for each group was evaluated in terms of Student's *t* test, and using Snedecor's tables we have calculated the probability that the differences could have occurred by chance.

Table 2 presents the results of these comparisons. Castration followed by no treatment resulted in an average increase in lateny which would be expected to occur by chance less than 1 time in 100 and is therefore clearly significant in the statistical sense. Apparently 50 micrograms of testosterone propionate per day was adequate to maintain latency scores at preoperative levels. Lower doses were associated with latencies which, while shorter than would be expected without any replacement therapy, were significantly longer than normal. One hundred micrograms of androgen per twenty-four hours produced a decrease in latency scores which was significant at the 6 per cent level of confidence, and under the influence of 500 micrograms of hormone per day castrated male rats showed latencies that were definitely shorter than normal.

Data collected during the second post operative period corroborate these conclusions. Males in Group II were given 25 micrograms during the first and 75 micrograms during the second post operative period. The increase in dosage occasioned a shortening of the average log latency which was significant at below the 1 per cent level. In fact, under the influence of 75 micrograms per day these animals showed latencies shorter than they had displayed before castration (significant at the 8 per cent level).

Males in Groups III and V received 50 and 500 micrograms of androgen respectively during the first post operative period and no hormone in the second post operative period. The withdrawal of hormone was followed by increases in log latencies which were highly significant in both cases.

Changes in the copulatory response

The rat's copulatory response includes five distinct elements. (a) The male mounts the receptive female from the rear, clasping his forelegs around her sides in the lumbar region. (b) The male's forelegs are pressed downward and backward and then brought forward again in a series of very rapid "palpation" movements which help stimulate the female to elevate the perineum and thus expose the genitalia. (c) Concomitantly with the forelimb palpations, the male executes a series of short pelvic thrusts which bring the penis into contact with the genitalia of the female. (d) After several such preliminary pelvic movements the male accomplishes the brief intromission by means of a single, deep thrust. (e) Insertion usually is maintained for only a fraction of a second and the male dismounts abruptly with a vigorous backward lunge which often carries him half a foot or more away from the female. This pattern is clearly recognizable under ordinary circumstances. Ejaculation does not occur with each copulation, and when it does take place the behavior is different as subsequent description will reveal.

Per cent of each group copulating in each test: The first noticeable post operational change in sexual behavior was the complete disappearance of the copulatory response from the behavior of some males in some tests. We shall first consider the results on a "present-or-absent" basis, paying no attention to the frequency of copulations during tests in which such behavior occurred.

The trends represented in figure 2 are based upon 3-point moving averages. Results of tests 1 and 2 are not shown, but the percentage of copulators increased for all groups in the first few preoperative tests and then tended to stabilize. This probably reflects gradual adaptation to the testing situation as well as some general "conditioning" effect. In the interest of legibility we have refrained from reproducing the curves for Groups III and IV which received 50 and 100 micrograms per day respectively. The performance of Group IV males was very similar to that of Group V and the curve for Group III falls midway between those of Groups II and IV.

During the first post operative period the records of the several groups fell in a definite order which corresponded to the magnitude of hormone dosages. Because of inter-group differences in preoperative performance, comparisons between behavior in the first and second periods of the experiment must be drawn

47

with considerable caution. However, comparing each group with itself in the two periods we may tentatively conclude that daily injections of 50, 100, or 500 micrograms sufficed to maintain the proportion of copulators at preoperative levels if not to increase it somewhat.

Administration of 25 micrograms per day resulted in a decrease in the per cent of a group copulating in each test. Control injections of sesame oil had no recognizable effect, and the proportion of copulators in Group I progressively decreased until only approximately 10 per cent of the animals were responding in any given test. It should be stated that this did not reflect the persistence of copulatory behavior in a single member of the group. It was not always the same individual who copulated in successive post operative tests. Various males executed the

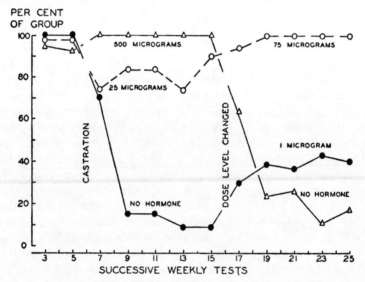

FIG. 2. PROPORTIONS OF THREE EXPERIMENTAL GROUPS SHOWING THE
COPULATORY RESPONSE

response at different times, and the post castrational decline in behavior was not a regularly progressive change in individual animals although it was for the group as a whole. For example, one male in Group I exhibited no sexual activity for the first seven weeks after castration and then, in the eighth postoperative test, displayed 3 complete and 20 incomplete copulations. Another rat's scores were negative until the fifth test after castration at which time copulation and ejaculation occurred. Similar irregularities appeared in the individual records of males in Groups III and V during the second post operative period when androgen treatment was discontinued.

Hormone dosages were changed nine weeks after castration and subsequent alterations in behavior fully supported the foregoing conclusions. An increase from 25 to 75 micrograms in the daily dosage for Group II occasioned a prompt rise in the proportion of copulators per test, and within three weeks the record of this group was equal to its own preoperative scores. Groups III and V were deprived of androgen in the second post operative period and the change in their performance was comparable to that of Group I during the first period after castration when plain oil was injected. The scores for Group I during the second post operative period show that as little as 1 microgram of testosterone propionate per day was sufficient to increase the percentage of copulators.

Frequency of copulations in positive tests: Thus far we have paid no attention to the number of times an animal copulated in a particular test. As long as the response occurred once the individual was scored as positive. We are now ready to consider the effects of castration and subsequent androgen administration upon copulatory frequency during those tests in which such behavior appeared.

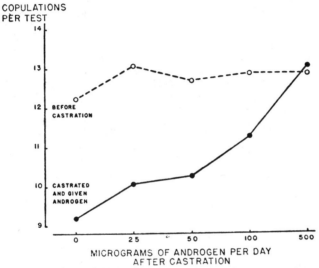

FIG. 3. AVERAGE FREQUENCY OF COPULATIONS PER TEST BEFORE AND AFTER CASTRATION

The average copulation frequency tended to increase in all groups during the first 2 or 3 preoperative tests and then to level off at approximately 12 to 13 copulations per test. During the last 4 tests before operation no major changes occurred in the group averages. Following castration the scores for all groups declined progressively. The drop from preoperative levels was quite marked in Groups II and III which received 25 and 50 micrograms of androgen respectively, and somewhat less pronounced in Groups IV and V receiving 100 and 500 micrograms. The average scores for Group I (no hormone) were extremely variable, due to the small number of males copulating, but obviously tended to decrease sharply.

In order to determine the statistical significance of these changes we have compared the average copulatory frequencies for each group during the last 3 preoperative tests with the mean scores during tests 10 to 15 which took place 4 to 9 weeks after castration. The comparisons are graphically represented in figure 3.

It is plain that in every group save one the average frequency of copulation was lower after castration, and the extent of the decrease was inversely proportional to the amount of androgen administered. Only those males which received 500 micrograms per day maintained their behavior at preoperative levels.

Application of the t test to differences between pre- and post operative means revealed that the difference was not statistically significant in the case of Group I, presumably because of the small N (5) and the high test-to-test variability shown by these untreated castrates. The decreases were significant, however, in the case of Groups II, III and IV (at the 1 per cent level for II and III and the 3 per cent level for IV).

During the second post operative period when Groups III and V were deprived of hormone the average frequency of copulations per positive test decreased markedly for both groups, but because relatively few males copulated at all the group averages were too variable to justify further statistical analysis. Group I males which had been receiving plain oil during the first post operative period showed no increase in copulation frequency when they were given 1 microgram of androgen per day although, as pointed out earlier, this treatment did produce an increase in the proportion of the group displaying the copulatory response at least once in a test.

During the first period after operation males in Group II were given 25 micrograms of hormone per day and copulated much less frequently during positive tests than they had before castration. In the second post operative period when the dosage was increased to 75 micrograms copulation frequency gradually increased until the group was performing as well as it had prior to operation. During the last 5 tests under the influence of 75 micrograms these animals copulated an average of 13.5 times per positive test which was slightly higher than their preoperative average. The average increase consequent to raised hormone dosage amounted to 3.4 copulations per rat per test and this difference proved to be significant at below the 1 per cent level of confidence.

It seems paradoxical that 100 micrograms of testosterone propionate per day was insufficient to maintain normal copulatory frequency during the first 9 weeks after castration in the case of Group IV and that 75 micrograms constituted an adequate dosage at a later stage in the experiment for Group II. It is possible that the long period of treatment at a relatively low dose level somehow sensitized the males in Group II so that the increase to 75 micrograms was peculiarly effective. At the present state of knowledge it seems useless to speculate further.

Changes in the incomplete copulatory response

The male rat's incomplete copulatory response consists of mounting the female and displaying pelvic thrusts and palpating movements of the forelimbs. There is no intromission, and when he dismounts the male slides weakly off the female's back instead of throwing himself backward with the vigorous lunge which terminates the complete copulatory response. The difference between complete and incomplete copulations usually is quite obvious to the trained observer although in perhaps 5 per cent of the cases there may be some question as to the occurrence of intromission.

Per cent of each group showing incomplete copulations in each test. Castration without androgen replacement resulted in a decrease in the number of animals exhibiting the incomplete copulatory response. Groups I, III and V were injected with plain oil during either the first or the second post operative period and in all groups the proportion of tests in which this response occurred decreased by approximately 50 per cent from precastrate levels.

Castrates receiving 1, 25, or 50 micrograms of testosterone propionate per day showed incomplete copulatory reactions in about the same proportion of tests as they had before the operation. A daily dose of 100 micrograms caused some increase in the number of tests in which such behavior appeared; and 500 micrograms produced an even more marked rise. These results are summarized in table 3.

TABLE 3

Proportion of tests in which incomplete copulation occurred

| GROUP | N | AVERAGE PERCENTAGE OF TESTS POSITIVE FOR GROUP* | | | MICROGRAMS OF ANDROGEN PER DAY |
		Normal (tests 4–6)	Castrate (tests 10–15)	Difference	
I	11	48	46†	−2	1
II	10	53	62	+9	25
III	10	40	45	+5	50
IV	10	47	65	+18	100
V	11	54	82	+28	500
			(tests 20–25)		
I	11	48	21‡	−27	0
III	10	40	15	−25	0
V	11	54	30	−24	0

* A test is "positive" if the response in question occurs at least once.
† Refers to tests 20–25 for this one group.
‡ Refers to tests 10–15 for this one group.

It should be held in mind that we are considering here merely the proportion of tests in which incomplete copulation occurred one or more times. The question as to the number of responses executed per test is dealt with below.

Frequency of incomplete copulations in positive tests. There was a marked tendency for untreated castrates or those receiving small amounts of androgen to display an increase in the average frequency of incomplete copulations during those tests in which such behavior appeared. Mean values presented in table 4 show that when they were injected with bland oil animals in Groups I, III and V displayed more of these responses in positive tests than they had prior to castration. Administration of 1 or 25 micrograms per day was accompanied by a statistically significant rise in the average number of incomplete copulations per positive test (Groups I and II). Castrates receiving 50, 100, or 500 micrograms during the first post operative period exhibited no significant change in the average frequency of incomplete copulations. Furthermore, when the dosage for males in Group II was raised from 25 to 75 micrograms per day the average frequency of this response decreased from 5.2 to 2.7 per test.

The data shown in tables 3 and 4 suggest that lack of testis hormone caused castrated male rats to show incomplete copulation in fewer tests but to execute the response a greater number of times during the tests in which it did occur. Administration of 1 or 25 micrograms of testosterone propionate per day induced no marked change in the proportion of tests in which incomplete copulations appeared, but the frequency of responses per positive test was increased by these dosages. This particular reaction was comparatively unaffected by castration if 50 micrograms of androgen were injected daily after the operation. Finally, large doses of 100 or 500 micrograms produced an increase in the number of tests in

TABLE 4

*Average number of incomplete copulations per positive test**

GROUP	N	NORMAL (TESTS 4–6)	CASTRATE (TESTS 10–15)	DIFFERENCE	PROBABILITY	MICROGRAMS OF ANDROGEN PER DAY
I	7	1.7	9.4†	+7.7	<.01	1
II	10	1.8	5.2	+3.4	.02	25
III	8	2.1	2.5	+0.4	.25	50
IV	9	3.2	2.7	−0.5	.32	100
V	8	2.9	3.7	+0.8	.35	500
			(TESTS 20–25)			
I	5	2.1	8.7‡	+6.6	.05	0
III	5	2.4	13.8	+11.4	.01	0
V	5	2.0	4.9	+2.9	.07	0

* A positive test is one in which the response in question occurred at least once.
† Refers to tests 20–25 for this one group.
‡ Refers to tests 10–15 for this one group.

which the behavior was present but did not change the frequency of incomplete copulations per test.

Changes in the ejaculatory response

When a male rat ejaculates the event is clearly reflected in the overt behavior. At first the pattern progresses in the manner described for the copulatory response, but when insertion is achieved the male does not release the female immediately. Instead he maintains the mating clasp with his forelegs and prolongs intromission, pressing tightly against the female's hind quarters. After several seconds the clasp is relaxed and the male raises his forelimbs rather slowly, rearing upwards and backwards and often coming to rest in a semi-sitting position. Occasionally instead of releasing the female in this manner the male may grip her more tightly and fall slowly to one side, pulling her with him. In either event the difference from a copulation without ejaculation is clear-cut and unmistakable.

In the present report the terms "ejaculation" or "ejaculatory pattern" refer exclusively to this sequence of overt responses and not to the occurrence of emission. Under certain circumstances male rats may display all of the outward signs of ejaculation even though they are incapable of emitting seminal fluid (e.g. after removal of the accessory sex glands), but in our records an animal was credited with an ejaculation each time the behavioral pattern appeared regardless of the presence or absence of a vaginal plug.

Analysis of ejaculatory frequency in normal animals. Before discussing the effects of castration and androgen treatment upon the ejaculatory response it will be profitable to consider briefly several aspects of this element in the sexual performance of the normal male. The data to be described bear directly upon

TABLE 5

Comparison of average scores of rats that ejaculated once with those that ejaculated twice in the sixth preoperative test

ITEM FOR COMPARISON	17 MALES THAT EJACULATED TWICE	29 MALES THAT EJACULATED ONCE	DIFFERENCE	PROBABILITY
Copulations preceding first ejaculation	8.5	12.2	+3.7	< .01
Seconds per copulation before first ejaculation	18.6	31.3	+12.7	< .01
Seconds to attain first ejaculation after mating began	152.4	349.2	+196.8	.05
Seconds latency	24.9	36.5	+11.6	†
Seconds recovery after first ejaculation	261.5	270.7*	+9.2	†

* Only 8 of the 29 males in this group recovered sufficiently to resume copulating before the test was terminated.

† Not significant.

questions of individual differences in sexual ability and possess considerably significance in relation to problems of sexual impotence.

The data upon which the following analysis is based were obtained during the last preoperative test for all rats. During this test a few animals failed to copulate, 5 copulated without ejaculating, 29 ejaculated once and 17 ejaculated twice. Disregarding those males which did not copulate or copulated without ejaculating we have attempted to discover such differences as might have existed between the 17 rats that ejaculated twice and the 29 males that did so only once.

The first and most obvious possibility was that a difference in general health was involved, but a comparison of body weights revealed no significant difference. Males ejaculating once weighed an average of 316 grams and the average weight of those ejaculating twice was 327 grams with a great deal of overlap between the two distributions.

Other possible explanations were analyzed and the results are summarized in table 5. Rats that ejaculated twice in a time-limited test were animals that reached the first ejaculation after fewer intromissions than did those individuals who ejaculated only once in the same period. "Double-ejaculators" allowed less time to elapse between preejaculatory intromissions with the result that the average number of seconds per copulation was significantly lower, and, as a natural consequence of these two differences, "double ejaculators" achieved their first ejaculation earlier in the test than did "single ejaculators".

Males which were going to ejaculate twice tended to initiate sexual activity with less delay than rats that were going to ejaculate once, but this difference in latency scores was not statistically significant. Following the first ejaculation the "double ejaculators" resumed their copulatory activity after an average delay of about 4½ minutes. Eight "single ejaculators" recovered in approximately the same length of time but failed to ejaculate again even though they completed as many copulations after the first ejaculation as did the other males. Twenty-one "single ejaculators" showed no copulatory behavior after the initial ejaculation. In their case ejaculation occurred so late in the test that not enough time remained for recovery and the renewal of sexual activity.

TABLE 6

Comparison of activity preceding first and the second ejaculations for 17 rats that ejaculated twice during the sixth preoperative test

ITEM FOR COMPARISON	FIRST EJACULATION	SECOND EJACULATION	DIFFERENCE	PROBABILITY
Copulations preceding ejaculation	8.5	4.6	−3.9	< .01
Seconds preceding ejaculation after mating begins......................	152.4	97.7	−54.7	< .01
Seconds per copulation................	18.6	21.5	+2.9	†

† Not significant.

One additional item of interest was seen in the fact that when a second ejaculation occurred it seemed to do so after fewer copulations and in less time than had been necessary for the first ejaculation. These conclusions are based upon data summarized in table 6.

Per cent of each group ejaculating in successive tests. Most of the sexually-active male rats of our strain will ejaculate at least once in the majority of a series of weekly tests, but not every male will ejaculate on every test. Figure 4, based on a 3-point moving average, shows the proportion of Groups I, II and V which achieved ejaculation at least once in successive tests. The curves for Groups III and IV fall between those of II and V but have been omitted so that the figure can be read easily.

There was an obvious relationship between the amount of androgen administered and the proportion of a group ejaculating. Comparison of figures 2 and 4 reveals that more hormone was necessary to evoke ejaculation than to call forth copulation.

Frequency of ejaculation in positive tests. Having seen the effects of castration and hormone treatment upon the presence or absence of the ejaculatory response, we are now in a position to consider the frequency of ejaculations during those tests in which this reaction occurred at least once. Most rats which continued to ejaculate under the influence of very small amounts of androgen tended to do so only once within the 10-minute test. For example, before castration males in Group I ejaculated an average of 1.7 times per positive test. When these animals were castrated and injected with 1 microgram of testosterone propionate per 24 hours ejaculation occasionally occurred, but the average frequency was 1.0 per positive test. The difference between the means was significant at the 3 per cent level.

Animals in Group II also displayed a drop in the frequency of ejaculations when they were castrated and given daily injections of 25 micrograms, but the change was slight and the difference between the pre- and post operative means

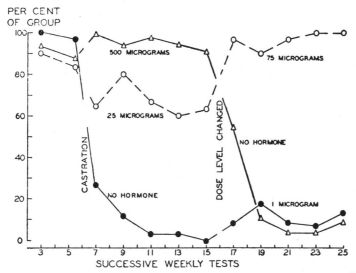

FIG. 4. PROPORTIONS OF THREE EXPERIMENTAL GROUPS EJACULATING AT LEAST ONCE DURING THE TEST

not statistically significant. Rats receiving 50 or 100 micrograms of androgen per day tended to ejaculate more frequently after castration although the increases were not marked. However, males in Group V ejaculated an average of 1.4 times per positive test before castration and 1.7 times after operation while receiving 500 micrograms of hormone per day. This rise in ejaculatory frequency was highly significant ($<.01$).

None of the animals ejaculated more than twice in the same 10-minute test and changes in ejaculatory frequency during positive tests can therefore be illustrated by showing the proportions of the various groups which ejaculated twice in each test. This scheme has been followed in preparing figure 5 which is based upon a 3-point moving average. Curves for groups III and IV are not shown but they fitted in with the trends revealed by the other three groups. Group IV males received 100 micrograms and their scores fell approximately midway between the 25 and 500 microgram groups. Group III with 50 micrograms showed two ejaculations about as frequently as rats receiving 25 micrograms.

There was some tendency for the percentage of a group ejaculating twice to increase in successive tests before operation. After castration if no androgen was given the ability to ejaculate twice in the same test was soon lost by all males. Administration of 25 or 50 micrograms of hormone per day maintained the proportion of "double ejaculators" at levels roughly comparable to those attained in the last preoperative tests. Higher dosages, particularly 500 micrograms per

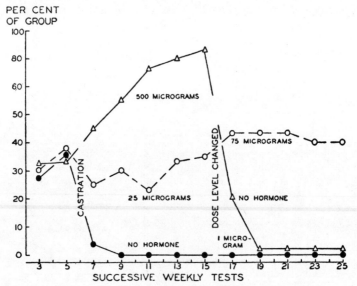

FIG. 5. PROPORTIONS OF THREE EXPERIMENTAL GROUPS EJACULATING TWICE
DURING THE TEST

day, brought about a striking increase in the number of males capable of attaining two ejaculations within the time-limited test.

Number of copulations preceding ejaculation. Normal male rats display considerable individual variation with respect to the number of intromissions which are necessary to evoke ejaculation, and at the same time most individuals show a remarkable constancy in this function from one test to another. Some males regularly achieve the first ejaculation in each test after 5, 6 or 7 copulations whereas others just as reliably take 13, 14 or 15 intromissions to attain the ejaculatory peak.

Table 7 shows the effects of castration and subsequent androgen therapy on this particular aspect of the male's sexual behavior. Animals receiving 1 microgram per day or no hormone at all ejaculated so rarely that there are no reliable data concerning the number of copulations necessary for ejaculation in such cases. However, the values incorporated in table 7 strongly suggest that castrated rats receiving 25, 50 or 100 micrograms of testosterone propionate per day tended to ejaculate after fewer copulations than they had before operation. In contrast, males injected with 500 micrograms per day showed no significant change in the frequency of copulations preceding the first ejaculation.

These mean scores suggest that relatively large amounts of androgen were necessary to maintain the preejaculatory copulation score at normal levels, and the performance of males in Group II during the three experimental periods supported this conclusion. Prior to operation these rats ejaculated after an average of 11.0 copulations. After castration while they were receiving 25 micrograms of testosterone propionate per day the mean number of copulations preceding ejaculation fell to 9.3. Then, when the daily dosage was increased to 75 micrograms the score rose again to an average value of 11.1. The increase occasioned by the higher dose level was highly significant (p = <.01).[3]

Time necessary to achieve ejaculation. The fact that castrated males receiving less than 500 micrograms of androgen per day tended to ejaculate after fewer intromissions during the first postoperative period than they had before gonad-

TABLE 7

Average number of copulations preceding the first ejaculation

GROUP	N	NORMAL (TESTS 4-6)	CASTRATE (TESTS 10-15)	DIFFERENCE	PROBABILITY	MICROGRAMS OF ANDROGEN PER DAY
II	10	11.0	9.3	−1.7	.05	25
III	10	11.1	8.5	−2.6	<.01	50
IV	10	11.7	8.7	−3.0	<.01	100
V	11	9.9	9.2	−0.7	.25	500

ectomy was an unexpected finding, and we proceeded to determine whether these rats ejaculated earlier in the tests after castration.

Basing our calculations upon the number of seconds intervening between the first sexual mount (complete or incomplete copulation) and the occurrence of the first ejaculation we found that rats receiving plain oil or 1 microgram of androgen invariably took longer to ejaculate than they had in preoperative tests. Group II males ejaculated in an average of 265 seconds before operation and 281 seconds after castration while they were receiving 25 micrograms of hormone per day. When the dosage was raised to 75 micrograms the average time needed to achieve ejaculation decreased to 257 seconds. Rats receiving 50 or 100 micrograms per day ejaculated more quickly in post operative tests than they had before castration; and under the influence of 500 micrograms of androgen per 24 hours castrated males displayed the most marked decrease in the number of seconds elapsing before ejaculation (255 seconds as normal and 211 as castrate, significant at the 5 per cent level).

Castrates receiving small doses of androgen tended to ejaculate after fewer

[3] Here, as at several other points in this report, it appears that Group II males responded more strongly to 75 micrograms than did Group IV animals to 100 micrograms. We are unable to explain this phenomenon but call attention to the fact that rats in Group II received 25 micrograms per day for 9 weeks before the higher concentration was employed.

copulations than normal, but it took them longer to reach the point of ejaculating. This suggests that the delay between copulations must have increased when the androgen level was low. Gonadectomized males injected with very large amounts of hormone ejaculated sooner than they had before operation, but the number of preejaculatory intromissions was unchanged. Here a shortened intercopulatory interval is indicated.

These indications are borne out by values presented in table 8 which shows the mean recovery period in seconds for each group before and after castration. Post operative changes in the amount of time elapsing between copulations were directly proportional to the amount of androgen given daily to the castrated animals. Only very large doses were sufficient to maintain copulation at normal speeds. The significance of all the differences shown in table 8 cannot be demonstrated statistically, but the regularity of progression in the Difference column argues against the belief that this is a chance effect. For Groups II, III, IV and

TABLE 8

Average number of seconds intervening between copulations which precede the first ejaculation

GROUP	N	NORMAL (TESTS 4–6)	CASTRATE (TESTS 10–15)	DIFFERENCE	MICROGRAMS OF ANDROGEN PER DAY
I	3	18.3	32.6	+14.3	1
II	10	24.1	30.2	+6.1	25
III	10	25.3	30.9	+5.6	50
IV	10	26.0	29.0	+3.0	100
V	11	25.7	22.9	−2.8	500
III	2	26.0	39.8	+13.8	0
V	2	21.7	44.8	+23.1	0

V in which the N was large enough to permit application of the *t* test the probabilities were .07, .02, .05 and .11 respectively. Finally, in Group II the mean intercopulatory interval increased by 6.1 seconds while the castrates were receiving 25 micrograms of androgen per day, and then decreased by almost the same amount when the dosage was raised to 75 micrograms.

Number of incomplete copulations preceding ejaculation. We have already seen that the frequency of incomplete copulations increased in castrated males given relatively small amounts of androgen. In the case of these animals if we combine the incomplete and the complete copulations preceding the first ejaculation we arrive at the following tentative conclusions. Copulations with intromission occurred less frequently after castration and in their place there appeared more of the incomplete copulatory responses. The increase in incomplete reactions equalled or exceeded the decrease in complete ones with the result that under the influence of low hormone doses gonadectomized males actually mounted the female at least as frequently to achieve ejaculation as they had before operation or as did other castrates which received larger amounts of androgen.

Changes in the recovery period after ejaculation

In male rats of our experimental strain ejaculation is followed by several minutes of relative inactivity during which the animal appears refractory to sexual arousal. The total refractory period can be thought of as including an "absolute" and a "relative" phase. During the absolute phase the male seems incapable of copulatory performance. No amount of stimulation will evoke mounting behavior and very little attention is paid to the female. During the relative phase of the refractory period sexual responsiveness is gradually regained. This period may be shortened if the female spontaneously resumes the exhibition of heat responses. It seems as though the copulatory threshold becomes progressively lower and can be crossed before it reaches normal levels if the stimulation is sufficiently intense.

We have defined the post ejaculatory refractory period as the number of seconds intervening between an ejaculation and the occurrence of the next complete or incomplete copulation. Inspection of the experimental records revealed that in the first test after castration the length of this period was greatly increased unless at least 75 micrograms of testosterone propionate were supplied each day. If no androgen was injected the refractory period was very long in tests given one week after gonadectomy or cessation of hormone treatment. In such cases, however, ejaculation was soon eradicated and consequently there are not enough data on duration of the refractory period to permit statistical treatment. A similar situation obtains in the case of Group I rats which received 1 microgram per day during the second period after operation.

Males in Groups II to V ejaculated with sufficient frequency after operation to permit quantitative analysis of changes in the post ejaculatory refractory period. Prior to gonadectomy the average recovery periods for these 4 groups varied from 251 to 280 seconds. During the first postoperative period the group averages ranged from 229 to 309 seconds. The longest mean refractory period was that of Group II whose members received 25 micrograms of hormone per 24 hours, and the shortest was that of rats in Group V which were given 500 micrograms daily. Animals treated with 50 micrograms showed a slight increase in average recovery time and those receiving 100 micrograms a minor decrease. The average times for Group II in the three periods of experimentation were as follows: 281 seconds before operation, 309 seconds after castration while receiving 25 micrograms per day and 256 seconds after castration while receiving 75 micrograms per day.

In order to measure the statistical significance of changes in the post ejaculatory refractory period the raw scores were transformed into logarithms and differences between the pre- and postoperative averages for each group were evaluated by means of the t test. The results appear in table 9. It will be seen that the increased mean refractory periods shown by animals receiving 25 or 50 micrograms of androgen per day were statistically significant. The shortened recovery times for Groups IV and V were not significant although of course the possibility of a true difference is not disproven.

When the dosage for Group II was increased from 25 to 75 micrograms the mean log refractory period decreased from 2.49 to 2.41 and the change was significant at the 2 per cent level.

Morphological changes and general health

Condition of seminal vesicles: Within 24 hours after the final test all animals were sacrificed and samples of seminal vesicle tissue were removed from several members of each group for sectioning and histological study. It was regarded as undesirable to subject the rats to abdominal operation at the conclusion of the first postoperative period and therefore no information is available concerning

TABLE 9

Average duration of recovery period after first ejaculation (in log seconds)

GROUP	N	NORMAL (TESTS 4–6)	CASTRATED (TESTS 10–15)	DIFFERENCE	PROBABILITY	MICROGRAMS OF ANDROGEN PER DAY
II	5	2.40	2.49	+.09	<.01	25
III	7	2.39	2.45	+.06	.02	50
IV	6	2.42	2.37	−.05	.14	100
V	11	2.38	2.34	−.04	.13	500

TABLE 10

Average body weights

GROUP	N	NORMAL WEEKS 4–6	CASTRATE WEEKS		PER CENT INCREASE OVER PREOPERATIVE WEIGHT		MICROGRAMS OF ANDROGEN PER DAY	
			10–15	20–25	First postoperative period	Second postoperative period	First postoperative period	Second postoperative period
I	11	300	344	350	15	2	0	1
II	10	297	339	367	14	8	25	75
III	10	306	366	376	19	3	50	0
IV	10	318	369	382	16	—	100	—
V	11	304	349	375	15	7	500	0

the response of accessory glands to hormone dosages used during the first 9 weeks after castration.

Males in Groups III and V were given 50 and 500 micrograms per day respectively in the first period and plain sesame oil during the second period after castration. At the end of the second period the seminal vesicles were typical of the long-term castrate. Gross size was markedly reduced and there was no evidence of secretory activity. Much of the epithelium was totally desquamated and such cells as remained covering the connective tissue were very low with quite small, spherical nuclei.

Group I males received no hormone for 9 weeks after operation and then were give 1 microgram per day for the next 10 weeks. This amount of testosterone propionate exerted no observable effect upon the seminal vesicles, and the cytological picture in these animals was exactly the same as that described for Groups III and V after 10 weeks without androgen. This finding is particularly interesting in view of the fact that 1 microgram per day did produce a distinct increase in some aspects of the sexual behavior.

Daily injections of 75 micrograms of hormone maintained normal seminal vesicles in males of Group II. In their gross and microscopic aspects the accessories of these animals were indistinguishable from those of intact rats.

General health: The health of the experimental males was checked by observation and by recording bodily weight once each week. There was no indication that either the operation or the hormone treatment had any adverse effects. All animals gained weight during the first 9 weeks after castration and group averages increased from 14 to 19 per cent with no apparent relationship between magnitude of increase and size of hormone dosage. During the second postoperative period 41 rats gained a little more weight and 11 lost an average of 1 to 15 grams. These minor decreases were shown by some rats in every group and probably reflect the fact t'· it the animals were getting to an age at which the weight curve tends to level off. Average weights are shown in table 10.

DISCUSSION

In interpreting the foregoing results it is necessary to hold in mind certain limitations of the conditions under which they were obtained. First, the experimental population was a highly selected group drawn from an inbred strain. Second, our own results show that the behavioral effects of a given concentration of androgen tend to vary depending upon previous hormonal treatment. Castrated rats which have never been given androgen show an improvement in some aspects of their sexual performance when they are injected with 1 microgram per day; yet five times this amount has no appreciable effect upon the behavior of castrates previously treated with 100 micrograms. Finally, although it is probable that other forms of androgen would facilitate sexual performance in the castrated male, it is entirely possible that the magnitude of such an effect is a function of the form of the hormone and the method of its administration.

Within these limits the results of our experiment are reasonably consistent and clear cut. It has long been known that sexual behavior can be maintained in castrated rats by the administration of androgen but it can now be added that if the androgen is testosterone propionate, and if it is administered in the form of daily injections, the amount needed to hold performance at or near preoperative levels is approximately 50 to 75 micrograms. This is considerably less than the amounts employed in most of the earlier studies. We have also obtained new evidence concerning the effects of hormonal concentrations falling well above or below this "maintenance level."

In a very general way it is correct to state that the strength of the "sex drive" in a castrated male rat is roughly proportional to the amount of exogenous testicular hormone which is injected. But any such generalization must be followed by the qualification that "sex drive" is a meaningful concept only when it is operationally defined, and that when this is done the strength of the drive tends to vary according to the behavioral criterion selected to measure it. For example, if we choose the frequency of intromissions per 10 minutes as *the* criterion our results indicate that sex drive decreases after castration unless more than 100 micrograms of androgen are given each day. In contrast it would be entirely permissible to state that sex drive will be measured in terms of the number of intromissions necessary to cause a sexual climax, or the speed of sexual arousal as measured by initial latency scores. In either instance the minimum androgen concentration sufficient to maintain normal sex drive would be considerably below 100 micrograms.

We are impressed with the need for an unambiguous, operational definition of sex drive,—a definition based upon actual mating performance rather than some less direct type of behavior such as the tendency to cross an electrically charged grid to reach the receptive female. Present data are insufficient to establish the point but they contain some suggestion that initial latency, time to achieve the first ejaculation, and duration of the post ejaculatory recovery period may be positively correlated. If such a correlation were high enough it would be entirely feasible to devise a combined score which would incorporate these measures and would serve as a valid and reliable index to sex drive in the individual male. When this can be done it seems likely that there will prove to be a fairly high positive relationship between sex drive and concentration of exogenous androgen administered to the castrated male.

A word of caution should be addressed to the reader who is unfamiliar with endocrinological data. The results herein reported *do not* signify that individual differences in the sexual responsiveness and potency of *unoperated* male rats are due to differences in levels of endogenous hormone. This particular problem cannot be approached until there is available some sensitive and reliable method of determining the amount of hormone present in the blood. Nothing of this sort can be done as yet with an animal as small as the rat. There are many potential sources of variability in sexual excitability and differences between intact males are quite possibly entirely independent of any hormonal basis.

The results of this experiment tell us nothing with respect to the locus or specific nature of hormonal action. They are, however, harmonious with the general hypothesis that one important function of the hormone is to lower the threshold in those nervous mechanisms specifically concerned with the mediation of sexual responses. For instance, the reduced initial latencies characteristic of castrated males receiving high doses of androgen suggest that such animals reach the threshold of arousal necessary to mounting activity with less preliminary stimulation than do other males whose blood contains less male hormone. Similarly, the abbreviated period of sexual refractoriness after an ejaculation is suggestive of a lowered threshold and a consequent reduction in the duration or intensity of the liminal stimulus.

SUMMARY

Fifty-two male rats were observed in six weekly mating tests with receptive females and then castrated. Sex tests were continued after operation while the males received daily injections of testosterone propionate. The amount of androgen administered to different groups varied from 1 to 500 micrograms per day. A control group was treated with plain sesame oil.

The amount of hormone necessary to maintain sexual performance at or near preoperative levels varied somewhat depending upon the behavioral criterion selected as a measure. In the main, however, 50 to 75 micrograms per day represented a maintenance dose.

Castrated rats receiving no hormone and those injected with less than 50 micrograms were less likely to show any sexual responses toward the female than were normal animals or other castrates given higher concentrations of male hormone. When they did display copulatory reactions the low dose castrates were slow in initiating sexual contact, and mounting activity was apt to be preceded by relatively protracted periods of inattention or investigation of the estrous female. Such mating behavior as did occur consisted to a large extent of copulatory attempts which failed because of lack of intromission. Successful intromissions often were widely spaced in time. Ejaculation rarely was achieved but when it did appear it usually occurred after fewer intromissions and was followed by a longer period of sexual inactivity than is the rule in normal rats or in castrates receiving larger amounts of hormone.

Castrated male rats injected with 100 or 500 micrograms of testosterone propionate per day exhibited sexual behavior that was equal or superior to that shown prior to operation. They were more likely to copulate in every test than they had been before gonadectomy. At the same time the occurrence of incomplete copulations increased. Castrates receiving these larger doses of hormone tended to initiate sexual relations after shorter delays than they had before operation. The frequency of intromissions in a 10-minute test was not increased but multiple ejaculations occurred more frequently, and the first sexual climax was reached at an earlier point in the test. Finally, the castrates supplied with large amounts of androgen tended to recover more rapidly from the sexually depressing effects of an ejaculation.

REFERENCES

1. BEACH, F. A.: *Hormones and Behavior*, New York: Paul B. Hoeber, Inc., 1948.
2. MOORE, C. R. AND PRICE, D.: Some effects of testosterone and testosterone propionate in the rat. *Anat. Rec.*, 1938, **71**, 59–78.
3. SHAPIRO, H. A. Effect of testosterone propionate upon mating. *Nature*, 1937, **139**, 588–589.
4. STONE, C. P.: Copulatory activity in adult male rats following castration and injections of testosterone propionate. *Endocrinol.*, 1939, **24**, 165–174.

Reprinted from *Endocrinology*, **51**, 237–248 (1952)

DIFFERENTIAL REACTIVITY OF INDIVIDUALS AND THE RESPONSE OF THE MALE GUINEA PIG TO TESTOSTERONE PROPIONATE[1]

JEROME A. GRUNT[2] AND WILLIAM C. YOUNG

From the Department of Anatomy, University of Kansas, Lawrence, Kansas

MANY investigators have directed attention to the close relationship between the reactivity of tissues, sometimes referred to as sensitivity or responsiveness, and the degree of response to hormonal stimulation. The literature in which the subject is discussed has never been reviewed completely, but a few references will indicate something of the extent to which differences in reactivity influence endocrine function.

Smith and Engle (1927) reported that the response of immature rats and mice to pituitary transplants increased with age; that in older animals fewer transplants were required to produce precocious sexual maturity. Bradbury (1944) and McCormack and Elden (1945) found an inherent seasonal variation in the sensitivity of the rabbit to pituitary extracts. Albright, Burnett, Smith, and Parson (1942) presented evidence that in certain clinical cases of idiopathic hypoparathyroidism the disturbance was not a lack of hormone, rather a resistance to it. Selye and Albert (1942) described a differential reactivity of the adrenal gland. They found that estradiol administered to immature male rats in large doses was not followed by the adrenal cortical hypertrophy that was produced in the adult, in fact, it tended to decrease the size of the tissue. Sprague (1951) writes that in many conditions of seeming adrenal insufficiency, no deficiency of the adrenal hormones exists. Apparently, there had been an alteration in the manner in which the target organs were responding to the hormones.

The importance of the relationship of reactivity for tissue response is not restricted to mammals. Lillie (1932) showed that different feather tracts of the brown leghorn fowl have inherently different growth rates, and exhibit correspondingly different thresholds of reaction to the female hormone. The response of the oviducts in young New Hampshire Red chicks to injections of stilbestrol was shown to depend on the level of folic

Received for publication May 19, 1952.

[1] This investigation was supported in part by a research grant from the National Institutes of Health, Public Health Service, and in part by a grant from the University of Kansas Research Fund.

[2] Submitted in partial fulfillment of requirements for the degree of Doctor of Philosophy at the University of Kansas.

acid intake (Hertz, 1945). Williams (1942), early in his work on the metamorphosis of the native silkworm moth, concluded that the ultimate result is determined by the ability of the tissues to react.

Early reports of differences in the reactivity to gonadal hormones and an enumeration of some of the factors which might influence reactivity have been reviewed (Young, 1941; Beach, 1948), but there are more recent additional studies. The seminal vesicles of the rat are most sensitive to stimulation by androgens at 40 to 60 days of age, a time when this tissue is differentiating most rapidly (Hooker, 1942). Price (1944) indicated that at 26 days there is a marked increase in the reactivity of the uterus and seminal vesicle of gonadectomized rats to testosterone propionate. A similar change was found in the male and female prostate on the thirty-sixth day. Hamilton (1948) found that no correlation exists between certain secondary sexual characteristics of the human male, such as baldness and the presence of auricular hair, and the titers of excreted steroids. He concluded that the action of the hormones in endocrine-dependent states is controlled by genetic, aging, environmental, and other factors. Lyman and Dempsey (1951) working with hibernating, castrated, male hamsters showed that following the injection of testosterone, the seminal vesicles of the animals which returned to hibernation revealed little enlargement or histological change, whereas those of the animals which remained awake showed an increase in weight and histological alteration. They concluded that it is not the amount of circulating hormone which determines the condition of the affected organ, rather the transient condition of the organ.

It will be apparent that most of these differences in reactivity have been associated with age, season, and nutrition. Other factors however may also be involved. Regulation of the reactivity of tissues by endocrine substances, such as the thyroid hormone, has been suggested (Smelser, 1939; Salmon, 1941; Langham and Gustavson, 1947; and others). Another example is the priming effect of an initial dose of estrogen on the sensitivity of the vaginal mucosa (Kahnt and Doisy, 1928). Differences in reactivity may have a genetic basis. Bates, Riddle, and Lahr (1941) report that two hereditary strains of chicks showed differential thyroid responsiveness to thyrotrophin. Chicks from one source required four times as much hormone to produce an amount of stimulation equal to that obtained in chicks from the other.

In few if any of the studies cited above were the investigators concerned primarily with reactivity. Most information came as a by-product of experiments directed toward other problems. Lacking was a study of the extent and importance of differences in the reactivity of tissues for hormonal action in animals homogeneous with respect to age, diet, the season of experimentation, methods of handling, and general endocrine balance as judged by the condition of the individuals at the beginning of the investigation.

MATERIALS AND METHODS

Reproductive behavior of the male is particularly well adapted for investigations of this type. The end points are definite. The strength of sex drive displayed by individual male guinea pigs is extremely variable from animal to animal, but for individuals quite constant from test to test. Long.range experiments requiring comparatively few animals are possible. Individual animals can be used repeatedly giving continuous information throughout the experimental period with a minimum of operational trauma. The importance of this last point can be illustrated briefly. In this investigation each animal was given 41 tests, consequently much more information was obtained from single animals than would have been possible had procedures been employed that involved sacrifice of the animals following each test.

Sexual behavior patterns of the adult, male guinea pig and their development have been described (Avery, 1925; Louttit, 1927, 1929; Sollenberger and Hamilton, 1939; Seward, 1940; Young and Grunt, 1951; and Webster and Young, 1951), but since an understanding of what follows depends on some familiarity with the behavior, a brief description is given.

When a female guinea pig in heat is placed with a male the following sequence of behavior is usually seen. The male follows the female and there is frequently a generalized sniffing and licking, or pulling of the hair with the teeth. The term *nibbling* is used to describe these actions. Shortly afterwards, the male centers his attention on the anogenital region; *nuzzling* is used for this localized sniffing and licking. In the next type of action the male mounts the female and executes a series of pelvic movements. Most often this occurs posteriorly; frequently, however, in his excitement the male mounts elsewhere. In either case, the action is called *mounting*. In the display of the complete pattern terminating in ejaculation, *intromission* with pelvic thrusts usually follows mounting. In those cases in which ejaculation occurs, the activity is brought to an abrupt end with a grasping of the female's back, and a convulsive drawing in of the flanks. This position is held for a number of seconds following which the male dismounts and licks his penis. If watched, he will be seen to drag his pudendum along the floor of the cage. Usually the time required to achieve ejaculation is less than 10 minutes. More frequently than not, ejaculation is followed by a period of quiescence which lasts at least an hour (Grunt and Young, 1952). This sequence of behavior is not invariable. At times the lower degrees of the behavior pattern are omitted and mounting with intromission or even ejaculation takes place almost immediately. At other times there is little sexual behavior beyond nuzzling or mounting.

Although there is wide variation in the age at which these elements of the sexual behavior pattern are first displayed, the approximate age at which each component appears is known. Generally, nibbling is first seen about the fourteenth day, nuzzling about the seventeenth day, mounting about the forty-fifth day, intromission about the fifty-fourth day, and ejaculation about the sixty-fourth day (Avery, 1925; Louttit, 1929; and Webster and Young, 1951). All this information was utilized in the development of the method (Young and Grunt, 1951) by which the strength of sex drive was measured in this investigation.

When determining the strength of sex drive a female displaying good heat responses (Young, Dempsey, Hagquist, and Boling, 1937) was placed in the cage occupied by a male. The amounts of the five measures of behavior, nibbling, nuzzling, mounting, intromission and ejaculation, displayed by the male during a 10-minute period were recorded. Each minute of the test was divided into four 15-second periods, and a measure was recorded only once during each period in which it was shown. Intromission and ejaculation, being more discrete and of shorter duration, were recorded only in the period during which the measure of behavior began.

In determining the score for each test, the number of 15-second periods was counted during which each measure of behavior was displayed. Only the highest measure of behavior displayed during each period was used in the calculations. Therefore, the maximum number of measures in each test was 40. Since, however, ejaculation usually is achieved in less than 10 minutes, the number of measures was less. The number of 15-second periods during which a given measure of behavior was displayed was then divided by the duration of the test. The quotient represents the number of times per minute when that measure of behavior was the highest degree shown. Each quotient was then multiplied by a factor arbitrarily chosen as representing the importance of the measure; 1 for nibbling, 2 for nuzzling, 3 for mounting, 4 for intromission, and 5 for ejaculation. These values were chosen because, as was indicated above, the elements of the behavior pattern appear in this order during maturation. The sum of the products was taken as the score of the test. When this procedure is followed, an animal that mates within the first 15 seconds is scored 20.0 while one that does nothing for the entire 10 minutes of the test is scored 0.0. A male having an average of 8.50 or higher for 10 tests is considered *high drive*. If the score averages between 6.25 and 8.50, he is considered *medium drive;* if below 6.25, *low drive*.

In the first of two experiments, each of 39 male guinea pigs, 80 to 150 days of age, was isolated in a cage 24 inches square. Beginning approximately seven days later each male was given tests at weekly intervals to determine the strength of his sex drive. After 10 tests all but 10 of the animals, the intact controls, were castrated. Their average scores ranged from 2.2 to 11.1. For the following 31 weeks both experimental and control animals were tested weekly. In all 1599 tests were made. During the first 16 of the 31 weeks no therapy was given. This was the period during which sexual activity decreased from the precastrational level to a constant low level, which is referred to as the *base line*. During the following 15 weeks, each of 22 of the castrated males was injected with testosterone propionate.[3] The androgen was administered daily in doses of 25γ per 100 grams body weight for the first 10 weeks of this period, and 50γ per 100 grams body weight for the last 5 weeks. The volume of injected sesame oil containing dissolved hormone was kept constant for each 100 grams body weight. Seven of the 29 castrates were given sesame oil and served as the castrate control group.

At the end of the experiment the scores for each week were averaged. In a first analysis of the scores the animals were divided into intact controls, castrate controls and castrates given testosterone propionate, thus revealing the effects of castration and subsequent androgen therapy. A second type of analysis was suggested by the great differences between individuals. For this analysis the animals were divided into high, medium, and low drive groups, thus relating the performance of each group during the period of therapy to the strength of drive prior to castration.

In the second experiment 31 animals were used. On the basis of the scores of 10 tests prior to castration, these males were divided into four balanced groups whose scores averaged 7.8. Each group contained 7 or 8 animals. Following castration, tests were made at weekly intervals for 31 weeks. During the first 16 weeks no androgen was given,

---

[3] Testosterone propionate was supplied by the courtesy of Ciba Pharmaceutical Products, Inc.

and the scores decreased to the base line. At the beginning of the seventeenth week following castration and for the remaining 15 weeks, the animals in each of the four groups were given daily injections of testosterone propionate in doses of 12.5, 25, 50, or 100γ per 100 grams body weight. These amounts were used in order that the effects of a variable dosage could be compared. In all, 1271 tests were made.

Following the last test, the weekly scores were averaged. In a first analysis, the averages for each of the four groups were plotted in order to show the effects of castration and therapy with these amounts of androgen. In a second analysis the performance was related to the strength of drive prior to castration.

After the last test, the animals were sacrificed and portions of the endocrine and reproductive systems were fixed for histochemical and histological study. The results of the microscopic findings will be reported elsewhere.

Throughout the work the temperature was maintained between 70° and 75° F. The diet always included rabbit pellets, oats, green vegetables, alfalfa hay, and water. All the animals were healthy, and showed little if any effect of the operation or daily injections.

RESULTS

The results obtained from the experiment involving observation of males prior to castration, during the period of castration without therapy, and while testosterone propionate was being administered in the same amount per 100 grams body weight to each animal are summarized in figure 1.

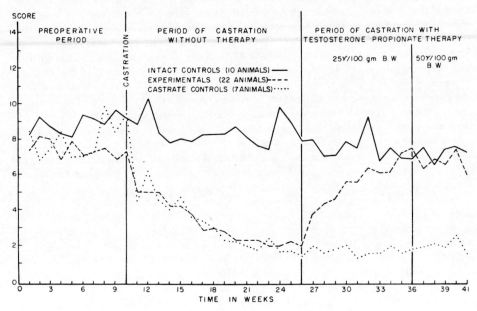

FIG. 1. Strength of sex drive in male guinea pigs before and after castration and following therapy with testosterone propionate daily.

During the 41 weeks of the experiment the intact controls showed a very slow decline in sexual activity. By the end of the period approxi-

mately 18% of the original drive had been lost. This loss manifested itself
mainly in a decline in the frequency of ejaculations which was possibly
associated with aging. Removal of the gonads in the animals to be studied
experimentally was followed by a gradual decline in sexual activity. The
base line was reached within 14 weeks, during which time more than 75%
of their original drive was lost. The activity at this level consisted basically
of nibbling and the score achieved was approximately 2.0. The activity of
the animals which received no treatment save the daily injection of sesame
oil remained at this level during the remainder of the experiment. When
the experimental animals were given testosterone propionate beginning

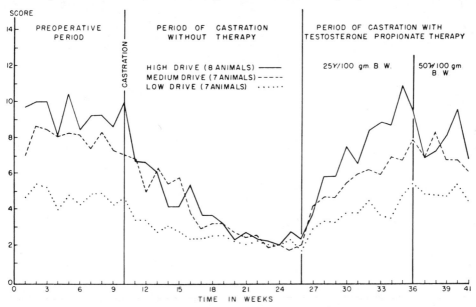

FIG. 2. The effect of castration and therapy with testosterone propionate daily on the
sex drive of high, medium, and low drive male guinea pigs.

the seventeenth week following castration, their sexual activity increased.
By the tenth week after the start of therapy, they returned to a level of
drive similar to that of the intact controls, and only slightly lower than
that seen prior to castration.

It will be recalled that the 22 castrated animals given testosterone propionate were divided into three groups according to the strength of their
sex drive prior to castration. The average scores of these groups before
castration, during the period of castration without therapy, and after
androgen treatment was begun are shown in figure 2. Almost immediately
after the beginning of testosterone propionate therapy a redistribution of
the animals into the groups seen prior to castration became apparent. The

high drive animals showed the highest amount of sexual activity, the low drive animals the lowest, and in general, the activity of the medium drive animals was between the two. Not once during the period of androgen therapy did the score of the low drive group equal that of the medium or high drive groups. In all three groups the strength of drive following androgen administration was comparable with that prior to castration.

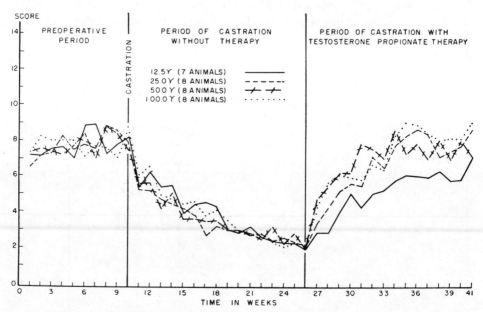

FIG. 3. Strength of sex drive in male guinea pigs before and after castration and following therapy with different quantities of testosterone propionate daily.

The significance of the differences between the groups was tested by the "t" test. In doing so the scores of the tests for the thirty-fourth, thirty-fifth, and thirty-sixth weeks were used. This was the end of the period during which therapy consisted of 25γ testosterone propionate per 100 grams body weight. It was found that the differences between the low drive group and the other two were significant at the 2% level or better. Because of the variability within the medium drive group, the difference between this group and the high drive group was not significant, nevertheless, the trend was apparent. Similar comparisons were made for the thirty-ninth, fortieth, and forty-first tests. This was the end of the period during which therapy consisted of 50γ testosterone propionate per 100 grams body weight. Although the same trend manifests itself, the increase in variability was sufficient to decrease the significance of the differences. Only that between the high and low drive animals remained significant, but at the

3% level. The correlations of the data from all the castrates given androgen, when compared with the average of the precastrational scores, were found to be significant at the 1% level or better.

With few exceptions, the results of the experiment in which four different doses of androgen were administered were comparable with those obtained when each animal received the same amount (figure 3). As before, sexual drive of the four groups decreased to the base line after castration and, following the injection of testosterone propionate, there was an increase in sexual behavior. The only conspicuous deviation from the rule

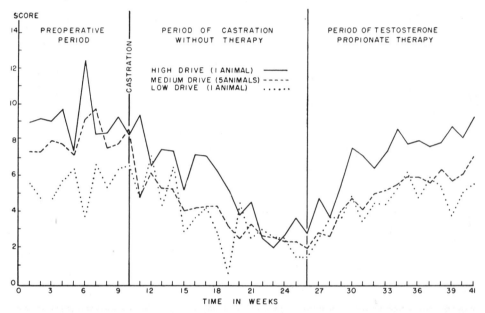

FIG. 4. The effect of castration and therapy with 12.5γ testosterone propionate daily on the sex drive of high, medium, and low drive male guinea pigs.

that individuals return to the level displayed prior to castration, was that the rate of increase in the sexual activity of the group given daily injections of 12.5γ testosterone propionate per 100 grams body weight was slower than in the other animals. Never during the 15 weeks of therapy did this group reach its precastrational average, although at the end of the experiment there were indications of a return to this level. The low scores were due basically to the failure to achieve intromission and ejaculation.

By the eighth week after the beginning of testosterone propionate therapy, the sexual drive of the animals in the other three groups had returned to the precastration level. With minor exceptions, they continued to display sexual activity of this degree for the remainder of the experiment. The behavior of the animals given 100γ testosterone propionate was comparable with that of the others, except that, starting 9 weeks after the beginning of therapy, the scores always equalled or surpassed by a slight degree, the highest scores of the other groups. The differences were tested for significance. Using the means of the last five tests, it was found that only the differences between the group given 12.5γ testosterone propionate and each of the other groups were significant at the 5% level. None of the other

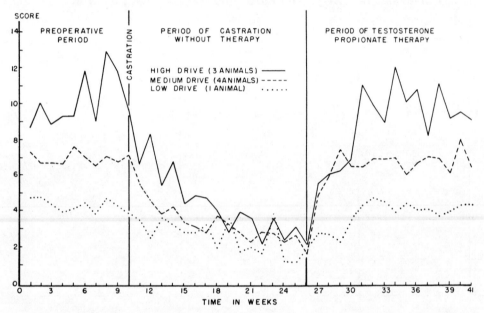

FIG. 5. The effect of castration and therapy with 50.0γ testosterone propionate daily on the sex drive of high, medium, and low drive male guinea pigs.

differences was significant. It is seen, therefore, that once the threshold was reached, 4-fold differences in dosage did not cause significant variation in sexual behavior.

In the first part of the experiment it was shown that when male guinea pigs are injected with a given dosage of androgen, the high drive animals returned to a high level of activity, and the low drive animals returned to a low level. It became of interest therefore, to see whether this would also be true when the range of experimental dosages was varied as in the second experiment. Even though the number of animals in each group was less, comparable data were obtained (figures 4, 5, and 6). The strength of drive following castration and androgen therapy was related to that prior to castration. The response of the eight animals receiving 25γ testosterone propionate are not summarized here because it was similar to that of the animals used in the first part of the experiment (figure 2).

DISCUSSION

Although the importance of the reactivity of a tissue or organ for its responsiveness to hormonal stimulation has been widely noted, this study is believed to be one of the first, if not the first, in which adult animals, homogeneous with respect to age, diet, the season of experimentation, method of handling and caging, have been used. The degree of variation in the hormonally induced reaction that was studied is of interest. More so,

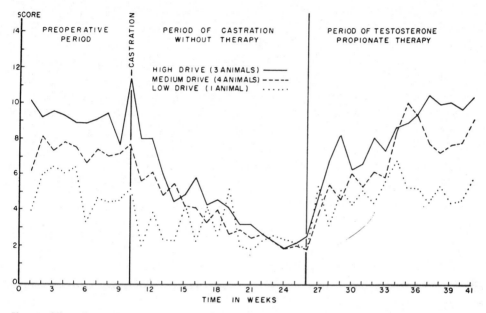

Fig. 6. The effect of castration and therapy with 100γ testosterone propionate daily on the sex drive of high, medium, and low drive male guinea pigs.

however, is the demonstration of the faithfulness with which the strength of the precastrational sex drive of individuals was restored by hormonal treatment following castration, and, in the second part of the experiment, the demonstration that within the limits of a 4- to 8-fold increase above the threshold, variation in strength of drive was attributable to the reactivity of the tissues on which the hormone was acting rather than to variation in the amount of hormone. Increased emphasis is thereby given to the principle that the response to a hormone is influenced by the character of the soma.

The reader will want to know whether this suggestion is consistent with other data that are reported by those who have followed the effects of androgens on adults. Hamilton's (1948) demonstration that testosterone propionate will not cause alopecia in some eunuchs, particularly those having relatives that are not bald, is similar in principle. The point was also made by Dempsey (1951) when he called attention to the observation that certain patients with interstitial cell tumors, although producing great amounts of androgen, fail to exhibit any of the symptoms usually associated with high levels of testosterone. The results from two other studies, on the other hand, indicate that the response may be proportional to the amount of hormone. Gordon and Fields (1942) showed that large amounts of androgen, when given to hypogonadal individuals often cause excessive penis growth, priapism, and "over strong sex interest." Beach and Holz-Tucker (1948) indicated that castrated male rats receiving 100 to 500γ testosterone propionate per day exhibited sexual behavior that was equal or superior to that shown prior to gonadectomy. Species differences may account for the lack of agreement, but even this suggestion is premature because no two studies are more than remotely comparable. Additional work must be done before the principle that is postulated can be fully evaluated. Also of importance is the full meaning of reactivity in terms of what is taking place within the cells that are being stimulated by hormones. No approach to the problem was made during this study, nor is any suggested in the literature.

SUMMARY

The strength of sex drive displayed by individual male guinea pigs is extremely variable from animal to animal, consequently the tissues mediating this behavior are well adapted for a study of the relationship between tissue reactivity and the response to testosterone propionate. Male guinea pigs, homogeneous with respect to age, diet, the season of experimentation, methods of handling, and general endocrine balance as judged by the conditions of the individuals at the beginning of the investigation, were used. The administration of 12.5, 25, 50 or 100γ of the androgen per 100 grams body weight daily led to the demonstration that the strength of sex drive displayed by individuals during therapy is similar to that shown prior to castration. With androgen therapy, animals that were characteristically high drive prior to gonadectomy returned to the high level, while the animals that were low drive returned to the low level. It is postulated from the results obtained when the tissues mediating sexual behavior in the male guinea pig were used as the target organ, that much of the difference between individuals is attributable to the reactivity of the tissues rather than to differences in the amount of hormone. The strength of the reaction to the androgen was not altered significantly by a 4- to 8-fold increase in dosage.

REFERENCES

ALBRIGHT, F., C. H. BURNETT, P. H. SMITH AND W. PARSON: *Endocrinology* **30**: 922 1942.

AVERY, G. T.: *J. Comp. Psychol.* **5**: 373. 1925.

BATES, R. W., O. RIDDLE AND E. L. LAHR: *Endocrinology* **29**: 492. 1941.

BEACH, F. A.: *Anat. Rec.* **92**: 289. 1945.

BEACH, F. A. AND A. M. HOLZ-TUCKER: *J. Comp. and Physiol. Psychol.* **42**: 433. 1948

BRADBURY, J. T.: *Endocrinology* **35**: 317. 1944.

DEMPSEY, E. W.: Personal communication, 1951.

GORDON, M. B. AND E. M. FIELDS: *J. Clin. Endocrinol.* **2**: 531. 1942.

GRUNT, J. A. AND W. C. YOUNG: *J. Comp. and Physiol. Psychol.* In press. 1952.

HAMILTON, J. B.: *Rec. Adv. Hor. Res.* **3**: 257. 1948.

HERTZ, R.: *Endocrinology* **37**: 1. 1945.

HOOKER, C. W.: *Endocrinology* **30**: 77. 1942.

KAHNT, L. C. AND E. A. DOISY: *Endocrinology* **12**: 760. 1928.

LANGHAM, W. AND R. G. GUSTAVSON: *Am. J. Physiol.* **150**: 760. 1947.

LILLIE, F. R.: *Amer. Nat.* **66**: 171. 1932.

LOUTTIT, C. M.: *J. Comp. Psychol.* **7**: 247. 1927.

LOUTTIT, C. M.: *J. Comp. Psychol.* **9**: 293. 1929.

LYMAN, C. P. AND E. W. DEMPSEY: *Endocrinology* **49**: 647. 1951.

McCORMACK, G. AND C. A. ELDEN: *Endocrinology* **37**: 297. 1945.

PRICE, D.: *Physiol. Zool.* **17**: 377. 1944.

SALMON, T. N.: *Endocrinology* **29**: 291. 1941.

SELYE, H. AND S. ALBERT: *Proc. Soc. Exp. Biol. and Med.* **50**: 159. 1942.

SEWARD, J. P.: *J. Comp. Psychol.* **30**: 435. 1940.

SMELSER, G. K.: *Anat. Rec.* **73**: 273. 1939.

SMITH, P. E. AND E. T. ENGLE: *Am. J. Anat.* **40**: 159. 1927.

SOLLENBERGER, R. T. AND J. B. HAMILTON: *J. Comp. Psychol.* **28**: 81. 1939.

SPRAGUE, R. G.: *Am. J. Med.* **10**: 567. 1951.

WEBSTER, R. C. AND W. C. YOUNG: *Fertility and Sterility* **2**: 175. 1951.

WILLIAMS, C. M.: *Biol. Bull.* **82**: 347. 1942.

YOUNG, W. C.: *Quart. Rev. Biol.* **16**: 135, 311. 1941.

YOUNG, W. C., E. W. DEMPSEY, C. W. HAGQUIST AND J. L. BOLING: *J. Lab. and Clin. Med.* **23**: 300. 1937.

YOUNG, W. C. AND J. A. GRUNT: *J. Comp. and Physiol. Psychol.* **44**: 492. 1951.

Reprinted from *Endocrinology*, **57**, 139–146 (1955)

6

DEVELOPMENT OF SEXUAL BEHAVIOR IN MALE GUINEA PIGS FROM GENETICALLY DIFFERENT STOCKS UNDER CONTROLLED CONDITIONS OF ANDROGEN TREATMENT AND CAGING[1]

WALTER RISS,[2] ELLIOT S. VALENSTEIN, JACQUELINE SINKS AND WILLIAM C. YOUNG

Department of Anatomy, University of Kansas, Lawrence, Kansas

THE differential responsiveness of male guinea pigs to testosterone propionate (t.p.) as measured by their sexual behavior (Grunt and Young, 1952) is attributed in part to the influence of early experience on the organization of sexual behavior patterns (Valenstein, Riss and Young 1955; Valenstein and Young, 1955). These patterns differ however in males from genetically different strains (Valenstein, Riss and Young, 1954). It appeared likely therefore that genetic as well as experiential factors contribute to the organization of these patterns and therefore to the responsiveness to t.p. The point was checked in the present experiment in which the responses of different genetic strains to t.p. were studied and in which the influence of early experience was controlled. Opportunity was also provided to ascertain if the testes are necessary for the organization of sexual behavior patterns (Valenstein, Riss and Young, 1955) and to compare the schedules of development of sexual behavior in genetically different strains.

METHODS

Briefly, males from 2 strains were compared in the rate of development of sexual behavior when reared under equivalent conditions and given equivalent quantities of t.p. Each of 47 males (the isolated males) was caged with a lactating female until day 25

Received January 10, 1955.

[1] This investigation was aided in part by research grant M504(c) from the National Institute of Mental Health of the National Institutes of Health, Public Health Service, and in part by grant 44 from the University of Kansas Research Fund.

[2] Public Health Service Research Fellow of the National Institute of Mental Health, present address: Department of Anatomy, State University of New York, State University Medical Center at New York City, Brooklyn 2, N. Y.

and then isolated. Each of 51 males (the socially raised males), weaned on day 25, was caged with two females of the same age from the day of birth. In each of the conditions of rearing, approximately half the animals were from a genetically heterogeneous stock and half from the highly inbred Strain 2. This history of Strain 2 is given elsewhere (Riss, 1955). The isolated and socially raised males were subdivided into four groups containing approximately equal numbers of males from the two strains: (1) intact males, (2) males castrated within 2 days after birth and injected daily with 25 μg. of t.p.[3] per 100 gm. body weight beginning the day after castration, (3) males castrated and injected as above except that 500 μg. of t.p. were given, and (4) untreated castrates. The t.p. dissolved in sesame oil was injected intraperitoneally.

Rate of development of sexual behavior was measured in two ways: by recording the age at which a particular kind of response first appeared, i.e., mounting, and by using an index of the rate at which all responses were displayed, i.e., the sexual behavior score.

Animals were given a ten-minute test at each week of age. In such direct tests of sexual behavior the male was given an estrous female of nearly equal size. If the male was caged with females, they were removed. The type and number of responses were recorded as they occurred. A more complete description of the technique of testing and scoring is given elsewhere (Grunt and Young, 1952; Valenstein, Riss and Young, 1954).

Tests were terminated at the 17th week of age or earlier if a male had completed 6 tests after the appearance of his first ejaculation. After the test in which ejaculation first appeared, the females with which a socially-reared male had been living were not returned to him. If no ejaculation occurred by day 90 the females were removed from him at that time.

The advantage of using both rate and kind of response lies in the fact that it provides a means of distinguishing between the particular responses which an animal is able or stimulated to display and the rate at which they are shown. It is recognized that scores, which reflect rate of response, are comparable only if animals show evidence of being able to display the same responses. Such responses as intromissions and ejaculation may not be shown by animals for at least 2 reasons: 1) Failure of the male to have learned to mount and maneuver a female properly. This occurs when males have not lived with other animals for a sufficient length of time (Valenstein, Riss and Young, 1955). 2) Inadequate development of the structures which make intromissions possible, i.e., the penis.

Precautions were taken to assure that comparisons of score were made only on animals who were potentially capable of intromission within the 17-week testing period. The precautions were the following: 1) Gross measurements were made of the penes and fresh seminal vesicles from a number of the castrates treated with 25 and 500 μg. of t.p. and from intact controls at the end of 17 weeks. The length of the seminal vesicles was measured from the bifurcation to the distal ends of the horns. The weight was that of the entire organ. The penis, with glans retracted, was measured from its distal end to the insertion of the most distal striate muscle. 2) Since Strain 2 males generally do not learn to mount properly after living the first 25 days with only a lactating female (Valenstein, Riss and Young, 1955), animals living with female companions were included. Ninety days, the age chosen for separating the females from the males, is beyond the length of time found to be adequate for intact males (Valenstein, Riss and Young, 1955). It was expected therefore that a 17-week testing period was sufficiently long to demonstrate any retardation due to other variables.

To test whether the process of learning to mount and maneuver the female occurs independently of the presence of testes, untreated castrated males raised in the social

---

[3] Testosterone propionate was supplied by Ciba Pharmaceutical Products, Inc. Hereafter the amount will be given without reference to 100 gm. body weight per day.

situation were tested until day 70 or, in the case of two Strain 2 males, until day 90. The males were tested as the others and observed particularly for mounting and maneuvering of the female. At the stated ages, the females were removed and the males were injected with 500 μg. of t.p. for 7 weeks. The tests of sexual behavior were continued during this period. It was expected that if the males had learned to mount and maneuver properly, intromission and ejaculation would occur automatically as the necessary structures became sufficiently developed.

RESULTS

The data are uniform in establishing a distinction between the males from the heterogeneous stock and Strain 2. The latter develop equivalent sexual responses at later ages (Table 1) and finally attain scores (rates of

TABLE 1. AVERAGE AGE IN WEEKS AT FIRST APPEARANCE OF MEASURES OF SEXUAL BEHAVIOR AND PROPORTION OF ANIMALS EXHIBITING THE BEHAVIOR UNDER CONTROLLED CONDITIONS OF TESTOSTERONE PROPIONATE TREATMENT AND CAGING

Group	Heterogenous stock				Strain 2			
	Socially raised		Isolated		Socially raised		Isolated	
	Age	Pro-portion	Age	Pro-portion	Age	Pro-portion	Age	Pro-portion
Mounting								
500 μg.t.p.	4.4	6/6	3.0	7/7	10.4	6/7	7.0	4/7
Intact	4.7	7/7	4.7	5/7	10.7	7/7	15.7	2/7
25 μg.t.p.	6.9	6/6	6.4	6/7	9.7	3/6	9.6	3/7
Untreated Castrates	5.0	5/6	3.9	2/2	11.0	3/5	—	0/2
Intromissions								
500 μg.t.p.	6.1	6/6	4.4	6/7	11.1	6/7	10.0	2/7
Intact	7.3	7/7	6.7	5/7	11.3	7/7	—	0/7
25 μg.t.p.	13.1	2/6	11.1	2/7	—	0/6	15.0	1/7
Untreated Castrates	—	0/6	—	0/2	—	0/5	—	0/2
Ejaculations								
500 μg.t.p.	8.4	6/6	8.0	6/7	9.9	4/7	13.0	1/7
Intact	7.9	7/7	7.1	5/7	11.7	7/7	—	0/7
25 μg.t.p.	11.1	1/6	15.7	2/7	—	0/6	—	0/7
Untreated Castrates	—	0/6	—	0/2	—	0/5	—	0/2

sexual responses) which are less than those attained by the heterogeneous males (Fig. 1). Daily injections of 500 μg. of t.p. did not eliminate the differences.

The influence of the genetic factor on the schedule of development of sexual behavior is revealed by a comparison of the intact males raised in the social situation. Each measure of sexual behavior was displayed earlier by the heterogeneous than by the Strain 2 males: mounting with pelvic thrusts at 4.4 and 10.4 weeks of age, intromission at 7.3 and 11.3 weeks and ejaculation at 7.9 and 11.7 weeks. Data obtained in earlier studies were not controlled with respect to the genetic and experiential factors, but the average ages at the time of the first mounting, intromission and ejaculation were given as 45 days (Louttit, 1929), 54 days and 64 days (Webster and

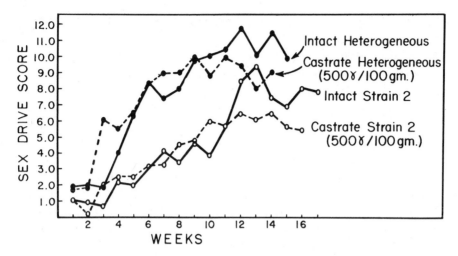

FIG. 1. Average sex drive scores of males raised in the social
situation at each week of age.

Young, 1951), respectively. One male Avery (1925) observed with a receptive female is said to have copulated on day 35.

With respect to the rate of sexual behavior only intact animals and
animals treated with 500 μg. t.p. which were socially reared are shown in
Figure 1 because they are the only groups whose scores are comparable.
They, as groups, have met the stated criteria of sufficient structure and
learning to show complete sexual responses, including intromission, within
the 17-week period. Isolated males generally showed insufficient learning
(Table 1) and animals treated with 25 μg. t.p. showed retarded structure
(Table 2). A comparison of penes and seminal vesicles reveals the average
weight and length of these structures were significantly greater (P <.01)
in males receiving the larger quantity of hormone than in those receiving
the smaller. In the males receiving 500 μg. of t.p., the measurements
approached those in the intact animals; seminal vesicle- and penis-length
were not significantly different but seminal vesicle-weight was less
(P <.01). The reason for the latter difference is not clear to the writers.
However, the obtained difference may have contributed to what appears

TABLE 2. AVERAGE WEIGHT AND LENGTH OF SEMINAL VESICLES AND AVERAGE PENIS-
LENGTH IN INTACT AND TESTOSTERONE PROPIONATE TREATED CASTRATE MALES*

	Weight—Seminal vesicle (gm./100 gm. body wt.)			Length—Seminal vesicle (cm./100 gm. body wt.)			Length—Penis (cm./100 gm. body wt.)		
	Av.	S.D.	N	Av.	S.D.	N	Av.	S.D.	N
Intact	.43	.08	6	1.47	.19	6	.57	.05	6
500 μg.t.p.	.25	.11	9	1.41	.25	9	.55	.06	8
25 μg.t.p.	.04	.02	10	.73	.22	11	.44	.04	9

* Data of males from Strain 2 and the heterogeneous stock were combined since there were
no significant differences.

to be a decrease in the performance towards the end of the testing period (Fig. 1).

Animals given 500 μg. were precocious in the expression of sexual vigor following treatment with the larger quantity of androgen. At 3 to 4 weeks of age the sexual behavior scores of the injected heterogeneous and Strain 2 males in the isolated situation were significantly higher than those of their untreated, intact controls (P <.01 and P <.001). A similar trend is seen when the injected heterogeneous males raised in the social situation and their untreated intact controls are compared. The increased score at 3 and 4 weeks may be spoken of as increased sexual vigor because no Strain 2 males mounted at that time and only some of the hetrogeneous were beginning to mount (Table 1). The score at that age, therefore, reflected an increased rate of nuzzling and abortive mounting (hurdling or leaning on the female) more than an earlier maturity of behavior.

Two other socially reared animals showed sexual responses earlier than the others in their group. They, one heterogeneous and one Strain 2 male receiving 25 μg. of t.p., were found to have testicular fragments. Their record of behavior is not included in the data summarized in Table 1, and penis- and seminal vesicle-length were not obtained, but both males exhibited mature behavior considerably earlier than any of the other animals in the 25 μg. group and earlier than most of the animals in the 500 μg. and intact groups. The heterogeneous male displayed mounts and thrusts, intromissions and ejaculations at 3, 6 and 6 weeks; the Strain 2 male at 4, 7 and 7 weeks. The performance of these animals suggests that the testicular fragments in which there was hyperplasia of interstitial cells similar to that described by Lipschutz (1950) were secreting androgen.

The suggestion made elsewhere (Valenstein, Riss and Young, 1955) that development of the capacity for the complete copulatory act depends on contact with other animals is confirmed. Without exception, in the four groups in which sufficient hormone was present, more socially raised than isolated males attained complete copulation (Table 1). The results are more striking in the Strain 2 males, but had the more rapidly developing hetrogeneous males been isolated on day 10 instead of on day 25, the differences within this stock would have been greater (Valenstein, Riss and Young, 1955).

The conclusion that the effectiveness of t.p. is influenced by the experiential background of the animal (Valenstein and Young, 1955) is also confirmed. With regard to the latter point, it should be noted that t.p. alone did not substitute for living with other animals (Table 1). The data on the socially reared, untreated castrates which were later treated with hormone provides further support for the conclusion. Most untreated castrates in the heterogeneous stock displayed mounting proficiency before hormonal treatment, although the amount of mounting was significantly less than in their intact controls. Subsequently, when they received daily injections of 500 μg. of t.p., the amount of mounting increased, approaching that exhibited by the controls. Five of the 6 went on to achieve ejaculation from 4 to 6 weeks later. The one animal that did not copulate after treatment had never demonstrated mounting. The performance of the castrated Strain 2 males raised with females and later given androgen seems to depend on the time they were isolated. Three that were isolated and injected with the larger quantity of t.p. beginning on day 70 did not exhibit ejaculation throughout 7 weeks of treatment, although 1 male mounted occasionally. On the other hand, the 2 that were isolated on day 90 ejaculated after 4 to 5 weeks of treatment. This is not surprising in view of the fact that Strain 2 males do not consistently display mounting before day 70 (Table 1).

DISCUSSION

Grunt and Young (1952) demonstrated that the characteristic level of sexual performance of the male guinea pig depends on the responsiveness to t.p. rather than on the quantity of the androgen. It has since been shown (Valenstein, Riss and Young, 1955; Valenstein and Young, 1955) that the experiential background of an animal can be an important factor in determining the responsiveness to a hormone. The present study has demonstrated the importance of the genetic background. Regardless of the quantity of t.p. administered, the genetically heterogeneous males performed at a higher level than the highly inbred Strain 2 animals and exhibited the various components of the sexual act at earlier ages.

The data suggest the role of t.p. to be that of an activator rather than a direct organizer of sexual behavior. The organization is dependent on variables associated with the strains and upon opportunity to learn the techniques of mounting and maneuvering a female. Its role as an activator is suggested by the better performance of the males treated with 500 μg. t.p. when they were 3 to 4 weeks of age. The performance was better at that time largely because the experimental males showed an increase in the rate of sexual response. It is not clear from the present data, however, that the age at which particular responses appear is set ahead by large quantities of t.p.

The problem of precocity will require further study. When androgen was injected into prepubertal rats in attempts to induce sexual behavior precociously, the median age of the first copulation (pelvic thrusts with intromissions but not ejaculations) was set ahead about 20 days (Stone, 1940; Beach, 1942). Two of the four males studied by Beach displayed ejaculation approximately 30 days earlier than the average age of the untreated males. These reports led us to expect that treatment of prepubertal male guinea pigs with sufficiently large amounts of t.p. would be followed by a precocious display of more mature sexual responses. This expectation was partially realized for the lower measures of sexual behavior, but not consistently for intromission and not at all for ejaculation. The difference in the response to treatment may be attributable to the circumstance that the rats used by Stone and Beach were intact whereas the guinea pigs used in this study were castrated. This suggestion is supported by the fact that considerable precocity of behavior was displayed by the two male guinea pigs containing fragments of what appeared to be active testicular tissue. The need of the young male for large quantities of androgen, whatever its origin, could be accounted for by the low sensitivity of its tissues to androgen (Hooker, 1937, 1942; Price and Ortiz, 1944). This hypothesis would be supported by the fact that, in their behavior and in the growth of penes and seminal vesicles, the response of the castrated male guinea pigs receiving 500 μg. of t.p. was much nearer that of the intact males than was the response of the males receiving the smaller quantity of hormone.

Still to be explained is the fact that the young males which received 25 μg. of t.p. did not acquire the sensitivity to this hormone that is characteristic of the adult (Grunt and Young, 1952). Sexual behavior was at a low level and the penes and seminal vesicles were small, even at 17 weeks of age. The males treated by Grunt and Young were injected subcutaneously whereas those used in this experiment were injected intraperitoneally. Reports of the relative effectiveness of the two routes of injection are conflicting (Deanesly and Parkes, 1937; Rubinstein, 1944; Leathem, 1948), consequently, until this uncertainty can be resolved, the possibility that the difference in responsiveness of pre- and post-pubertally castrate animals to 25 μg. of t.p. was due to the manner of injection cannot be excluded.

Another possible explanation comes from a comparison of our results with those reported by Beach and Holz (1946) who castrated a group of rats the day of birth and injected them with t.p. beginning approximately 5 months later. These rats did not execute the complete copulatory pattern as frequently as males which had been castrated when they were older and ejaculation was achieved by only 1 of 9 males. The general failure to ejaculate was attributed to a retardation of penis development which was not overcome by subsequent treatment with t.p. In our experiments in which the seminal vesicles were also studied, both structures were retarded in males castrated within 2 days after birth, but the response was better when 500 μg. rather than 25 μg. of t.p. were injected; probably therefore the young guinea pig requires more than 25 μg. of t.p. for the growth of the penis and seminal vesicles. A comparable suggestion was made by Beach and Holz who stressed the importance of testis secretion during a critical period of penile growth in the rat.

SUMMARY AND CONCLUSIONS

In an experiment in which the influence of contact with other animals was controlled, an effort was made to determine (1) if the responsiveness of castrated male guinea pigs to injected androgen as measured by sexual behavior is associated with the genetic background, (2) if genetically determined differences can be overcome by hormonal administration, and (3) if testicular hormone is necessary for development of the capacity for sexual behavior.

Regardless of the quantity of t.p. administered up to 500 μg. per 100 gm. body weight per day, or the social condition of rearing, the genetically heterogeneous males developed faster and displayed a higher level of sexual behavior than the inbred males. It is concluded, therefore, that factors peculiar to the strains modify the responsiveness to this hormone. The sexual behavior of animals castrated at birth and receiving no hormonal therapy provided evidence that the presence of the testes is not necessary for the organization of sexual behavior. However, relatively large quantities of testicular hormone seem necessary for the growth of the penes and seminal vesicles of young animals.

The administration of 500 μg. of t.p. resulted in precocity in mounting, but neither the time of the first ejaculation nor the ultimate level of sexual behavior was similarly affected. Schedules of development of sexual behavior of males raised with female cage mates and in isolation are given.

REFERENCES

AVERY, G. T.: *J. Comp. Psychol.* **5**: 373. 1925.

BEACH, F. A.: *J. Comp. Psychol.* **34**: 285. 1942.

BEACH, F. A. AND A. MARIE HOLZ: *J. Exp. Zool.* **101**: 91. 1946.

DEANESLY, R. AND A. S. PARKES: *Proc. Roy. Soc.* London, s.B., **124**: 279. 1937.

GRUNT, J. A. AND W. C. YOUNG: *Endocrinology* **52**: 237. 1952.

HOOKER, C. W.: *Endocrinology* **21**: 655. 1937.

HOOKER, C. W.: *Endocrinology* **30**: 77. 1942.

LEATHEM, J. H.: *Proc. Soc. Exper. Biol. & Med.* **68**: 92. 1948.

LIPSCHUTZ, A.: *Steroid Hormones and Tumors*, Williams and Wilkins Co., Baltimore, 1950.

LOUTTIT, C. M.: *J. Comp. Psychol.* **9**: 293. 1929.

PRICE, D. AND E. ORTIZ: *Endocrinology* **34**: 215. 1944.

RISS, W.: *Am. J. Physiol.* **180**: 530. 1955.

RISS, W. AND W. C. YOUNG: *Endocrinology* **54**: 232. 1954.

RUBENSTEIN, H. S.: *J. Urology* **51**: 88. 1944.

STONE, C. P.: *Endocrinology* **26**: 511. 1940.

VALENSTEIN, E. S., W. RISS AND W. C. YOUNG: *J. Comp. & Physiol. Psychol.* **47**: 162. 1954.

VALENSTEIN, E. S., W. RISS AND W. C. YOUNG: *J. Comp. & Physiol. Psychol.*, in press 1955.

VALENSTEIN, E. S. AND W. C. YOUNG: *Endocrinology*, **56**: 173. 1955.

WEBSTER, R. C. AND W. C. YOUNG: *Fertil. and Steril.* **2**: 175. 1951.

Editor's Comments on Papers 7 Through 11

The following selections were chosen primarily because they demonstrate the application of a variety of interesting techniques to questions of the site of action of androgens in the control of male behavior. Using methods such as the production of lesions (Brookhart and Dey, Paper 7; Schreiner and Kling, Paper 8; and Hart, Paper 11) and localized chemical (Fisher, Paper 9) or electrical stimulation (Vaughan and Fisher, Paper 10), the hypothalamus, the limbic system, and the spinal cord have been implicated as tissues sensitive to testosterone. The small number of animals responding and the difficulty of replicating the effects of both chemical and electrical stimulation have caused particular criticism of the Fisher and Vaughan and Fisher papers. Nonetheless, current research continues to suggest roles for the spinal cord, hypothalamus, amygdala, and associated neural pathways in the stimulation and inhibition of male behavior.

Reprinted by permission of the American Physiological Society from *Amer. J. Physiol.*, **133**, 551–554 (1941)

7

REDUCTION OF SEXUAL BEHAVIOR IN MALE GUINEA PIGS BY HYPOTHALAMIC LESIONS[1]

J. M. BROOKHART AND F. L. DEY

From the Institute of Neurology, Northwestern University Medical School, Chicago, Ill.

Accepted for publication April 4, 1941

Lesions in the ventral portion of the hypothalamus of the female guinea pig may bring about disturbances of the reproductive cycle (Dey, Fisher, Berry and Ranson, 1940; Dey, 1941). In addition to the ovarian disturbances which occur in some animals, all animals of the group showed a complete absence of mating behavior and copulatory reflexes. Subsequent studies have suggested that the interference with copulatory behavior is not secondary to ovarian or anterior pituitary hormonal imbalances, but is the result of the destruction of elements of the central nervous system which are necessary to the integration of the behavior pattern (Brookhart, Dey and Ranson, 1940; Brookhart, Dey and Ranson, in press). It is the purpose of this communication to make a report of the effects of similar lesions in the hypothalamus upon the mating behavior of male guinea pigs.

Nine male guinea pigs weighing between 600 and 800 grams were used in this experiment. The lesions were placed in the hypothalamus by means of the Horsley-Clarke instrument bearing a unipolar electrode. In 4 of the animals the lesions had previously been placed by Dr. R. Gaupp in the course of studies on diabetes insipidus and no data on the preoperative sexual behavior are available. The preoperative sexual behavior of the remaining 5 animals was observed by placing them singly in cages with one or more spayed females in induced estrus, and was seen to be normal. Similar behavior tests were made on each animal 2 to 4 weeks postoperatively, each animal being given 3 or more observation periods of at least an hour in duration. In order to obviate the possibility that any deficiency in sexual behavior was due to the strangeness of the surroundings or lack of sexual experience, each of the operated males was placed in a cage with three normal females for a sufficient length of time to allow all of the females to go through two sexual cycles. The presence or lack of pregnancy at the end of the period was regarded as an indication of the sexual potency of the males.

[1] Aided by grants from the Rockefeller Foundation and from the Committee for Research in Problems of Sex, National Research Council.

At the end of the experimental period, which lasted 6 months, each of the animals was subjected to the Batelli (1922) electrical ejaculation test, and the ejaculate was examined for sperm.   At the time of sacrifice, smears of the epididymides were examined for motile sperm.   Portions of the testes and seminal vesicles were fixed, sectioned and stained with hematoxylin and eosin.   The hypothalami were fixed and sectioned, and alternate sections were stained with cresyl violet and Weil stains.

When a normal male guinea pig is placed with receptive females, he begins an immediate round of investigation accompanied by purring, treading, ruffling of the hair on the back of the neck and shoulders, dilatation of the para-anal pouches, and sniffing of the genitalia of the female. A short period of this courtship activity appears to be necessary for the full

TABLE 1

NUM-BER	ACTIVITY TESTS	SEMEN	SPERMA-TOZOA	TESTES	SEMINAL VESICLES	ES-TROUS PE-RIODS	PREG-NAN-CIES
1*	No activity	Scanty	Normal	Normal	Normal	6	0
2*	No activity	Normal	Normal	Normal	Normal	6	0
5	No activity	Normal	Normal	Normal	Normal	5	2
6	No activity	Normal	Normal	Normal	Normal	6	0
7	Treading, purring and sniffing only	Normal	Normal	Normal	Normal	5	1
8	Few mild mountings	Normal	Normal	Normal	Normal	6	0
9	Few mild mountings	Normal	Normal	Normal	Normal	6	0

Total number of estrous periods—40.   Pregnancies—3

3*	Normal behavior	Normal	Normal	Normal	Normal	5	2
4*	Normal behavior	Normal	Normal	Normal	Normal	5	2

Total number of estrous periods—10.   Pregnancies—4

* Animals operated by Doctor Gaupp.   No preoperative data.

development of sexual excitement.   The act of sniffing and rubbing the hair on the back of the female in the wrong direction finally culminates in mounting and copulatory thrusts.

The results of the various observations on the operated males are summarized in table 1.   In 2 of the animals which were not operated by us, copulatory behavior was normal.   Three of the animals exhibited courtship behavior of varying degrees of intensity, one failing to mount while the other two mounted a few times but executed no copulatory movements. The remaining 4 animals showed absolutely no interest in the estrual females with which they were placed.

Each of the males was caged with 3 normal females throughout the period of 2 full cycles.   Depending on whether or not they impregnated one of the females during the first cycle, each male was with females through 5 or 6 estrous periods and thus had 5 or 6 opportunities to cause pregnancy. Out of a total of 50 opportunities which were offered the 9 males only 7 pregnancies resulted, and no seminal remnants were noted on the genitalia of any of the females which did not become pregnant. The 2 males which showed normal behavior in the preliminary tests accounted for 4 of the pregnancies, impregnation occurring in 4 out of a total of 10 estrous periods. One of the animals which showed mild courtship behavior had 5 opportunities and only impregnated one of the females in his cage. The remaining 2 pregnancies occurred in the case of one of the animals which previously had shown no interest in the females. The remaining 5 animals failed to impregnate any of the females in their cages. Thus, the animals which showed reduced sexual activity while under observation produced only 3 pregnancies out of a total of 40 opportunities.

In all cases but one, the semen which was obtained upon electrical ejaculation was normal in volume and coagulated almost immediately after ejaculation. The seminal vesicles of the one animal which delivered only a scanty amount of semen were seen to be distended with secretion when the animal was autopsied immediately after the attempted ejaculation. The seminal and epididymal smears showed numerous motile sperm in all cases. Microscopic examination of the testes indicated active spermatogenesis and lack of atrophy in the seminiferous tubules. The interstitial tissue was normal in all animals. The seminal vesicle epithelium was high, and negative Golgi images distal to the nuclei were evident in all cases.

With the exceptions of animals 3 and 4, the lesions were similar in location to those already described for the sterile female animals (Dey, 1941). The lesions occurred bilaterally near the ventral border of the hypothalamus between the optic chiasma and the stalk of the pituitary. In the two exceptions which showed normal sexual activity the lesions were similarly located. However, the lesion in one of the animals was unilaterally placed, while that of the other involved only the most superficial portions of the basal surface of the hypothalamus. No significant differences could be noted between the lesions of animals 5 and 7 and the animals which failed to impregnate any of the females in their cages.

The proximity of the lesions in these animals to the anterior pituitary raises the possibility that the reduced sexual behavior was a result of altered anterior pituitary function. However, the immediate postoperative reduction in sexual activity which has been observed argues against the possibility that such a reduction is the result of a lack of testicular hormone, since postpubertal castration causes only a gradual reduction in sexual drive (Ball, 1937; Stone, 1939). In addition, the condition of the seminal vesicle epithelium, the motility of the sperm, and the production of a copious coagulable ejaculate in these animals may be taken as an indication of a continued production of the testicular hormone (Moore, 1928; Moore and Gallagher, 1930; Moore, Hughes and Gallagher, 1930). The continuation of active spermatogenesis, the lack of atrophy of the seminiferous tubules, and the continued secretion of testicular hormone, in turn, point to a continuation of normal gonadotropic function on the part of the anterior pituitary. It is therefore suggested that the behavioral deficiency in these animals is the result of destruction of elements of the central nervous system rather than gonadotropic or androgenic hormonal imbalance.

SUMMARY

Lesions which are properly placed in the hypothalamus may abolish or greatly reduce sexual activity of male guinea pigs. This decrease in activity is manifested by a lack of interest in estrous females and by the failure of operated males, with a few exceptions, to impregnate normal estrous females.

The continuation of active spermatogenesis, and the normal maintenance of the seminiferous tubules and of the seminal vesicles indicate normal gonadotropic function on the part of the hypophysis. It is therefore suggested that the behavioral deficiency is the result of a destruction of elements of the central nervous system which are necessary for the mating reactions.

REFERENCES

BALL, J. J. Comp. Psychol. **24**: 135, 1937.
BATELLI, F. C. R. Soc. de Phys. et d'hist. Natur. de Geneve **39**: 73, 1922.
BROOKHART, J. M., F. L. DEY AND S. W. RANSON. Proc. Soc. Exper. Biol. Med. **44**: 61, 1940.
          Endocrinology **28**: 561, 1941.
DEY, F. L. Am. J. Anat., in press.
DEY, F. L., C. FISHER, C. M. BERRY AND S. W. RANSON. This Journal **129**: 39, 1940.
MOORE, C. R. J. Exper. Zool. **50**: 455, 1928.
MOORE, C. R. AND T. F. GALLAGHER. Am. J. Anat. **45**: 39, 1930.
MOORE, C. R., W. HUGHES AND T. F. GALLAGHER. Am. J. Anat. **45**: 109, 1930.
STONE, C. P. Endocrinology **24**: 165, 1939.

Reprinted by permission of the American Physiological Society from *Amer. J. Physiol.*, **184**, 486–490 (1956)

Rhinencephalon and Behavior

LEON SCHREINER[1] AND ARTHUR KLING[2]

8

From the Department of Neurophysiology, Army Medical Service Graduate School, Walter Reed Army Medical Center, Washington, D. C.

ABSTRACT

Lesions of the rhinencephalon, primarily restricted to the amygdaloid complex, modify aggressive behavior of lynxes (*Lynx rufus*), agoutis (*Dasyprocta agouti*), monkeys (*Macacus rhesus*) and domestic cats toward relative docility and precipitate a state of chronic hypersexuality. Relative docility was characterized by failure of the experimental animals to exhibit aggressive behavior, fear, or escape activity in the presence of threatening situations which precipitated such behavior in their preoperative periods. Hypersexuality was exhibited by marked increases in copulatory activity with males and females of their own and other animal species. It is concluded that the rhinencephalon and its diencephalic connections, in association with endocrine systems, are important regulators of emotional and sexual behavior of rodents, carnivores and primates.

RECENT interest in the rhinencephalon has resulted from Papez' suggestion that this structure and its diencephalic connections serve to regulate emotional behavior in addition to well known functions of olfaction (1). In association with exaggerated oral and sexual activity, changes in affect toward relative docility have been described in various species of animals after injury of the rhinencephalon (2–10). With few exceptions, such studies have been carried out on the domestic cat and dog or animals (rat and monkey) which tend to become more placid when actively trained or merely housed in the laboratory.

In contrast to the normal behavior of such animals is the intrepid savageness of the adult lynx (*Lynx rufus*) which is thought to be permanently resistant to acquisition of docile behavior through captivity (11). Similarly, the caged agouti (*Dasyprocta agouti*—a large Central American rodent) remains remarkably hostile toward man for long periods of time. A comparison of the postoperative behavior of these animals with that of the cat and monkey with comparable lesions of the rhinencephalon

affords an opportunity to observe effects of such injury upon three different attitudinal states; namely, savageness (lynx), hostility (agouti and monkey) and tameness (cat).

METHODS

One female and two male agoutis; two male lynxes; twelve male cats; and four male rhesus monkeys, all adults, were used. The lynxes were obtained from an animal supplier within a few days of being taken into captivity and were individually housed in steel cages containing small feeding doors and movable rear walls. The agoutis were housed in the laboratory approximately 1 year, and the cats and monkeys were observed for periods of 1–6 months prior to placement of lesions.

Preoperative observations including general cage behavior, reactions to routine feeding, care and handling by the experimenters, and activities in the presence of other animals were repeatedly made for several days in the case of the lynxes and for several weeks or months on the other animals.

Lesions in the amygdaloid complex of the rhinencephalon were made by aspiration, utilizing the approach and surgical technique described in previous communications (5, 6). In their postoperative periods, all animals were observed for periods of 3 months to more than 1 year. Finally, they were killed, and the brain of each animal was sectioned and stained for histologic evaluation of the lesions.

RESULTS

In presenting our results, which are necessarily of a qualitative nature, behavioral changes observed in the cats and monkeys will

---

Received for publication July 18, 1955.

[1] Present address: Mayo Foundation, Rochester, Minn.

[2] Present address: Univ. of Utrecht, Netherlands.

FIG. 1. Photographs of sections taken through greatest extent of lesions which destroyed the amygdaloid complex bilaterally in behaviorally typical *monkey 2 (left)* and *cat 54 (right)*. Diagrams illustrate extent of lesions in each animal, respectively, with *stippled areas below solid lines* representing lesions on the left side in each brain and *stippled areas below broken lines* representing those on the right side. Abbreviations as follows: *B*, basal amygdaloid nucleus; *C*, cortical amygdaloid nucleus; *G*, globus pallidus; *I*, internal capsule; *L*, lateral amygdaloid nucleus; *M*, medial amygdaloid nucleus; *OC*, optic chiasm; *P*, putamen; *R*, rhinal sulcus; *T*, optic tract; *TH*, thalamus; *V*, third ventricle.

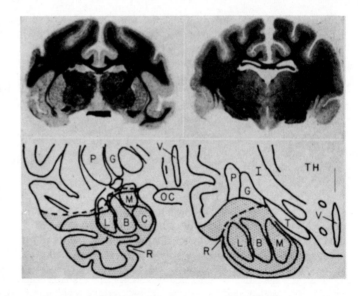

be briefly described since findings in these species have been amply reported (2, 3, 5–7, 9, 10). To our knowledge, however, there is no published information concerning the effects of rhinencephalic injury upon more refractory aggression as displayed by lynxes and agoutis. Observations on these animals will, therefore, be given in greater detail.

Preoperative Behavior. The lynxes were observed for 1 week prior to surgery. Attempts to transfer them from their shipping crates to permanent cages gave considerable information as to their basic attitudes toward the observers. The animals were extremely savage, carried out repeated attacks with bared teeth and claws and conveyed further evidence of unfriendliness through elicitation of penetrating hisses and growls. Transfer was finally accomplished after each animal was anesthetized. In the succeeding preoperative observation periods, the only direct contact with the caged lynxes was through small feeding doors. When approached, they crouched in the rear of the cages, vocalized savagely and occasionally sprang forward to bite or claw the cage doors. Similar responses ensued when animals of other species were placed before their cages. Only when fasted for a day or two would the lynxes eat in the presence of the observers.

The agoutis, having a far less dangerous potential, were less difficult to manage. These rodents, which are comparable in size to an adult rabbit (fig. 2), were individually housed in small wire cages, and when approached by the observers they would retreat, piloerect, and vocalize in a peculiar grunting fashion. Further attempts at handling resulted in attacks with teeth and claws upon the gloved hand. Because of their long curved claws, agoutis are quite clumsy on smooth surfaces and in such instances were easily handled although they did carry out effective maneuvers of defense and attack.

The monkeys were observed for periods as long as 3 months prior to surgery. Although they became accustomed to the presence of the observers, they remained aggressive and rather difficult to manage. The domestic cats, on the other hand, were friendly toward the observers and displayed typical patterns of feline defensive and aggressive behavior when confronted by dogs and other animals.

Postoperative Behavior. After placement of rhinencephalic lesions limited largely to the amygdaloid nuclei (figs. 1 and 2), all animals became docile and easy to handle. This was most strikingly demonstrated by the agoutis and lynxes. During the 1st postoperative month, the lynxes were repeatedly released from their cages and allowed to wander freely about the laboratory and hallways. They reacted favorably to petting and were frequently fed from the observer's hand (fig. 2). In addition, they were carried about the laboratory by the experimenters and were cuffed or otherwise roughly handled without retaliation. Their behavior during periods of free association with other animals (cats and monkeys) was one of friendliness with no evidence of hostility (6).

The agoutis displayed similar postoperative changes in affective behavior. For the duration of their survivals of 4–6 months, they could be removed from their cages and handled with ease. The grunts, piloerection and acts of hostility which they so prominently displayed in the preoperative periods were no longer apparent. The preparations responded favorably to stroking and petting and on numerous occasions would rise on their hind legs to paw, sniff and lick the observer's arm (fig. 2). Their conduct in the presence of operated animals of other species did not include hostility or fright. In contrast, unoperated agoutis which were housed in adjacent cages remained difficult to manage and continued to display hostility toward other animals.

Like the agoutis and lynxes, the monkeys underwent behavioral changes in the direction of decreased fear, escape activity and aggressiveness toward the observers and could be easily caught and otherwise handled when loose in the laboratory. Their behavior in the presence of other animal species was characterized mainly by friendly investigation. The domestic cats with similar lesions remained docile and tolerated considerable noxious and threatening stimulation before conveying evidence of anger. Their friendliness was markedly demonstrated in the presence of dogs and

other animals which they approached, licked, rubbed and attempted copulation (fig. 3) with no outward evidence of fear or hostility. Such activities are rarely, if ever, demonstrated by unoperated alley cats.

Associated with the reduction in aggression was the delayed appearance of hypersexual behavior which became manifest within 30–60 days after placement of the lesions. This was most pronounced in the cats which invariably mounted, took the neck grasp and carried out pelvic thrusts upon dogs, monkeys, agoutis and chickens (fig. 3). Male preparations of the other species also showed similar, but less diverse and intense, patterns of overt sexual behavior. Thus, the lynxes and agoutis were seen to make attempts to copulate with operated male cats as well as with male members of their respective species.

After the hypersexual behavior of the preparations had reached a stable plateau, their friendliness toward the experimenters became less pronounced. In the succeeding observation periods, it was, therefore, impossible to evaluate their individual state of aggressiveness as a separate entity since an emotional background of increased sexual activity was present at all times.

In the succeeding observation periods of $2\frac{1}{2}$ months, the lynxes were no longer released from their cages. This precaution was not taken because either preparation again reverted to his original state of savageness, but because it was deemed unsafe to explore limits of toler-

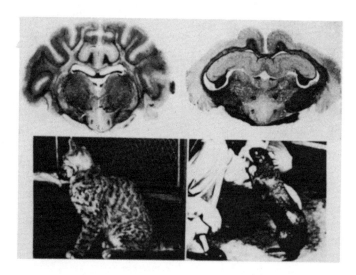

FIG. 2. Photographs of sections taken through brain of *lynx 2* (*left*) and *agouti 1* (*right*). Amygdaloid complex was destroyed bilaterally sparing the hippocampus in these and other animals of the series. Lower photos illustrate some aspects of postoperative behavior of lynx (*left*) and agouti (*right*) as further described in text.

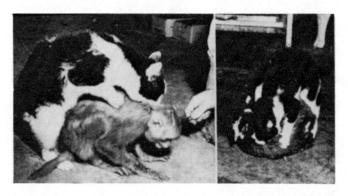

FIG. 3. Illustrations of post-operative hypersexual activity of male *cat 31* as repeatedly observed for duration of survival of *1* year.

ance toward the experimenters of an animal weighing 35 pounds and having the potential ability to inflict severe injury. For the duration of their survivals, the lynxes, when approached, would move to the front of their cages, carry out treading movements with their forelegs, lick the cage doors, and elicit high pitched noises or low growls which were not accompanied by outward evidence of anger. In this period, it was difficult to keep the animals to the rear of their cages sufficiently long to place food and water through the feeding doors. In such instances, it was necessary to chase them away by gently prodding them with a broomstick which they grasped with their teeth or paws in a manner which appeared to the observers as inquisitive or playful in nature. From all aspects of their postoperative behavior, it was concluded that the level of aggression displayed by the lynxes remained far below that exhibited by them in their preoperative periods.

While the appearance of hypersexual behavior in the agoutis and monkeys was also followed by a reduction in friendliness toward the observers, their levels of aggression also remained far below those present preoperatively. In these animals, as well as in the cats, the absence of aggressive behavior toward each other remained unaltered, and on this basis it is concluded that the relative docility displayed by them was a durable postoperative finding.

DISCUSSION

After comparable lesions of the rhinencephalon primarily restricted to the amygdaloid complex (figs. 1 and 2), the behavior of the various species of animals utilized in this study,

each with markedly different preoperative temperament, was altered toward the side of increased docility. This change was manifested by the animals toward the experimenters (fig. 2) as well as toward each other during periods of free association. In the latter situation, the absence of aggression (6) contrasted sharply with that exhibited by the animals in their preoperative periods.

The extent to which animals will tame under laboratory conditions varies with the species. This factor, then, played a role in assessing the postoperative behavior of the cats and to a lesser extent of the monkeys. The lynxes, however, showed no tendency to tame preoperatively and, while not having a comparable dangerous potential, the agoutis remained hostile after long periods of observation. It is concluded, therefore, that the postoperative change in behavior toward docility exhibited by these two species resulted directly from injury of the rhinencephalon with no essential contribution from factors of laboratory conditioning.

The specific identification and mode of function of neural mechanisms contributing to the postoperative behavior of the preparations are not clearly defined. The relative docility cannot be attributed to altered functions of the diencephalon and cerebral cortex resulting from interruption of the hippocampus-fornix system (12) since these structures in our animals remained relatively undisturbed. However, with the recent description of direct connections between the amygdaloid nuclei and the hypothalamus (13), it is suggested that interruption of this neural system withdraws influences which normally facilitate rage-provoking mechanisms of the hypothalamus.

The delayed appearance of hypersexual behavior, however, cannot be entirely considered a direct consequence of the lesions since such behavior in amygdalectomized cats is abolished by castration (14). The expression of hypersexual behavior appears, therefore, to be dependent upon release from suppression of neuroendocrine mechanisms including the hypothalamus, hypophysis and brain stem which, in association with peripheral structures, serve to influence sexual behavior (14, 15–17).

It is concluded that the rhinencephalon, in addition to olfactory functions, serves as an important regulator of affective and sexual behavior of rodents, carnivores and primates.

REFERENCES

1. Papez, J. W. *Arch. Neurol. & Psychiat.* 38: 725, 1937.
2. Kluver, H. and P. C. Bucy. *J. Psychol.* 5: 33, 1938.
3. Kluver, H. and P. C. Bucy. *Arch. Neurol. & Psychiat.* 42: 979, 1939.
4. Fulton, J. F. *Frontal Lobotomy and Affective Behavior.* New York: Norton, 1951.
5. Schreiner, L. and A. Kling. *J. Neurophysiol.* 16: 643, 1953.
6. Schreiner, L. and A. Kling. Medical Film, U. S. Army M. F. 8-7935.
7. Anand, B. K. and J. R. Brobeck. *J. Neurophysiol.* 15: 421, 1952.
8. Hillarp, N. A., H. Olivecrona and W. Silfverskiold. *Experientia.* x/5: 224, 1954.
9. Pribram, K. H. and M. H. Bagshaw. *J. Comp. Neurol.* 99: 347, 1953.
10. Rosvold, H. E., A. F. Mirsky and K. H. Pribram. *J. Comp. Physiol. Psychol.* 47: 173, 1954.
11. Personal communication.
12. Rothfield, L. and P. J. Harman. *J. Comp. Neurol.* 101: 265, 1954.
13. Adey, W. R. and M. Meyer. *Brain* 75: 358, 1952.
14. Schreiner, L. and A. Kling. *Arch. Neurol. & Psychiat.* 72: 180, 1954.
15. Brookhart, J. M. and F. L. Dey. *Am. J. Physiol.* 133: 551, 1941.
16. Smith, P. E. and S. L. Leonard. *Proc. Soc. Exper. Biol. & Med.* 30: 1250, 1933.
17. Dempsey, E. W. and D. McK. Rioch. *J. Neurophysiol.* 2: 9, 1939.

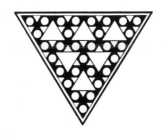

Reprinted from *Science*, **124**, 228–229 (1956)

9

Maternal and Sexual Behavior Induced by Intracranial Chemical Stimulation

Alan E. Fisher

A technique permitting chemical or electric stimulation, or both, of restricted brain areas in unanesthetized rats, and electroencephalographic (EEG) recording from these areas, has been developed and found to be of value (1).

Implants are prepared as follows. Two Tygon-insulated copper or silver wires (0.1 mm in diameter) are baked along the outside of No. 22 hypodermic tubing extending from 2 to 9 mm below the base of a plastic holder (2). The wires lead from contact points on the holder and terminate at opposite sides of the end of the shaft as a bipolar stimulating-and-recording electrode. The implant shaft is permanently inserted in the brain while the anesthetized rat is held in a stereotaxic instrument. Four holes in the base of the holder permit rigid attachment to the skull with jeweler's screws.

Two or more days later, rats are placed in 3- by 3- by 2.5-ft boxes for stimulation testing. A small clip connects the implant to light overhead leads from a 0- to 12-v, 60 cy/sec stimulator, or to an EEG machine. The clip also contains a No. 30 metal cannula that penetrates the implant shaft to the depth of the electrode tips or lower. Seven feet of PE 10 polyethylene tubing (0.024 in. in diameter) leads from the cannula to a microsyringe which can release a minimum of 0.0001 cm^3 of solution into the brain. All overhead leads are intertwined and spring-mounted, permitting repeated, controlled stimulation of a freely moving animal. Behavioral tests with a series of external stimulus objects are given before, during, and after chemical stimulation, and placebo solutions and nonhormonal neural excitants are used for control brain injections during initial tests or retests.

The technique is first being used to test whether there are "primary drive centers" under hormonal influence or control

(3). Thus far, maternal and sexual behavior have been elicited from separate brain loci in a series of males during stimulation with sodium testosterone sulfate in 0.09-percent saline, and, subject to verification, a mixture of a pure salt of estrone and a suspension of progesterone has induced heat behavior in two females (4). Maternal behavior elicited during chemical stimulation includes nest building and a persistent retrieving and grooming of litters of young. All aspects of mating behavior have been induced or accentuated. Attendant high-drive states are suggested by an exaggerated speed, compulsiveness, and frequency of all overt responses during positive test periods.

Although individual animals respond positively for up to 4 test days, duplication of effect from animal to animal has imposed difficult problems. Of 130 male operates tested with the testosterone salt, five have shown complete maternal response, 14 have shown nesting behavior only, and six have shown exaggerated sexual response. Histological data suggest that locus of immediate action is a critical factor, with slight variation in placement leading to incomplete, confounded, or diffuse drive states, or negative results. Initial findings tentatively implicate the medial preoptic area in maternal behavior and the lateral preoptic area in sexual behavior.

A variety of effects have been noted during testosterone stimulation at adjacent loci. Those seen in five or more cases include respiratory changes, diffuse hyperactivation, long-lasting exploratory-like behavior, repetitive localized muscular response, digging, leaping, and seizures.

The following list includes other significant points of departure from chemical- or electric-stimulation data reported in the literature. Since positive test responses completely transcend control be-

94

havioral data, two single cases of possible theoretical significance are also briefly described.

1) Elicited behavior, whether specific or diffuse, commonly continues without decrement for more than 90 minutes following chemical stimulation.

2) An entire hierarchy of related responses can be brought to the threshold of activation, with adequate stimulus objects insuring integrated behavior.

3) A segment of a response hierarchy may occur alone. Excessive nest-building is often seen, and one aspect of the nesting pattern, "pick up paper and push under body," continued rapidly for more than 1 hour in three cases.

4) Specific behavior has occurred in the absence of appropriate external stimuli. One male continuously "retrieved" his tail when stimulated with testosterone and then repeatedly retrieved a female in heat. When pups and paper were supplied, however, the male built a nest and retrieved and groomed the young, neglecting the objects to which he previously reacted.

5) In one case, maternal and sexual drives were activated simultaneously. The testosterone-treated male reacted to stimuli related to each drive and to a degree never shown in control tests. Double activation was most convincingly illustrated when a female (not in heat) and newborn rat pups were presented. The male attempted copulation twice while a pup he was retrieving to a nest was still in his mouth. Shaft placement was adjacent to both areas previously implicated in sexual and maternal response.

All effects have followed injection of minute amounts of solution, containing from 0.003 to 0.05 mg of the testosterone salt. In this connection, it must be emphasized that the possibility remains that causative factors other than hormonal properties may be operating. Thus far, however, control testing with neural excitants, physiological saline, and electric stimulation has failed to produce or perpetuate these complex behavior patterns.

Initial EEG data are promising. Records from six testosterone-treated "maternal" or "nesting" males have shown single spiking lawfully spaced in a normal record rather than the general spiking seen after picrotoxin or metrazol injection. Selective chemical action seems probable. Also, testosterone-induced spiking occurs before, not during, elicited overt behavior. Correlation of EEG changes with brain stimulation and with elicited response presents technical problems but could become a powerful tool.

The early data suggest other implications and further applications.

1) A neurophysiological definition of drive seems within reach. The role of hormones in eliciting behavior should be clarified as well as the organization of neural circuits that mediate or integrate primary drives. Present data favor a "neural center" theory, but control studies are needed.

2) Responses analogous to symptoms of mental dysfunction often occur during chemical stimulation. These include "obsessive-compulsive acts," tics, diffuse excitation, and states of hypo- and hypersensitivity to sensory stimuli. Further work may establish tie-ins between shifts in chemical balance in the central nervous system and certain forms of mental dysfunction.

3) Males having no adult contact with females, young, or paper have responded to chemical stimulation with integrated maternal behavior on the first trial. The data are pertinent to the problem of whether innate, centrally organized sensory-motor connections exist for complex response systems.

4) Testosterone has ostensibly elicited both sexual and maternal patterns. The findings may reflect multiple properties for the hormone. Limited progesterone-like activity has been proposed for the androgens by Selye, and progesterone has been linked to maternal response.

5) Chemical stimulation has elicited long-lasting, integrated behavior that was free of lapse or interference. The data strongly suggest selective chemical action within the central nervous system. Further work may demonstrate that differential sensitivity to specific physiological change by functionally organized areas of the nervous system is a basic principle of neural function.

In summary, integrated, long-lasting drive states have been induced by direct chemical stimulation of brain loci. Further work with techniques of this type could well lead to breakthroughs in the study of pharmaceutical action, brain organization and function, and the dynamics of behavior.

ALAN E. FISHER
Department of Psychology,
University of Wisconsin, Madison

References and Notes

1. I initiated the work as a public health post-doctoral fellow at McGill University with D. O. Hebb as sponsor. Work continues at the University of Wisconsin during the second fellowship year with H. F. Harlow as sponsor.
2. The Lucite holder is identical with that described by J. Olds and P. Milner, *J. Comp. Physiol. and Psychol.* 47, 419 (1954).
3. For example, F. A. Beach, *Psychosomat. Med.* 4, 173 (1942).
4. Hormone salts supplied by Ayerst, McKenna, and Harrison, Ltd., Montreal, Canada.

17 May 1956

Reprinted from *Science*, **137**, 758–760 (1962)

10

Male Sexual Behavior Induced by Intracranial Electrical Stimulation

Eva Vaughan
Alan E. Fisher

Abstract. Electrical brain stimulation in the anterior dorsolateral hypothalamus produced a marked increase in sexual capacity in some male rats. Several measures of sexual behavior, including the length of the postejaculatory refractory period, were significantly affected.

The importance of the role of certain areas of the anterior hypothalamus in the mediation of male sexual behavior has been indicated by studies of ablation, chemical stimulation, and intracranial self-stimulation (*1*). The purpose of our investigation was to determine whether changes in the sexual behavior of male rats could be produced by electrical stimulation of the hypothalamus. Electrodes were permanently implanted in 30 adult male rats. After recovery the animals were tested with estrous female rats. Mounts without intromission, with intromission, and with ejaculation, with and without stimulation, were recorded. A short-pulse stimulator (Grass) delivered alternating positive and negative square waves.

Three of the subjects showed the following behavior, to varying degrees, during electrical stimulation in the 8- to 10-ma range. Within seconds after the onset of stimulation, mounting began and continued at a high rate; it stopped immediately on termination of stimulation. The grooming behavior which ordinarily follows an intromis- sion was notably absent in all three subjects. After intromission the male's behavior remained oriented toward the female, even if no mounting occurred for some time. The postejaculatory period was significantly shortened in two of the subjects and not affected in the third. Penile erection was virtually constant during stimulation. Normal sexual behavior in the absence of stimulation was sometimes observed, but it disappeared after 5 to 10 minutes of stimulation.

The major effects were especially marked in one animal, rat C. Its median postejaculatory period was 27 seconds, as compared with the normal average of more than 5 minutes. This rat had as many as four ejaculations during one 5-minute stimulation period. We observed that the ejaculations were not merely "behavior" but were physiological ones: four sperm plugs were found on the cage floor after the first 5-minute stimulation period on one of the test days.

, After preliminary studies to determine the general characteristics of the effect, two of the subjects, rats B and C, were subjected to further tests of specific variables. These included the duration of and progressive changes in sexual behavior under stimulation, the effect of variation of the stimulus animal, and changes in sexual responses as a function of variation in the intensity of electrical stimulation.

To determine changes in the effect over time, rat C was tested with an estrous female for 7½ hours, during which time the stimulation was turned alternately on and off every 5 minutes. The results are shown in Fig. 1. During the 220 minutes of stimulation, the subject had 174 mounts without intromission, 81 mounts with intromission, and 45 ejaculations. The following changes occurred over time: (i) the number of intromissions and ejaculations declined rapidly and the number of mounts without intromission declined more slowly; (ii) the latency of the first response and first ejaculation for each stimulation interval increased throughout; (iii) postejaculatory refractory periods increased in length; (iv) the number of intromissions for each ejaculatory sequence decreased; and (v) the ratio of mounts without intromission to mounts with intromission increased throughout. Very similar results were obtained in a test with rat B, in which the frequency of stimulation was somewhat different, but the total stimulation time was the same. Neither of these animals had ceased responding when the tests were terminated.

The effect of a change in the stimulus animal was investigated by testing rat C for 60 minutes (stimulation on 5 minutes, off 5 minutes) with an ovariectomized female. The subject had 15 mounts without intromission and no other sexual responses. This may be compared with the same animal's score of 28 mounts without intromission, 36 with intromission, and 19 ejaculations during the first 60 minutes of the previously described test. On the other hand, the subject without stimulation did not respond at all to the ovariecto-

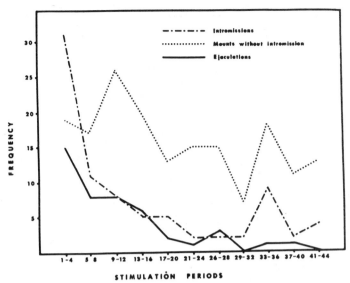

Fig. 1. Number of mounts without intromission, intromissions, and ejaculations of rat C in 44 successive 5-minute stimulation periods.

mized female. Thus, the stimulation produced some increase in sexual behavior toward a nonestrous female, but far less than toward an estrous one.

How important is the intensity of the electrical current? An intensity of 5 ma in rat B produced somewhat heightened activity and caused no increase in sexual behavior, 8-ma intensity produced a marked increase. In rat C, three intensities—4, 6, and 8 ma—were systematically varied (8 ma produced the optimal effect in this animal). The amount of sexual behavior was significantly affected, an increase in all three categories of sexual response occurring with an increase in the level of stimulation. The results were almost entirely accounted for by behavioral differences occurring at the 8-ma level; the difference under 4 and 6 ma was negligible. Furthermore, at both 4 and 6 ma, the subject showed grooming and resting responses, while at 8 ma such activity was absent. Ten milliamperes produced disorganized leaping or seizures in both animals.

An attempt was made to assess the "drive value" of the stimulation in rat C by requiring it to learn to run to the right side of a T-maze, while under electrical stimulation, for the opportunity to copulate with an estrous female. Its rate of learning was compared with that of two control subjects (without stimulation) under the same conditions. When the number of trials required to

learn was computed the control rats were significantly superior.

We doubt that these results could be attributed to a low drive level under electrical stimulation. Other factors, such as aversive side effects of high-intensity stimulation or motor effects which interfere with performance could be involved. Even more probable, an unusually intense state of drive, indicated in these animals by both objective and subjective assessment, may in and of itself include aversive components which would preclude clear cut results in a test of motivated learning.

Histologies of rats B and C showed that the electrode tip was in the lateral anterior hypothalamus. In one case the tip impinged on the upper boundaries of the lateral preoptic area. More exact localization, as well as specific differentiation between positive and negative electrode placements, was not possible, thus indicating that the area involved may be quite circumscribed.

The extremely short latency of the first sexual response after onset of stimulation and the immediate termination of sexual behavior with the termination of stimulation indicate that the effect is mediated neurally rather than humorally. Apparently the electrode tip rested in an excitatory area for male sexual behavior. The stimulation of this area not only increases sexual response level but also elimi-

97

nates or reduces the effects of certain inhibitory factors—evidenced by the lack of postintromission grooming and the marked curtailment of the postejaculatory refractory period. Exactly how the mediation of the effect takes place cannot be determined from this study, but several possibilities may be noted. Stimulation of the excitatory area may: (i) raise its firing level to a point where it overcomes normal inhibitory influences; (ii) block the action of an inhibitory mechanism directly; (iii) lower the firing threshold of the excitatory area; or (iv) produce a combination of two or more of these effects.

The results of our experiment supplement and extend the experimental data obtained with different techniques (1) and provide additional evidence for an anterior hypothalamic integrating system within the sexual behavior circuit (2).

EVA VAUGHAN
ALAN E. FISHER
Department of Psychology,
University of Pittsburgh,
Pittsburgh 13, Pennsylvania

References and Notes

1. A. E. Fisher, *Reticular Formation of the Brain*, Henry Ford Hospital Symposium, H. H. Jasper *et al.*, Eds. (Little, Brown, Boston, 1959), pp. 252–54; ———, *Current Trends in Psychological Theory* (Univ. of Pittsburgh, Pittsburgh, Pa., 1960), pp. 70–86; C. H. Sawyer, *Handbook of Physiology: Neurophysiology*, sect. 1, vol. 2, H. W. Magoun, Ed. (Am. Physiol. Soc., Washington, D.C., 1960), pp. 1225–1240.
2. Hormone preparations used in this study to bring the females into estrus were generously supplied by Dr. Edward Henderson of the Schering Corp., Bloomfield, N.J. Acknowledgment is due Harry LeWinter for aid in implanting many of the animals used in this study; the research was supported by a grant (M1951) from the National Institute of Mental Health.

25 June 1962

Reprinted from *Science,* **155,** 1283–1284 (1967)

11

Testosterone Regulation of Sexual Reflexes in Spinal Male Rats

Abstract. Castrated male rats with complete midthoracic spinal transections were maintained on exogenous testosterone; they showed intermittent clusters of genital responses consisting of erections, quick flips, and long flips of the penis when gentle pressure was constantly applied to its base. The number of these genital responses per 30-minute test was markedly influenced by withdrawal or administration of testosterone.

The mating behavior of male rats declines rapidly subsequent to gonadectomy. All elements of the mating pattern, however, do not disappear at the same time. It is reported that the ejaculatory pattern is the first response to disappear (4 to 8 weeks for most animals), whereas appetitive responses such as investigation of the estrous female, mounting, and pelvic thrusting disappear more gradually (*1, 2*). My results suggest that the decline in the ejaculatory response is due to the effect of withdrawal of androgens on spinal elements. The more gradual decline of appetitive responses presumably reflects the effects of withdrawal of androgens on hypothalamic or other forebrain structures.

A preliminary study showed that when a male rat with its spinal cord transected is restrained on its back and the preputial sheath is held behind the glans penis, a series of clusters of genital responses occur intermittently. Each cluster usually consists of several brief erections of the glans penis, two or three quick dorsal flips of the glans, and one to three extended, long dorsal flips of the glans (Fig. 1). The long flips are accompanied by strong ventral flexion of the pelvis. The pattern and duration of the long-flip response closely resemble the pattern and duration of an ejaculatory response of the intact rat. I have analyzed these reflexes in detail elsewhere and have suggested that the interval between the onset of response clusters (2 to 3 minutes) represents a refractory period of spinal sexual responses and that the erections and quick flip responses that occur as spinal refractoriness dissipates represent gradations of the long-flip response (*3*).

I used adult, sexually naïve, Long-Evans male rats. The animals were castrated at 120 to 131 days of age and subsequently given daily subcutaneous injections of 0.2 mg of testosterone propionate in oil (*4*). Five days

after the rats were castrated a spinal transection was performed in the midthoracic region (between thoracic spinal nerves six to ten) while the animals were under barbiturate anesthesia. The spinal cord was exposed by means of a laminectomy. To be certain that the spinal cord was completely transected, I removed a segment of cord approximately 2 mm long by aspiration through a fine glass tube. The incised muscle and skin were pulled together with separate layers of simple interrupted sutures. The postoperative care consisted of expressing the urinary bladders two or three times per day and washing off any accumulated urine or feces. The flexion reflex was frequently monitored to assess the integrity and condition of the isolated spinal cord. Twelve rats which survived the surgery and postoperative period in satisfactory condition and which showed strong somatic reflexes were used as subjects.

It was judged from a preliminary study that spinal elements mediating sexual reflexes have recovered from spinal shock before 20 days. In this study tests for sexual reflexes were initiated on the 20th day subsequent to spinal transection. By this time the rats were 145- to 156-days old and had received daily injections of testosterone for 25 days. Each test for sexual reflexes consisted of placing the animal on its back in a glass cylinder for partial restraint, pushing the preputial sheath behind the glans penis and holding it in this manner for 30 minutes. The time of onset of each response cluster and the types of responses occurring within each response cluster were observed and recorded. All such 30-minute tests were conducted at 2-day intervals, and all animals were tested at 2-day intervals throughout the experiment. After the first four tests, exogenous testosterone was withdrawn from six of the rats, specified as group A, while the other six rats, specified as group B, were maintained on testosterone. The latter group thus served as a control for the possible influence of further recovery from spinal shock on sexual reflexes. Daily injections of testosterone (0.2 mg) were administered again to the animals in group A after the completion of four tests (8 days) conducted while the animals were off testosterone. Six 30-minute tests (12 days) were conducted on the animals in group A after they were again placed on testosterone. Exogenous testosterone was withdrawn from the animals in group B after the eighth 30-

minute test. Six 30-minute tests (12 days) were then conducted on these animals while they were off testosterone. The schedule of hormone administration is summarized in Fig. 2. One of the animals in group A became sick after the eighth test and was killed; therefore, calculations for group A on tests 9 through 14 represent the mean of five animals rather than six.

There was a pronounced decline in the total number of erections, quick flips, and long flips per 30-minute test when testosterone was withdrawn (Fig. 2). The decline appeared to be progressive. When data from the two groups were pooled for analysis of the effects of withdrawal of testosterone, the Wilcoxon matched-pairs signed-ranks test (*5*) showed that the number of erections, quick flips, and long flips for all 30-minute tests conducted while the subjects were off testosterone were significantly (*P* < .05) below the mean number of these responses for the four 30-minute tests conducted immediately before withdrawal of testosterone. In light of the possible confounding influ-

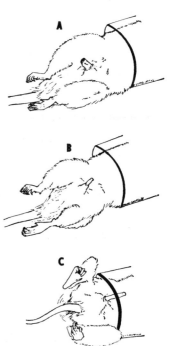

Fig. 1. Illustrations of an erection (A), quick flip (B), and long flip (C) which occur in a response cluster and which are evoked when gentle pressure is constantly applied to sides of the base of the penis. The rats are shown as they were restrained in a glass cylinder.

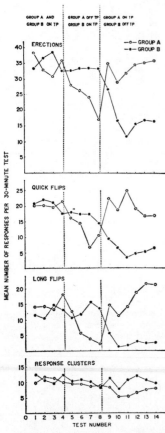

GROUP A AND	GROUP A OFF TP	GROUP A ON TP
GROUP B ON TP	GROUP B ON TP	GROUP B OFF TP

—o— GROUP A
—•— GROUP B

ERECTIONS

QUICK FLIPS

LONG FLIPS

RESPONSE CLUSTERS

MEAN NUMBER OF RESPONSES PER 30-MINUTE TEST

TEST NUMBER

Fig. 2. Influence of withdrawal and administration of testosterone on number of erections, quick flips, long flips, and response clusters per test. Tests were conducted at 2-day intervals, and all animals were tested at 2-day intervals throughout the experiment. There were six rats with transected spinal cords in each group. When hormone withdrawal is indicated there was no injection given on the day of the last test which was conducted while the animals were on testosterone (TP). When readministration of hormone is indicated, the first injection was given 48 hours before the first test which was conducted while the animals were on testosterone. When it was withdrawn from rats in group B, the number of long flips per test fell to approximately two per test (individual range of zero to nine) on tests 10 through 14.

ence on sexual reflexes of recovery from spinal shock, the most revealing effect of withdrawal and administration of testosterone is seen in the reversal of mean number of responses for the two groups, after the eighth test, when group A was placed on testosterone and group B was taken off testosterone. All animals showed the change char-
-acteristic of their respective groups, and the probability is less than .01 of this reversal occurring by chance. There was no appreciable effect on the number of response clusters per test (Fig. 2). Nor was there any detectable change in the latency to the first response cluster or in the intervals between response clusters which could be attributed to withdrawal or administration of testosterone. Thus there was a decline in the number of genital responses per cluster (and hence a decline in the duration of the cluster), but not in the timing mechanism controlling the onset of a response cluster.

It could be argued that the decline in sexual reflexes in the spinal animals following withdrawal of androgen is a reflection of a decreased sensitivity of genital sensory receptors, since it has been reported that genital papillae on the glans penis of the male rat decrease in size and number subsequent to castration (6). Two facts argue against this contention: (i) there is an occasional male rat which shows a complete ejaculatory pattern several months after castration (2); (ii) a complete mating response can be evoked in castrated male rats with hypothalamic implantation of testosterone in amounts too small to affect genital morphology (7).

The role that the sexual reflexes, which can be evoked from spinal rats, play in mating behavior of the intact male rat is uncertain. Assuming they have some function in copulation, it appears as though the decline of ejaculatory and, possibly, intromission behavior in the male rat following castration may be due to the influence of withdrawal of gonadal androgens on spinal neurons mediating the sexual reflexes. The more gradual decline in appetitive responses, such as investigation

of the female genitalia, mounting, and pelvic thrusting, is probably a reflection of the effect of withdrawal of gonadal androgens on hypothalamic or other forebrain structures. This is suggested by studies which show an abolishment of mating behavior in male rats (which could not be attributed to impairment of gonadal androgen output) following hypothalamic lesions (8) and by a study by Davidson showing a resumption of mating activity in castrated male rats caused by the implantation of testosterone into the hypothalamus (7).

BENJAMIN L. HART
Departments of Psychology and
*Anatomy, University of California,
Davis 95616*

References and Notes

1. C. P. Stone, *Endocrinology* 24, 165 (1939); F. A. Beach and A. Holz-Tucker, *J. Comp. Physiol. Psychol.* 42, 433 (1949).
2. J. M. Davidson, *Animal Behav.* 14, 266 (1966).
3. B. L. Hart, *J. Comp. Physiol. Psychol.*, in press.
4. Preliminary observations showed that in some male rats with transected spinal cords seminiferous tubules of the testicles undergo marked degeneration. Since there was some question whether interstitial cells (and hence androgen secretion) were also affected, all animals were castrated and administered a replacement dosage of androgen considered to be well above that required to maintain normal mating behavior of castrates. The daily dose of 0.2 mg of testosterone propionate is two to four times that estimated to be an adequate replacement dosage (1).
5. S. Siegel, *Nonparametric Statistics for the Behavioral Sciences* (McGraw-Hill, New York, 1956), pp. 75–83.
6. F. A. Beach and G. E. Levinson, *J. Exp. Zool.* 114, 159 (1950).
7. J. M. Davidson, *Endocrinology* 79, 783 (1966).
8. A. Soulairac and M. L. Soulairac, *Ann. Endocrinol.* 17, 731 (1956); K. Larsson and L. Heimer, *Nature* 202, 413 (1964).
9. Supported by grants FR-05457 and MH-12003 from the National Institutes of Health. I thank C. M. Haugen, G. M. DaVirro, and L. Farrell for technical assistance.

10 January 1967

Hormones and the Behavior of the Female

II

Editor's Comments on Papers 12 Through 18

Following the description of vaginal changes indicative of ovarian function, it became possible to study the cyclic behavior of intact females. The portions of the influential monograph by Long and Evans (Paper 12) that deal with the vaginal and behavioral changes indicative of the estrous cycle have been reproduced here. Owing in part to the difficulties associated with studying a cyclic animal, descriptive studies of female sexual behavior have coincided with or followed studies of female reproductive physiology. The studies of W. C. Young and his associates have provided many excellent examples of the type of parametric data essential to the interpretation of female cyclicity (see Paper 14). Goy and Young (Paper 17) subsequently dealt with questions of the genetic basis of individual differences in the hormone sensitivities of females.

Building on the contributions of Long and Evans, Allen and Doisy (Paper 13) described the first bioassays for the follicular hormones (which are now known as estrogens), as well as the partial isolation of an active follicular hormone. Although Allen and Doisy reported full estrous responses in rats treated with follicular hormone only, it was later shown by Dempsey, Hertz, and Young (Paper 15) for guinea pigs and Boling and Blandau (Paper 16) for rats that the ability of estrogens to induce behavioral estrus in these species was markedly enhanced by subsequent treatment with progesterone (a hormone produced in quantity by the corpus luteum). In females of both species, 36 to 48 hours of estrogen priming, followed by progesterone injection, induced, within 4 to 6 hours, behavioral receptivity essentially identical to that observed during natural estrus.

Recently the importance of progesterone to female behavior in the intact guinea pig has been further suggested by the observation of high levels of systemic progesterone coincident with the onset of behavioral estrus (Feder, Resko, and Goy, Paper 18). The relatively accurate hormone assay techniques now available should shortly provide new information regarding the relationship between the presence of hormones and the occurrence of behavior in intact females.

Reprinted from *Mem. Univ. California Press*, 6–7, 17–21, 32–33, 70–78, 82 (1922)
Originally published by the University of California Press and reprinted by permission of The
Regents of the University of California.

12

The Oestrous Cycle in the Rat and Its Associated Phenomena

J. A. LONG
H. M. EVANS

For somewhat over thirty years scattered papers in the literature have indicated that periodic changes can be recognized in the structure of the mucosa of the generative tract in mammals, but no systematic study of these phenomena with a single recent exception seems to have been attempted. We refer to the papers of Moreau, Lataste, Retterer, and Königstein. Yet even though this older literature made it seem extremely probable that these changes were associated with the sexual history of the animal and in particular with oestrus, and were hence definitely related to ovulation, it was clearly impossible to follow these changes in any one animal, or even to predict when they would recur, unless possibly advantage was taken of their relation to some outspoken event such as pregnancy. By the fortunate discovery that in the guinea pig these mucosal transformations are accompanied by the dehiscence of epithelial cells so that at times the lumen of the vagina has a characteristic cell content, it has been possible for Stockard and Papanicolaou to show us that we may discover with ease in the living animal the exact occurrence and progress of these cycles. When it has been proved, as Stockard and Papanicolaou have done for the guinea pig, and as we have been able to do with exactitude for the rat, that these cycles are correlated with the rhythmic discharge of ova from the ovary, it will be seen that we now have in our hands for the first time an accurate method for the detection of ovarian function in experimental animals. This fact promises important consequences, for it enables us to investigate disturbances of ovarian function which may be experimentally produced.

Before proceeding to such studies, it would be necessary to establish clearly all the characteristics of what we have called the normal oestrus or reproductive cycle in the animal form which we have chosen to investigate. The present monograph is devoted to that study. We may remark at once that the rat, while exhibiting certain fundamental similarities when compared with the guinea pig, has also striking differences. It is hoped that this inquiry will serve as a preliminary to establish some of the more fundamental phenomena which may be expected to characterize the various steps in the oestrous cycle throughout the higher mammalia. It has been carried on by us for a period of some four years and has been conducted with sufficient deliberation and repetition to lead us to cherish the hope that we have been able to establish reliable generalizations. It has involved inquiries into the sexual behavior of about a thousand individual animals, upon each of which, for the period of observation or experiment, accurate daily records have been kept. No one who has not experienced a similar self-imposed, long-continued, and meticulous responsibility will readily appreciate the amount of conscientious concern necessary to such tasks. Well over five hundred careful autopsy protocols have been founded on the study of complete serial sections of ovarian and representative sections of vaginal material.

Grateful acknowledgment is made of the grants awarded by the Board of Research of the University of California and the American Medical Association for the prosecution of this work.

II. LITERATURE

In view of the excellent work of Marshall on the *Physiology of Reproduction,* a comprehensive review of the very extensive literature on this subject is unnecessary. However, for convenience, a list is presented of the times of oestrus and ovulation which have been reported for various mammals by various observers:

Animal	Length of Diœstrous cycle	Ovulation	Authority	
CARNIVORA: Cat	14 days Fraction of gestation About 14 days	Dependent on copulation Only at copulation Only after copulation and independent of abortion Only after copulation Only after copulation Only after copulation	Coste (Hansen) Marshall & Jolly Ancel & Bouin Bouin & Ancel Winiwarter & Saimont Marshall Longley Van der Stricht, R. Hammond & Marshall	1847 1905 1909 1909 1909 1910 1910–11 1911 1914
Dog		Spontaneous Spontaneous at heat Spontaneous Spontaneous Spontaneous (?) at heat Spontaneous at œstrus Spontaneous and associated with rut Spontaneous Spontaneous Spontaneous at œstrus	Spallanzani (Heape) Bischoff Sir E. Millais Pierre Rossi Iwanoff Marshall & Jolly Ancel & Bouin Ancel & Bouin Marshall Hammond & Marshall	1786 1844–5 1884 1884 1900 1905 1908 1909 1910 1914
Lioness	3 weeks		Marshall & Jolly	1905
Otter	Month		Marshall & Jolly Marshall	1905 1910
MARSUPIALIA: Dasyurus Didelphys virginiana		Independent of copulation Spontaneous, after œstrus Spontaneous	Hill Hill & O'Donoghue Hartman	1910 1913 1916 1919
RODENTIA: Ferret		Usually only with copulation, although sometimes spontaneous at œstrus	Marshall	1904

* * * * * * *

V. THE NORMAL OESTROUS CYCLE

A. Changes in the Vagina

1. OBSERVATIONS ON THE LIVING ANIMAL

If the vaginal mucous membrane of rats in full reproductive vigor is examined daily by means of an appropriately small speculum, two strikingly different conditions will be encountered from time to time. These conditions, which alternate with each other, may be roughly described as the "moist, pinkish," and as the "dry, white" conditions, and either condition will usually be found to recur in four days. With the short cycle of four days, which we have found to be normal for the majority of young adults, it accordingly happens that during about one-half of this time the vagina has the moist, pinkish appearance characteristic of the dioestrous pause and during the other half the "dry" condition associated with a succession of events grouped together and related to oestrus proper. The moist, pinkish condition typifies the dioestrous pause, or the interval between oestrous changes. The dry, lusterless, or white condition is usually associated with a swelling or turgescence of the small radiating folds about the vaginal aperture (fig. 3, pl. I). In its incipiency it is always associated with the manifestation of oestrus, and toward its close with the occurrence of ovulation.

Vaginal Smears.

These stages succeed one another in orderly sequence, and each is characterized by a different histological make-up of the vaginal fluid. It is indeed not the least remarkable of the histological phenomena encountered in these studies that we should have so clearly marked a succession of the cell types which are thrown off within the vagina. Our study has given us a satisfactory explanation of the origin of these cells and has also in a sense made it clear why we should have this succession, but they have only served to make more precise our acquaintance with the astonishingly orderly and steplike character of the cellular dehiscence, by means of which there appears in the vaginal lumen at any one time only one type of epithelial cell. These studies have shown us that the dehiscence of the epithelium, which proceeds relatively speedily, also takes place at about the same rate over the entire mucosal surface, for were it otherwise we would not have the homogeneous cell picture which we actually find, but only confusion arising from the occurrence in some localities of an earlier, and in others of a later, succession of the same events.

We have chosen the stages disclosed by change in the content of the vaginal smear as of fundamental value in our study—as the stages with which to correlate changes in the histological structure of the internal organs of reproduction for two weighty reasons:

(1) In contrast to other phenomena (e.g., the appearance of external swelling and, in particular, the duration of oestrus), each of the stages marked by changes in the vaginal smear is approximately constant in length in all animals.

(2) The "vaginal stages" constitute the only reliable method of recognizing subdivisions of the oestrous cycle in the living animal.

We have, accordingly, made the histological characteristics of the vaginal smear the criteria for delimiting steps in the oestrous cycle and for this reason, before giving an account of the changes undergone by the generative apparatus in each stage, shall begin with a description of the successive histological pictures found at each stage of oestrus in the vaginal smear. (See table 1, p. 42.)

Dioestrous Interval.

During the interval, or dioestrous pause, which constitutes about half of the entire cycle and which possesses a relatively clean, moist, glistening, normally somewhat translucent, pinkish mucosa, the pipette, spatula, or other sampling instrument will succeed in withdrawing from the vaginal lumen a variable quantity of thin, somewhat stringy mucus in which are entangled leucocytes and small, irregularly shaped, free epithelial cells (figs. 4 and 5, pl. II). Both types of cell may occur in considerable numbers in some animals or may be both of them very sparse in others. The leucocytes are usually fairly abundant and are the characteristic small, polymorphonuclear elements which, as is well known, often have annular nuclei in the rat. The leucocytes are often distorted in one dimension so as to be slightly elongated within the "strings" of mucus. The epithelial cells are always single, never in groups.

Stage One.—Oestrus is inaugurated by the occurrence of a distinct stage which we have designated as "Stage One" (the stage corresponding to Heape's pro-oestrum). While characterized by a distinctive histological picture of the vaginal content, it may also often be detected macroscopically by the changed appearance of the surface of the vaginal mucosa as disclosed by the speculum; the mucosa is no longer moist, seems distinctly less transparent, and, while possessing a slight sheen, or luster, has a characteristic opaque look. In some animals a further external characteristic of Stage One may be found in an increasing turgescence of the small radiating folds about the vaginal aperture (fig. 3b, pl. I) as compared with the condition during the dioestrous interval (fig. 3a). To the naked eye, however, these appearances are not so distinctive that they may not be overlooked, but under no conditions will the incidence of Stage One escape the observer who takes the precaution to obtain a microscopic sample of the contents of the vaginal lumen. To a small drop of physiological saline, Ringer's, or Locke's solution, a narrow spatula may be applied after its withdrawal from the vagina where it has been in contact with the mucosal surfaces. This simple method of sampling the vaginal content is actually adequate for an infallible diagnosis even when studied with a lens which gives a magnification of but eighty diameters. The small translucent, jelly-like, adhesive mass withdrawn on the tip of the spatula is not easily dislodged into the drop of saline solution upon the slide. In it leucocytes have entirely disappeared and great numbers of small, round, nucleated epithelial cells of strikingly uniform appearance and size are now present (figs. 10 and 11, pl. II). Occasionally small sheets of these cells may be encountered. The microscopic picture is absolutely characteristic and unmistakable, for these particular cellular elements have occurred and will occur at no other time in the oestrous cycle. The manner of their formation as a peculiar layer at the surface of the vaginal epithelium will be disclosed later. Not the less striking and unexplained is their sudden dehiscence, a fact correlated with the equally sudden cessation of leucocytic migration. Leucocytes, indeed, are not encountered again until the close of the next two stages in oestrus. Stockard has found a similar behavior on the part of these cells during the pro-oestrum and oestrus in the guinea pig and it would appear that this peculiar lull, or sharp pause, in the penetration of the mucous membranes by leucocytes, which is otherwise going on constantly throughout the dioestrous pause, is a very fundamental characteristic of the histology of oestrus in the two rodents which have been investigated. The average duration of Stage One is twelve hours.

During Stage One females will usually not accept copulation, defending themselves against any aggression on the part of the males. It is hence proper to recognize this as the stage which is normally preparatory for oestrus. In some cases, however, at about the middle of the stage, and in still more cases toward the end of it, animals will mate, but even under these circumstances they do not usually show oestrous excitement in the characteristic manner typical of the following stage. As will be described at some length below, the acceptance of coitus can easily be tested by employing a small number of young accustomed males in a well illuminated, flat cage which can be instantly opened in order to interrupt the act.

Stage Two.—In Stage Two the macroscopic changes which we had noted as characterizing Stage One, notably the beginning swelling of the vaginal lips and the dry mucosa, are now increasingly evident. The speculum or spatula meets notably more resistance at attempted introduction into the vagina. An equally definite change takes place in the microscopic character of the vaginal smear, for the small, nucleated, and somewhat granular appearing epithelial cells characteristic of Stage One are rather suddenly replaced by large, thin, transparent, non-nucleated, scalelike elements—the cornified cells (figs. 21 and 22, pl. II). For this reason Stage Two may well be designated as the "cornified cell stage." The sample, while still scanty in amount, is opaque, whitish, and granular, the crumbling particles being easily tapped off into the drop of saline, where they have a tendency to float. There is still a singular absence of leucocytes, whose sudden disappearance was so marked a characteristic of Stage One. It is during the beginning of this stage that the female usually shows unmistakable signs of heat. If placed within a mating cage and, through some chance, not approached by the males, she will manifest what we have been led to recognize as oestrous excitement by quick, darting motions, or hops, with the back arched, with occasional quivering of the body and with a curious shaking of the ears. We may state that this behavior, never encountered at any time other than oestrus, need not, however, invariably be manifested by individuals in heat. A simple and unmistakable test of heat is furnished by the attitude of the female on the approach of the males. Mating is made possible by a characteristic flattening of the back, or slight opisthotonos. During this period there is usually a disagreeable odor attached to the vaginal secretion, which perhaps is a means of attracting and exciting the male.

Stage Three.—Stage Three, which may be equally well named "the late cornified cell stage," cannot be separated abruptly from Stage Two, the histological picture of the vaginal smear being identical although an exaggeration of that characterizing the preceding stage. Indeed, the accumulation of cornified, nonnucleated epithelial plates within the lumen of the vagina now proceeds so rapidly that easily visible masses of whitish, granular, or pasty substance always occur deep in the vagina near the cervix. These masses consist exclusively of enormous numbers of the elements in question without admixture with other cells. One who is familiar with the character of the "plug" (*bouchon vaginale*) which the male rat leaves in the vagina after coitus will at first easily mistake these white masses for fragments of the "plug". The most superficial microscopic examination, however, shows that they are made up of sheets or of single cornified elements whereas the male secretory product is amorphous. But besides this accumulation of cheesy substance, macroscopically evident within the vagina, Stage Three is further typified by a very important characteristic in the fact that animals in this stage will usually no longer accept coitus.[1] The average combined length of Stages Two and Three would appear to be about thirty hours. Throughout this period the vagina is dry.[2] The swelling of the vaginal orifice may also persist.

Stage Four.—Stage Four, the metoestrum or the leucocyte-cornified cell stage, is inaugurated by the appearance in the vaginal smear of leucocytes among the cornified cells (figs. 31 and 32, pl. II) and ends with the disappearance of the latter. The leucocytes cause a softening of the granular masses seen in Stage Three and convert them into a substance of a cheesy, creamy, and increasingly fluid consistency. Before the cornified cells completely vanish from the smear, epithelial cells reappear so that during a short interval all three cellular types are present (cornified, non-cornified, and leucocytes). This ushers in the beginning of the dioestrous pause, which may be recognized by the complete

---

[1] However, see the description of length of oestrus on page 33.

[2] Except for the very brief period of time which may immediately succeed sudden emptying of the uterus and drainage of its fluid into the vagina.

disappearance of cornified elements so that the smear again consists of leuco-cytes and epithelial cells. Stage Four is normally of about six hours duration and might well be known as the stage of transition to the resting condition. It is chiefly characterized by the sudden resumption of leucocytic migration.

It is interesting that in a certain proportion of animals the epithelial des-quamation of the mucosa may proceed somewhat further than to the cornified cell layer before leucocytes come in, and hence we may have the cheesy state, Stage Three, immediately succeeded by a few hours of a stage in which rather large, spindle-like, nucleated epithelial cells are shed. In such cases, however, vestiges of the cornified cell are always present and when leucocytes appear three cell types are actually found.

Since there is thus an invariable succession of cell types occurring in the vaginal smear at various times in the oestrous cycle, the stages in the cycle may with equal propriety be named from the cell content of the vaginal smear. Should we do this, we could designate them as the *stage of the sudden appear-ance of masses of uniform sized, nucleated epithelial cells dehisced from the sur-face, the stage of few large cornified cells, the stage of extremely abundant cor-nified cells, the stage of many leucocytes admixed with cornified cells,* and, finally, in the dioestrous pause, *the stage of leucocytes with scanty epithelial cells.*

2. RELATION BETWEEN OESTRUS AND OVULATION

The relation between oestrus and ovulation has a very special interest from the standpoint of the time of occurrence of fertilization. We therefore discuss it here and begin with a more accurate description of the time of appearance of oestrus and the variations in that time.

Usually oestrus is manifested for a period of from nine to twelve hours, beginning in the last part of Stage One and occupying most of Stage Two. Instances of the long duration of oestrus, of its early occurrence, or of its late occurrence are not unusual; nor must we fail to note another peculiarity, that is, the occasional occurrence of the manifestation of oestrus for only three hours or a similar very short interval. We have accumulated these results by the method already indicated, i.e., by offering individual females to a cage of males at the three-hour intervals when a record was kept of the vaginal smears.

* * * * * * *

Early and late exhibitions of oestrus have a special significance when brought into relation with ovulation. They might enable us to explain the variation found in the development of young embryos when the only age criterion is the copulation date. Table 2 indicates that ovulation may occur as early as eighteen and as late as thirty hours after the first appearance of cornified cells in the vaginal smear. Oestrus may be exhibited as early as three hours before the appearance of cornified cells or as late as twenty hours after. Did ovulation succeed oestrus by an invariable interval of time, the phenomenon of unequally developed embryos at the same copulation age would have to be explained by the unequal rate of ascent by the spermatozoa through the genital passages. This explanation suffers, however, first, from the belief in the limited viability of unfertilized eggs, and, secondly, from the fact that the enormous number of sperm ejaculated make it likely that the eggs are reached in an approximately uniform time.

The last objection gains weight in the case of those forms which, like the rat, have ovulation normally at a very considerable time interval after copulation. In the case of the rat almost a whole day separates these events. Long has demonstrated that sperm may reach the distal part of the oviduct in the mouse, a related form, within four hours. In view of all the facts, then, it is extremely probable that, in the rat, sperm will usually await the arrival of the ovum for a considerable interval of time, i.e., fertilization ensues immediately upon ovulation. Now it is very likely that ovulation does not necessarily follow oestrus in a uniform period of time, for there is reason for relating it to the general progress of the oestrous changes taken as a whole, an indication of the orderly progression of which we have found in the vaginal smear. In the case of very late exhibition of oestrus associated with precocious ovulation there might, then, be an actual transposition of these two events and a failure of fertilization through the inability of ova to continue to maintain their vitality until reached by sperm. It is evident that the cases of speediest fertilization following copulation will occur under conditions in which oestrus and ovulation are nearest together, providing sufficient time (at least four hours) elapses for the sperm to accomplish the uterine and tubal journey; and, conversely, the greatest delay in fertilization would occur in those instances in which oestrus and ovulation are separated as far as possible from each other. For example, if oestrus is manifested three hours before the occurrence of Stage One and ovulation thirty hours subsequent thereto, the sperm would have to wait thirty-three hours before fertilizing the eggs. Thus age differences in embryos dated from the hour of copulation could easily arise and could conceivably involve somewhat over a day.

* * * * * * *

C. COPULATION

1. NORMAL COPULATION AND THE FORMATION OF THE VAGINAL PLUG

From the preceding sections it will be seen that the length of the ovarian cycle may be influenced by various conditions. It is interesting that the act of copulation itself also exerts a similar though not so great an influence. An attempt to analyze this phenomenon has led to the discovery of a still more remarkable set of facts which can best be presented after a discussion of normal copulation.

The manner of copulation in rodents and especially in the rat has been described by earlier investigators: Lataste, Steinach, Kirkham. Yet it may not be amiss to present a more complete account of certain aspects of the subject. The condition of the female is of importance because, as has been pointed out earlier in the paper, copulation can take place only when the female is in heat, which, as we have shown, occurs typically in the transition from Stage One to Stage Two of the vaginal smear and in the early part of Stage Two. It will be recalled that at this time the lips of the external orifice of the vagina are slightly swollen and tend to make the opening more prominent; the vaginal mucosa is dryish, and covered with the thick cornified layer which, as Lataste long ago pointed out, probably serves as a protection. The cornification is accompanied by and possibly is the cause of a somewhat disagreeable odor which not unlikely is a means whereby the male is made aware of the condition of the female, for, as is common among animals, recognition may be chiefly carried on through the olfactory sense. The condition of the vaginal smear is sufficient to enable the investigator to determine in nearly all cases when the female will copulate and is of the greatest service in breeding animals· for purposes of embryological study, for it is only necessary to look through a colony to pick out those individuals which will usually copulate with little delay. In investigating the relation between the stages of the oestrous cycle and heat it was, of course, necessary to resort to the use of males, for many individuals show no signs that enable one to recognize the heat condition. Careful observation and considerable experience has enabled us, however, to detect oestrous quickly in most cases by the behavior of the female when placed with males. It might be mentioned parenthetically that the female rat, unlike other mammals, only rarely attempts to play the part of the male in riding other females. When such a female in heat is placed in a large cage in which males are kept she does not remain quietly in a corner, but is more or less active, constantly moving about and often keeping near the males. If she is not in heat she does not respond in any way to the male and after a short time is ignored. On the other hand, if in heat, her reactions are characteristic and consist in running about more or less intermittently with a curious hopping gait, stopping when a male succeeds in making an attempt, when the back is flattened in a characteristic way; indeed, the back is bent so that it becomes concave with the tail up and the head back. Under these circumstances and also frequently when she is hopping about, the head is shaken so that the ears quiver with a fine vibrating movement. It is an interesting fact that the position of slight opisthotonos taken by the female at coition may often be elicited by inserting rather firmly a speculum into the vagina while the animal is held in the hand, but the elicitation of this reaction has not resulted on account of contact of the speculum with the actual cervical canal, and unlike cervical stimulation it does not delay the appearance of the next oestrous cycle. This can not be considered a sign of heat, for in some individuals the animal will respond in this way at any time during the cycle. Nevertheless, it is usually characteristic only of the period of oestrus.

One of the most striking features of the copulatory act in rats is the great celerity with which attempts are made and actual insemination is accomplished by males, the movements of the posterior part of the body being very rapid. After each attempt, of which there may be very many before a fruitful coitus is consummated, the male almost invariably rolls back on to his hind quarters and licks his genitals. To an inexperienced person these attempts may be easily mistaken for a true copulation. But experience and vigilant observation will show that the latter differs from the former in being slightly more prolonged and in the male more often rising on his hind feet instead of promptly rolling backwards. He is usually quiet for a time afterwards, but as such inactivity often follows abortive coitus it is not a reliable sign. The only certain indication is the presence of a "plug" in the vagina of the female immediately after. Unless this is looked for at once it may be lost if the female is left with males for the reason that males may copulate as many as five times within a short period of time, and with the same female, one or two attempts being sufficient to dislodge a plug from the vagina. If the female is isolated the plug will usually remain in the vagina about twelve hours, occasionally as long as twenty-four hours. Of course, in any case, the actual finding of sperm in the vagina is the crucial test of insemination. But in our experience a plug is never unaccompanied by sperm.

The vaginal plug of rodents has been the subject of study by a number of investigators with regard to its structure, origin, use, and chemical composition. Although its occurrence was known before the time of Lataste and has been mentioned by many others since, he seems to have been the first to give it careful attention in his papers from 1882 to 1893. Lataste called the plug the *bouchon vaginale,* and was aware that most of it came from the seminal vesicle. He believed, however, that the outer part, or "envelope," was a contribution from the vagina of the substance which we now know to be the cornified layer which in some instances was so abundant as to give rise to a large mass, in consistency not unlike a plug. Lataste's conception of an outer "envelope" or coat formed by the vagina is easily explained by the fact that most plugs which are allowed to remain in position until they fall out do carry with them various portions of the cornified layer of the vagina, to which the plug is adhering tightly, the cornified layers being in the act of dehiscence at this time. Indeed, since one may occasionally even withdraw a more or less perfect cast of the vagina in the form of a sheet of cornified cells which constitutes the entire stratum corneum removed *en masse,* it is easy to believe that many of the vaginal plugs which Lataste examined must have contained an external coating with these cornified cells. Walker (1910) has shown conclusively that the plug is derived from the secretion of the seminal vesicle coagulated by the secretion of a coagulating gland (as the writers can confirm). As sperm are being expelled the secretions of the seminal vesicles and of the coagulating gland mix and evidently "set" the ejaculated mass, making it adhere tightly to any object with which it is in contact.

One will have no difficulty in seeing the formation of a plug that may be withdrawn from the penis of an animal killed by a blow on the head. On the tip of such a plug is a small mass of spermatozoa, of which use has been made in artificial insemination. It might also be noted that occasionally a plug is found attached to the hair of the female. All these plugs have the same general shape. It is apparent that at the moment of ejaculation the plug is moulded to fit the vaginal lumen, especially at the cervix, for the plug is constant in shape

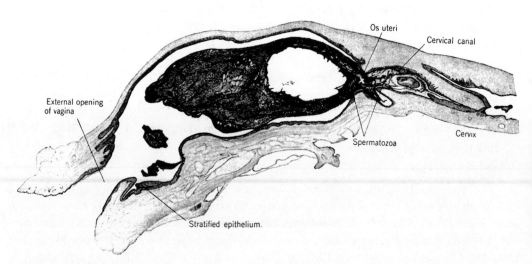

Fig. 84. Diagram of longitudinal section of vagina with plug *in situ* extending a short distance into the cervical canal. Rat 3580. × 6.

and has two little prolongations, each of which extends into one of the cervical canals (fig. 84). The plug adheres so closely to the cornified layer of the vaginal mucosa (fig. 85, pl. XI) that it is not easily distinguishable in section. The plug is naturally loosened so that it may be dislodged from the vagina when this cornified layer becomes stripped off spontaneously in every oestrous cycle, as described in an earlier section. In fact, the cornified layer may not only be considered a protection but also a means of insuring loss of the plug after it has performed its function. Sometimes two or even three plugs remain in the vagina at the same time.

The plug is not a solid body, but contains cavities, especially at its deeper end, these cavities being filled with spermatozoa. One of the obvious functions of the plug apparently is to confine spermatozoa between itself and the uterus as though to give abundant opportunity for them to make their way into the uterus.

Immediately or very soon after copulation the uterus is found to contain a considerable amount of dark reddish fluid in which there are enormous numbers of spermatozoa. The presence of this fluid has been noted by other investigators (Sobotta for the mouse and Königstein for the rat), but its relation to the fluid produced during Stage One has never before been understood. It will be recalled that in the first stage of the oestrous changes the uterus becomes distended with fluid. But this fluid, unlike that after copulation, is always, in normal animals, clear and colorless. That the bloody nature after copulation is not due to the spermatozoa is proved by the fact that after copulation by vasectomized males, which are unable to eject spermatozoa, the same bloody appearance is to be noted. It is probably caused by the act of copulation itself.

The male is capable of copulating and forming these plugs at any time of the day or night, but is most active at night. This last peculiarity may be the reason for a certain amount of apparent infertility in the case of females which are in heat for the first time in the morning hours and are past the heat period by night, when sexual activity in the male is at its height.

In the breeding of rats for embryos for any purpose except highly accurate studies on age, the observation of the plug is of the greatest value in saving time. It is only necessary to examine daily with the aid of a speculum the vaginae of mated females early in the morning or late at night for plugs from which to date the approximate age of embryos. Many animals can in this way be examined quickly. From the variations found in the development of embryos, even of the same copulation age, a fact which has been pointed out in earlier parts of this paper, it will be evident that some variation is unavoidable. But the inexactness in estimating the development of embryos dated from the finding of a plug is but slightly greater than if copulations be actually observed. Some plugs will prove to be infertile, of course, just as is the case with observed copulations. The following figures are of interest in this connection. One hundred and ninety-eight healthy young females were confined in several lots in each case with an almost equal number of healthy young males, their litter mates. During a fixed time interval we made daily observations with the vaginal speculum at an early hour each morning. Two hundred and forty-four gestations resulted during this time interval and three weeks thereafter, but during the same time interval only one hundred and fifty-two plugs had remained in the vagina long enough to be observed in the morning round of examinations, and of these one hundred and fifty-two plugs, one hundred and thirty-four, or slightly over 88 per cent, were followed by conception. It is evident that by this method alone, regardless of the fact that many plugs fall out during the night, a copulation record may be obtained for many pregnancies, and in almost 90 per cent of all cases where the plug has remained in situ until morning, gestation will follow. On the other hand, it is to be pointed out that in a colony handled in this way we will have almost as many pregnancies which are not preceded by the observation of a vaginal plug in situ the following morning. In our particular set of animals, of the two hundred and forty-four gestations resulting, in only one hundred and thirty-four cases was the plug seen the next morning, i.e., in only 54.8 per cent of the total gestations was it possible to obtain a record of the copulation by this method of observation. This figure might be higher if twice daily examinations were made.[4]

----

[4] This is a far more expeditious method of dating embryos than by actually observing copulation, and might well be an important factor in deciding whether to use rats or pigs for laboratory work in embryology.

2. INFERTILE COPULATIONS

Diseased females with normal males.

It has just been pointed out that a small percentage of copulations (or plugs) are sterile. It had been found early in this investigation that some of such females when left with males would copulate only at fairly long intervals without ever becoming pregnant. Such females turned out to be sterile because of an infection of the oviduct which occluded the lumen, but did not prevent ovulation in most cases. The reason for these long cycles is undoubtedly the same as for those which will be next reported.

TABLE 22.

Infertile females mated with normal males with
intervals between plugs.

Designation of Animals	Intervals in days between plugs in order of occurrence
158	7, 3, 3, 3, 3, 5, 6
159	13, 26, 14, 13, 13, 12, 12, 13
160	17, 13, 16, 15
161	14, 13, 13
197	12
198	10, 13
201	20, 13, 4
202	27
205	17, 23, 19, 16
206	15, 25, 13, 27, 12, 13, 23
208	13, 25, 13, 12, 14
209	4, 2, 17, 16, 18, 16, 18, 17, 40, 19
211	14, 10, 7
212	3, 3, 3, 4, 5
214	17, 16, 14, 15, 15, 15, 14, 14, 12, 4, 14
247	18, 17, 17
249	28, 13
294	17, 19
310	13
316	13, 13
339	5, 12, 4, 6, 7, 2, 2, 3, 3
342	7, 8, 4, 3
361	9, 15, 11, 4, 4, 3, 2, 4, 3, 4
386	6, 10, 8, 3, 5
393	9, 3, 7
394	23, 5, 10, 12, 13
395	9
396	14, 13, 14
404	18, 16
406	14, 12
407	12, 13
431	10
441	14
450	12, 12, 12
1188	12, 26, 13, 10

Normal females with vasectomized males.

In the course of a study of living eggs by one of us (Long, 1912), it was found that when females were allowed to mate with males rendered sterile by vasectomy, eggs could always be procured in the oviducts during the day on the morning of which a plug was found. Such males, from which a small piece of each vas deferens was resected, could copulate in a perfectly normal manner, but were impotent to inseminate a female. It seemed then as though the use of these males furnished an excellent method of studying the occurrence of ovulation since if copulation took place only at the time of ovulation the presence of the plug, uncomplicated by pregnancy, would be taken as an indication of ovulation. Females were accordingly placed each day with a new vasectomized male to give the maximum opportunity for coitus and careful daily observations made. The results of many of these cases are shown in table 23, from

TABLE 23.

Nine normal females mated with vasectomized males, with intervals between plugs.

Series Number	Intervals between plugs (in days) in chronological order
142	12, 24, 11
301	13, 13
305	19, 14, 14, 16
309	7, 4, 13, 13
410	12, 28
413	12, 16, 27, 15, 26, 16
422	14, 4, 15, 16, 16
423	8, 4, 4, 4, 12, 16, 16
440	10, 4, 4, 18, 15, 14, 14

which we would have to deduce an ovulation period that could not be harmonized with the earlier supposed cycle of ten days, nor with the true later discovered cycle of four to five days. This long copulation interval of healthy females with vasectomized males was in complete accord with the results on the copulation of sterile females possessing healthy ovaries but occluded oviducts. Furthermore, in both kinds of cases the vaginal smear showed a delay in the appearance of the next oestrus (table 24).

3. MECHANICAL STIMULATION OF CERVICAL CANAL

It was then that the suggestion offered itself that perhaps the secretions of the male accessory glands might be of importance in this connection. Previous studies had been made by other authors on the influence of prostate secretion on the activity of the spermatozoa and on fertility (Hirokawa, Iwanoff, Steinach, Walker), none on the influence on the female. Since it seemed possible that the plug itself, perhaps by being partly dissolved and absorbed, might exert an influence independent of the act of copulation, an attempt was made to mix

TABLE 24.

Length of œstrous cycles following infertile copulations, or copulations with vasectomized males,
as determined by vaginal smears.

Length of cycle in days	Number of cases	
3	1	
5	2	
6	1	
9	3	
10	5	
11	14	
12	8	
13	5	51 cases
14	4	average
15	5	13.1
16	3	days
17	2	
18	2	
19	3	

58 cases—average 12.4 days

the two secretions responsible for its formation in the vagina. Although a sort of plug was formed, but not successfully, no lengthening of the cycle followed. The only other possibility that suggested itself was the secretion of the prostate (table 25). Accordingly, the secretion obtained by triturating the bull's prostate gland, and also extracts of it in Ringer's solution, were injected into the uterus through the cervical canal by means of a glass syringe, using the method employed by one of us in earlier work (Long and Mark, 1911) with the result that the following cycle was longer than normal by several days, a very gratifying sequel. Controls consisting of injection of the same substance into the peritoneal cavity were negative. The same result followed, however, when Ringer's solution was injected into the uterus and in a few cases when the syringe was inserted but obstructed in some way so that the contents could not be expelled into the uterus. In fact, it was then discovered that merely inserting into the uterus a small glass rod had the same result! (table 26). In all of these cases the uterus was reached, of course, by traversing the cervical canal. As further experiments proved, the same results follow when the rod is inserted only about one millimeter into the cervical canal (table 26). This was done by using a slender brass rod with a small collar near the tip to prevent the rod being pushed in too far. It was apparent, therefore, that not only was it unnecessary to traverse the entire cervical canal but that the characteristic effect of cervical simulation could be elicited from that part of the canal known to possess the same stratified squamous epithelium as the vagina. The prolongation of the cycle follows only when the instrument is introduced during Stages One to Three (table 26), that is, during, and shortly before and after, the period of heat. This undoubtedly explains the action of the plug in prolonging the cycle since the plug fits tightly into the vagina and extends a short distance into the cervix.

This discovery of the profound influence on the length of the oestrous cycle of stimulating the tip of the cervix of the uterus by the mere momentary introduction into it of a slender rod is one of the most remarkable reactions known to us. The questions which naturally arise are what is the explanation of the mechanism of this striking phenomenon, and what is its significance? A partial answer to these questions can be given when dealing with the subjects of deciduoma formation and ovarian transplantation.

TABLE 25.

Effects of administering various substances *in utero* or intraperitoneally.

Figures in italics indicate the length of the period immediately following treatment, the other figures the length of periods preceding and following this.

Secretion of rat prostate. *In utero.*

Designation of animal	
2412	4, 4, *15*, 3, 4

Extract of rat prostate in Ringer's solution. *In utero.* About .2cc per dose.

2010	5, 4, *17*, 5, 8
2332	4, 4, *15*, 4, 6
2395	6, 5, *13*, 4, 4
2518	3, *20*, 5, 3

Extract of rat prostate in Ringer's solution. Intraperitoneally.

2383	5, 6, *5*, 6, 6	Given in middle of period. 1cc.
2395	4, 4, *6*, 5, 11	Given on second day of period. .2cc.
2412	4, 4, *4*, 3, 2	Given on second day of period. .6cc.
2557	5, 5, 6	Given on second day of period. 1.0cc.
2547	5, 5, *19*, 8, 3	.2cc.
2568	*4*, 5, 4	.2cc.
2897	*6*, 4, 5	.2cc.
3096	5; 6, *7*, 7, 7	.2cc.

.9% salt solution intraperitoneally. Doses of 4cc.

2208	5, 5, *8*, 3, 6
2383	5, 4, *5*, 6, 6
2502	5, 5, *5*
2537	5, 5, *10*, 11, 8
2571	9, 4, *4*, 4, 5

* * * * * * *

4. PROOF THAT COPULATION AND CERVICAL STIMULATION INDUCE A CONDITION OF "PSEUDO-PREGNANCY"

At first sight the remarkable effect of copulation or of cervical stimulation in delaying the appearance of the next oestrus and ovulation would appear to be unexplainable, for these procedures would appear thus only to express themselves some days after their occurrence and such a belated physiological response, or effect, would be difficult to understand. It would be easier to understand the delay of an immediately impending ovulation produced by nervous or other influences or the acceleration of that ovulation by direct nervous or vasomotor means, but, as we have shown, copulation or cervical stimulation seems not to influence the progression of oestrous changes and ovulation at the time, but only the time of appearance of the next oestrous events. *The entire matter becomes more intelligible, however, when we discover that the effect of copulation and cervical stimulation upon the reproductive system is immediate, although its manifestation is deferred. The stimulus produces a condition which justifies the designation "pseudo-pregnancy."*

Reprinted from *J. Amer. Med. Assn.*, **81**, 819–821 (1923)

13

AN OVARIAN HORMONE

PRELIMINARY REPORT ON ITS LOCALIZATION, EXTRACTION AND PARTIAL PURIFICATION, AND ACTION IN TEST ANIMALS *

EDGAR ALLEN, Ph.D.
AND
EDWARD A. DOISY, Ph.D.
ST. LOUIS

The fact that double ovariectomy abolishes the cyclic changes which normally occur in the genital tract of female mammals demonstrates that these changes are due to some influence from the ovaries. That this influence is hormonal in nature is indicated by the maintenance of cyclic changes by autotransplantation of the ovaries to other sites in the body.[1] Many attempts have been made to localize the seat of production of an internal secretion in definite ovarian structures. The follicles, the corpora lutea [2] developing in them after ovulation, and the interstitial tissue [3] have all been cited as possible sources, and many different ovarian preparations have been used clinically in the belief that therapeutic effects were being secured from accompanying hormones.

But there appears to be no conclusive evidence of either a definite localization of the hypothetic hormone or of the specific effect claimed for the commercial ovarian extracts in wide clinical use. The recent

7. Salmon, William: The Family Dictionary, London, 1705.
* From the Departments of Anatomy, Biologic Chemistry and Surgery, Washington University Medical School, St. Louis, with the assistance of Byron F. Francis, Harry V. Gibson, Leroy L. Robertson, Cleon E. Colgate and William B. Kountz.
1. Marshall, F. H. A.: The Physiology of Reproduction, 1922, p. 320.
2. Fraenkel, Ludwig: Die Function des Corpus Luteum, Arch. f. Gynäk. **68**: 438, 1903.
3. Marshall and Runciman: On the Ovarian Factor Concerned in the Recurrence of the Oestrous Cycle, J. Physiol. **49**: 17 (Dec. 22) 1914.

reviews of Frank[4] and of Novak[5] may be cited to illustrate the well founded skepticism concerning the activity of commercial preparations. The reason for this distrust is the absence of any suitable test for the activity of ovarian extracts. Without a practicable test, the search for any hormone is obviously a very haphazard and uncertain task. With a suitable test it has been relatively easy to establish the seat of production of the ovarian hormone of estrus, to observe its effect on animals, and to study methods for its extraction and preliminary purification.

THE TEST METHOD

In 1917, Stockard and Papanicolaou[6] described an exact method for following estrual changes in the living guinea-pig. This method has been applied to the correlation of estrual phenomena in the genital organs of the rat[7] and the mouse.[8] During the anabolic phase of the cycle in these rodents, the epithelium of the vagina grows to a considerable thickness and a cornified layer similar to that in the epidermis develops. During the catabolic phase, the outer layers of this epithelium degenerate and are removed by leukocytic action. These changes provide a definite succession of cell types in the vaginal lumen, each one characteristic of a certain phase of the cycle. Thus the microscopic examination of vaginal smears is a reliable indicator of the estrual condition of the living animal.

Since these cyclic phenomena in the genital tract cease after double ovariectomy,[9] their induction by the injection of ovarian extracts constitutes a positive test for the efficiency of those extracts. The estrual cycle of the mouse and the rat is of four to six days' duration, and this short period makes them especially suitable for test animals.

LOCALIZATION OF THE HORMONE OF ESTRUS

From the results of an earlier investigation by one of us[8] on the estrual cycle in the mouse, it was concluded that the corpora lutea of estrus[10] and the interstitial tissue could be excluded as possible sources of the hormone of estrus and that this was produced in the follicles. Among earlier workers, Frank[4] reported very briefly inducing uterine hyperemia in two normal virgin rabbits by the injection of liquor folliculi. Although his results appear to be positive, the ovaries were not removed from his test animals, and one cannot be quite sure that the changes described were not caused by the activity of the ovaries. Previously, Frank and several writers mentioned by him (especially Hermann[11]) had reported similar results from the injection of material obtained from the corpus luteum and the placenta (Hermann spayed some of his test animals). Consequently, this test does not seem to be specific for the hormone of estrus.

4. Frank, R. T.: The Ovary and the Endocrinologist, J. A. M. A. **78**:181 (Jan. 21) 1922.
5. Novak, Emil: An Appraisal of Ovarian Therapy, Endocrinology **6**: 599 (Sept.) 1922.
6. Stockard, C. R., and Papanicolaou, G. N.: Existence of a Typical Oestrous Cycle in the Guinea-Pig, with a Study of Its Histological and Physiological Changes, Am. J. Anat. **22**: 225 (Sept.) 1917.
7. Long, J. A., and Evans, H. M.: The Oestrous Cycle in the Rat and Its Associated Phenomena, Mem. Univ. California **6**: 1922.
8. Allen, Edgar: The Oestrous Cycle in the Mouse, Am. J. Anat. **30**: 297 (May) 1922; Racial and Familial Cyclic Inheritance and Other Evidence from the Mouse Concerning the Cause of Oestrous Phenomena, Am. J. Anat., to be published.
9. This has recently been checked in the rat (Long and Evans, Footnote 7) and in the mouse (Allen, Footnote 8).
10. The distinction is clearly drawn here between the corpora lutea of estrus and those of pregnancy. Only nonpregnant animals were referred to in the earlier paper.
11. Hermann, E.: Ueber eine wirksame Substanz im Eierstocke und in der Placenta, Monatschr. f. Geburtsh. u. Gynäk. **41**: 1 (Jan.) 1915.

PRELIMINARY EXPERIMENTS

Our first experiments were carried out with liquor folliculi from hog ovaries, which are a readily available source for the isolation of the contents of the follicles. Since the estrual cycle in the sow is of three weeks' duration,[12] and large follicles are present in the ovaries during only a part of this time, only one in every three or four ovaries is a profitable source of liquor folliculi.[13] A selection is made of ovaries containing follicles larger than 5 mm. in diameter. The follicular contents (liquor folliculi, follicle cells and occasional ova) are aspirated through a hypodermic needle into a suction bottle. At least 100 c.c. of readily workable material can be obtained from one pound of carefully selected ovaries.

In the first series of experiments, nine mice and rats were prepared for use as test animals by double ovariectomy. A week later they were given three injections at intervals of five hours of liquor folliculi aspirated from large follicles. These injections were made subcutaneously with the expectation that slow absorption from this region would be more closely comparable to the secretion of the hormone in normal animals. From forty to forty-eight hours after the first injection, all of the animals receiving liquor folliculi were in full estrus, as determined by microscopic examination of the smears. As a check on these results, the animals were killed and the uterus and vagina of each studied histologically. They were in a typical estrual condition as described for these rodents.[14]

Since the liquor aspirated from the follicles contains follicle cells and some ova, tests were next made with centrifugated liquor and Berkefeldt filtrate. Positive results in these experiments showed that the hormone is extracellular and present in the liquor folliculi. Realizing that liquor folliculi with its large protein content would be unsuitable for continued injections into patients and test animals, we began the preparation of extracts.

METHOD OF EXTRACTION

Fresh liquor folliculi is added to a double volume of 95 per cent. alcohol and allowed to stand until the proteins are coagulated (about twenty-four hours). The coagulum is filtered off. The filtrate, which is practically protein free, contains the active constituent. Further extraction of the coagulated protein with boiling alcohol yields an additional amount of the hormone. The alcohol is distilled off (the hormone being thermostable), and the residual aqueous suspension extracted with ether. The ether extract is evaporated, and the solids are dried in a vacuum desiccator. The residue is dissolved in a minimal quantity of ether, and a double volume of acetone is added. To insure completeness of separation, the solution and precipitation are repeated twice. The precipitate, which consists of lipoids (lecithin and cephalin), shows no activity in the test animal. The combined filtrates are evaporated, and the residue is dried. By boiling out the solid material with 95 per cent. alcohol, the active substance is obtained free from protein but contaminated with a little fatty material. The alcohol is evaporated off, and the minute

12. Corner, G. W.: The Ovarian Cycle in Swine, Science **53**: 420 (April 29) 1921.
13. This may be one reason for the conflicting reports of the therapeutic value of commercial extracts from ovaries unselected as to follicular development.
14. Stockard and Papanicolaou (Footnote 6). Long and Evans (Footnote 7). Allen (Footnote 8).

yield of oily residue taken up in purified corn oil or emulsified in dilute sodium carbonate. This constitutes a partially purified preparation of this follicular hormone, the subcutaneous injection of which produces no ill effects in our test animals.

A few points of chemical interest may be added. This hormone seems to be stable toward dilute alkali when boiled with it in alcoholic solution. The active substance resembles the neutral fats somewhat in that it is soluble in alcohol, acetone, chloroform and ether and it may be extracted from aqueous mediums by both ether and chloroform. Strongly acidic or basic groups seem to be absent, since it can be extracted with ether from either dilute acid or alkali. Most of our better preparations fail to give the biuret reaction, thus indicating the absence of protein and polypeptids.

ACTION IN TEST ANIMALS

Further animal experiments have confirmed our early results, and we are now in a position to make the following preliminary announcement concerning this follicular hormone and its action:

1. From one to three injections of this extract into spayed animals produce typical estrual hyperemia, growth, and hypersecretion in the genital tract and growth in the mammary glands. These changes include thickening and cornification in the vaginal walls, which constitutes a test easily followed in the living animal. After a time the effect of these injections wears off and typical degenerative changes set in. This hormone seems to be an efficient substitute for the endocrine function of the ovaries of the nonpregnant animal. It is probable that its alternate presence and absence in the circulation is sufficient to explain the mechanism of estrual phenomena in the genital tract in the absence of pregnancy.

2. While these spayed animals are in a condition of artificially induced estrus, they can be mated with normal vigorous males. They experience typical mating instincts, the spayed females taking the initiative in the courtship and showing no aversion to advances by the male. Successful copulation occurs, followed, as in normal animals, by the formation of typical vaginal plugs. Since these animals will copulate only when in estrus, the conclusion seems justified that this follicular hormone is the cause of estrual or mating instincts.

3. Several injections of active extracts were made into animals immediately after weaning, at an age of from 3 to 4 weeks. They became sexually mature [15] in from two to four days, at least twenty to forty days before the usual time of the attainment of puberty. These experiments were controlled by uninjected litter sisters which did not prematurely attain to puberty. From these experiments it is concluded that the attainment of sexual maturity, involving possibly the development of the secondary sexual characters, is brought about by a hormone from the follicles, although the consummation of their maturation is in some way restrained.[16]

4. So far we have obtained only negative results from our extracts of corpora lutea and commercial extracts of ovaries, corpora lutea, and ovarian residue from three of the largest firms manufacturing biologic products.

5. It is probable that this hormone is produced under the influence of maturing ova by their follicle cells. Since it is obtained from the ovaries of hogs and cattle and produces results in the mouse and rat, it is not species specific. It is probably produced in all ovaries as their ova mature, and therefore is probably common to all female animals.

The details of the experimental evidence for the foregoing statements will be published at an early date. Tests of the therapeutic value of this hormone are now being carried on.

———————

15. The attainment of sexual maturity is defined here as the opening of the vaginal orifice at the appearance of the first estrus.
16. Allen, Edgar: Ovogenesis During Sexual Maturity, Am. J. Anat. **31:** 441 (May) 1923. It is tentatively suggested that the development of the follicles in the immature ovary is restrained by the nutritional demands of rapid prepubertal body growth which limit the amount of nutriment necessary for the full development of the follicles and their contained ova.

Reprinted from *J. Comp. Psychol.*, **27**, 49–68 (1939)

14

SEXUAL BEHAVIOR AND SEXUAL RECEPTIVITY IN THE FEMALE GUINEA PIG[1]

WILLIAM C. YOUNG, EDWARD W. DEMPSEY, CARL W. HAGQUIST AND
JOHN L. BOLING

*Arnold Biological Laboratory, Brown University, Laboratories of Primate Biology
and Department of Anatomy, Yale University School of Medicine*

Received May 4, 1938

The behavior displayed shortly before and during the time the female guinea pig is sexually receptive has been described (17,[2] 18). Most important for the investigator is the straightening of the back and the elevation of the pudenda when the female is stroked on the back or mounted by a male because the period during which this response can be elicited is the period of sexual receptivity. Its length varies from 1 to more than 30 hours and averages 8 to 9 hours. Entirely different from the copulatory response, but usually associated with it, is the display of male-like mounting activity. This behavior is shown most commonly about the beginning of heat and its extent varies greatly. However, no correlation exists between the frequency of mounting and the length of heat. The latter may be unaccompanied by mounting activity, or cyclic periods of mounting activity may occur in the absence of heat. Obviously, therefore, mounting activity cannot be relied on for the identification of heat. For this purpose and for estimation of the time of ovulation which occurs near the end of heat, dependence must be placed on the copulatory response.

The reliability of the response is indicated by diverse observations and experiments. During more than 2000 experiences in mating guinea pigs, only rarely if ever did a female from which

---

[1] This investigation was supported by a grant from the Committee for Research in Problems of Sex, National Research Council.

[2] References to the observations of other investigators are cited in this article.

49

the response could be elicited refuse to accept the male (17). During a study of the ovarian condition in animals for which the behavior was known, the ovaries were removed at various stages in the cycle reckoned from the beginning of heat as determined by the copulatory response. In not one of 167 pairs of ovaries was any irregularity shown in the stage of follicular or corpus luteum development (19). In an experiment involving the artificial insemination of animals at varying intervals after ovulation, the percentage of sterile inseminations decreased with a regularity which would not have been probable had the method of estimating the time of ovulation been less reliable (2).

After the ease with which this behavior can be identified became evident, it seemed of interest to ascertain what hormones are responsible for its production and what factors are involved in the variations shown from animal to animal and from cycle to cycle in the same animal. It was found that although several estrogens induce heat in a certain percentage of spayed animals, all respond only when relatively small quantities of oestrin are followed by the injection of progesterone. The result led to the conclusion that in the normal female, sexual receptivity is caused by the synergistic action of oestrin and progesterone produced in the unruptured follicle (6).

The factors responsible for variations in the character of heat cannot be stated with the same certainty, but suggestive working hypotheses developed from a study of the ovarian condition supplemented by experimental procedures. As far as length of heat is concerned, the extent of general follicular development, the number of maturing follicles and rete cyst formation can be eliminated. On the other hand, extra-ovarian factors such as the sensitivity of the individual to hormonal action and the time required for the ovulation process to be completed are important. As far as the frequency of mounting activity is concerned, some factor not involved in the regulation of heat-length was assumed to play a part, because although the extent of general follicular development and rete cyst formation were eliminated, a direct relationship was found between this part of the behavioral complex and the number of maturing follicles (19).

While making these observations it was necessary to record the behavior of several hundred normal animals which were not used in the earlier part of the work.   Consequently, new data of interest were obtained.   In addition, consideration and study of the data resulted in the development of a point of view with respect to the measurement of sexual drive in the guinea pig which now seems appropriate for presentation.

The methods employed in making the observations are given in the articles cited above.   It is sufficient to state here that the data were accumulated during the continuous observation of 617 normal animals throughout a total of 2102 heat periods.   They bear on (*a*) the first heat periods, (*b*) the length of heat during consecutive reproductive cycles, (*c*) mounting activity, (*d*) cycle-length, (*e*) the split oestrus, and (*f*) the incidence of heat during each of the 24 hours of the day for this colony of animals.[3]

RESULTS AND DISCUSSION

The first heat periods

Data were collected from 54 animals (table 1).

The age at occurrence of the first heat varied from 33 to 134 days and averaged 67.8 days.   The time of first rupture of the vaginal closure membrane coincided with or preceded the first heat period.   The former varied from 33 to 111 days and averaged 58.2 days.   In the guinea pig as in the rat (13) a spread develops between the time of vaginal membrane rupture and the first heat when the latter is delayed.   Among the 29 animals in which the first heat occurred prior to the 67th day, the vaginal membrane ruptured 16 days or longer before heat in only three. Of the 25 animals in which the first heat occurred subsequent to the 67th day, the vaginal membrane ruptured 16 days or longer before heat in fourteen.

The length of each of the first four heat periods is shown for

³ We are deeply indebted to Messrs. Richard J. Blandau, Vincent J. Collins and Dr. Roy Hertz for assistance in observing the animals.   Likewise we wish to thank Dr. Erwin Schwenk of the Schering Corporation for his generous contributions of Progynon-B, Oreton, Oreton-B and Proluton, and Dr. Oliver Kamm of the Parke, Davis Company for his kindness in supplying Theelin.

TABLE 1

Summary of data from animals observed during first and subsequent heat periods

ANIMAL	AGE AT TIME OF 1ST HEAT	AGE AT TIME OF 1ST VAGINAL OPENING	NUMBER OF TIMES ANIMALS MOUNTED AT 1ST HEAT	LENGTH OF HEAT			
				1st	2nd	3rd	4th
	days	*days*		*hours*	*hours*	*hours*	*hours*
1736	33	33	5	5.0	6.0	6.5	9.0
1329	34	34	3	9.0			
1859	39	39	0	4.0	0.0	0.0	8.0
1385	41	39	7	7.5	5.5	6.0	6.0
1325	43	40	0	7.0	8.5	8.5	10.0
1375	43	42	0	9.0	10.5	9.0	8.0
1309	44	43	0	3.0	4.0	6.0	6.0
1323	45	44	4	7.5	6.5	6.5	8.5
1506	45	41	3	1.5	4.5	5.5	7.0
1327	47	44	1	4.5	10.5	12.0	12.5
1393	.47	45	2	7.5	8.0	12.5	10.0
1326	49	47	0	7.0	12.5	12.0	12.0
1396	49	39	0	9.0	10.0	11.0	6.5
1505	49	48	2	3.5	8.0	7.0	11.5
1514	50	48	6	9.5	6.5	12.0	16.5
1376	52	49	2	4.0	6.0	6.5	5.0
1377	52	48	12	8.0	9.5	7.0	3.5
1512	52	46	1	5.0	10.0	13.0	9.5
1300	53	36	0	5.5	4.0	5.5	13.0
1301	53	36	0	6.0	6.0	7.5	7.0
1330	55	55	21	10.0	10.5	9.0	11.5
1347	58	53	3	5.5	8.0	6.5	7.5
1331	60	38	0	8.0	16.5	7.5	4.5
1333	60	58	3	5.5	2.5	13.5	25.5
1306	63	55	29	11.5	11.0	14.0	8.5
1398	63	58	0	10.0	9.5	10.5	9.5
1397	64	58	0	6.5	8.0	9.5	
1503	64	56	0	5.5	9.0	6.5	11.0
1382	66	58	3	5.0	11.5	12.0	13.0
Averages when 1st heat was before 67th day........	50.8	45.9		6.6	8.0	8.7	9.7
1389	69	52	2	9.0	10.5	10.5	
1342	72	72	0	8.0	8.5	14.0	15.5
1510	72	70	0	2.0	5.5	6.0	6.0
1341	78	55	0	4.0	4.5	6.0	6.0
1378	80	63	0	4.5	10.0	3.0	
1386	80	80	0	6.0	6.0		
1392	80	69	0	4.5	8.5		
1308	82	31	1	5.5	8.5	8.5	7.5

TABLE 1—*Concluded*

ANIMAL	AGE AT TIME OF 1ST HEAT	AGE AT TIME OF 1ST VAGINAL OPENING	NUMBER OF TIMES ANIMALS MOUNTED AT 1ST HEAT	LENGTH OF HEAT			
				1st	2nd	3rd	4th
	days	*days*		*hours*	*hours*	*hours*	*hours*
1394	82	79	1	5.5	8.5		
1395	83	79	0	3.5	7.5		
1315	84	81	0	7.5	9.0	8.0	6.5
1318	84	65	0	7.5	4.5	5.0	10.5
1515	84	83	1	11.5	8.0	13.0	3.5
1317	85	68	2	4.5	5.0	9.5	0.0
1381	86	83	8	9.0	7.0	12.5	
1522	86	67	47	5.5	6.5	5.5	8.0
1517	87	83	0	11.5	10.0	10.5	14.5
1322	88	52	0	5.0	5.0	8.0	7.5
1310	88	70	0	6.5	3.0		
1313	92	64	0	3.0	7.0	8.0	9.0
1316	96	74	0	8.0	10.0	10.0	9.5
1387	97	70	0	5.5	9.5	7.5	
1303	104	93	0	3.5	0.0	0.0	0.0
1304	115	80	0	4.5	0.0	0.0	0.0
1311	134	111	0	4.5	8.0	10.0	7.0
Averages when 1st heat was after 67th day	87.5	71.9		6.0	6.8	7.8	6.9
Averages for all animals	67.8	58.2		6.3	7.4	8.3	8.6

most of the animals. The first two often are shorter than sub-sequent ones. A possible explanation is provided by supplementary observations. When newly spayed young animals are injected with oestrin and progesterone, the first heat period is shorter than subsequent heat periods produced by the same procedure (3). The result suggests that in animals which have not come into heat the sensitivity to hormonal action is lower, that the initial injection has a "priming" as well as a heat-stimulating action, and that the short first heat in many normal animals is attributable to the absence of hormonal action previous to this time.

The lower sensitivity of such animals to hormonal action is

also suggested by a comparison of follicular size at the beginning of heat in different groups of animals. In normal adults follicular size at the beginning of heat averages 619 × 10⁶ cu. μ (11). In animals killed at the beginning of the first heat after the early removal of the corpora lutea follicular volume averages 442 × × 10⁶ cu. μ (5). But in five young animals killed at the beginning of the first heat follicular volume averaged 761 × 10⁶ cu. μ. The result was surprising because no corpora lutea were present and it was expected the size would be similar to that found in the older animals from which the corpora lutea had been removed. Presumably, however, in animals coming into heat for the first time, either the hormones act over a longer period of time or more hormone is necessary before heat is produced.

The length of heat during consecutive reproductive cycles

When records of four or five consecutive heat periods were available, individual animals showed a tendency to have heat periods of rather constant length. Supplementary data have been obtained from 30 animals observed throughout 5 to 11 consecutive cycles during 1935–36 and 3 to 5 consecutive cycles during 1936–37 (table 2), from 24 animals observed throughout 7 to 9 consecutive cycles during 1935–36, but not during 1936–37 (table 3), and from 10 semi-spayed animals observed during 6 consecutive cycles or longer (table 4).

For the data shown in tables 2 and 3 a Pearson coefficient of correlation based on the odd-even heat periods was computed;[4] r was found to be 0.78. It is evident, therefore, that the tendency to consistency of heat length is shown over a relatively long period of the reproductive life. A similar analysis of the records of the semi-spayed animals would not have significance because so few animals were used, but casual inspection of table 4 indicates that the operation was without effect.

The factors responsible for the consistency of heat length are not known. Any specific ovarian condition can be eliminated because during the life of an individual the ovaries undergo changes which are not paralleled by changes in behavior (19).

[4] We are indebted to Dr. Frank K. Shuttleworth for this computation.

TABLE 2

CONSECUTIVE HEAT PERIODS DURING 1935-36 (LENGTH OF HEAT IN HOURS)

ANIMAL	1st	2nd	3rd	4th	5th	6th	7th	8th	9th	10th	11th	12th	1st	2nd	3rd	4th	5th
													1936-37				
912	10.5	10	9	0	11.5	11	14.5	10	7	6.5	9		7.5	7.5	12.5	5	6.5
960	6.5	8	8.5	0	6	9.5	5						7.5	11.5	7.5	12.5	
1010	11	10.5	8.5	10.5	0	>11	8.5	41.5					8.5	8.5	5.5	7.5	7
1014	10.5	>10.5	>5.5	10.5			>8.5	0	6	>12.5			>9.5	>8.5	5*	5.5	
1045	19.5	>20	7*	>31	26	24.5	26.5						>14	>10.5	18	11	
1046	8.5	0	0	>1.5	5.5	>5.5	>8.5	0	6				12.5	14	13	14	7
1055	5.5	10	8.5	6	5	7.5	7	0	0				5.5	>6.5	0	3.5	
1063	0	0	8.5	2	0	5	8.5	15	13.5				0	0	3.5	0	
1073	10	8.5	8.5	3	11	5	8.5	9	0	9	10		8.5	10	9.5	9	
1095	1*	7	5	4	6	8.5	10	9	0	5			8.5	9	0	4	>9.5
1120	13	8	7	>6.5	9	4	9	8	6	>7.5			>8.5	>7.5	8	7	
1180	0	13.5	3	>6.5	11	11	9	22.5	13	>4.5	26	17	7.5	11	10	9	
1251	12.5	12.5	>15.5	13.5	>14	5	22	17	11	24			12.5	15.5	>12.5	>11.5	
1274	16	6	14	7.5	9	9	10	12	6.5	8.5	9.5		10.5	8.5	9.5	8.5	
1306	11.5	11	14	8.5	12.5	10	3*	16	6.5				7.5	6	9.5	10.5	
1315	7.5	9	8	6.5	11	7.5	7.5	>7					6	6.5	5.5	5.5	
1317	0	4	5	9.5	0	7	8.5	>13					8	9	10.5	11.5	
1318	0	7.5	4.5	5	10.5	9	8	5					6	7.5	8	5	
1322	0	6.5	5	8	7.5	5	8	6.5					6.5	5.5	7	4.5	
1323	7	8.5	6.5	8.5	9.5	8	8.5						7.5	6.5	6	0	
1325	7	10.5	12	10	>6	13	14	11	16				9	6.5	13	6.5	
1327	4.5			>3.5*		7.5	20						5.5	9.5	2.5*	12	0
1330	10	10.5	9	>12.5	12	>8.5		3	6	2.5	9.5		8.5	7	7.5*	8	
1333	5.5	2.5	13.5	25.5	0	8.5		9.5	9	5			5*	20	7.5	6.5	>6.5
1393	7.5	8	12.5	10	8	>11	0	3	6		9.5		10.5	10	14	12.5	
1502	0	4.5	0	0	0	0.5	9.5	9.5	9	5			9	8	0	11.5	
1504	0	8	0	0	7	9.5	9	8	11		6		7.5	8	5	8.5	
1506	1.5	4.5	5.5	7	9	7.5	10						10.5	2.5	4.5	0	
1507	5	0	0	6	2	0		7					8	4*	12.5	8.5	7
1522	0	5.5	6.5	5.5	8	2	10	7					8	7	6.5	9	

* Split oestrus.

TABLE 3

ANIMAL	CONSECUTIVE HEAT PERIODS DURING 1935–36 (LENGTH OF HEAT IN HOURS)									
	1st	2nd	3rd	4th	5th	6th	7th	8th	9th	10th
918	4	>3	6.5	3.5	7.5	10	11.5	13.5	>5	>12.5
941	13	19	6	15	15	38.5	11	33		
1013	4.5	0	0	0	0	3	4	6		
1029	6	6.5	3	0	6.5	7.5	8	0	5	7
1032	7	6	2.5	4	0	4.5	6	6	>8	8.5
1066	0	7	>9	10	12	4.5*	3*	2*	2.5*	2.5*
						>9	>8.5	14.5	11	>4.5
1104	8.5	12	>7	8	7.5	12	7	11		
1105	4	10.5	0	7	7	10	5.5	5		
1116	>6	5	0	0	5.5	4	0	4.5	0	8.5
1144	9.5	8.5	9.5	9	9.5	8	6.5	9.5	10.5	
1185	7	4.5	10	8.5	11	0	4.5	7.5	11.5	
1197	6	8	8.5	8	>2.5	6.5	11.5	>10.5	24	0
1222	9.5	4*	15	1.5*						
		9.5		1	22	17.5	12	13	8.5	12
				>11						
1230	0	0	0	0	2	0	5.5	3.5	0	7.5
1232	7.5	6	9	>10.5	8	5*	1*			
						8.5	3.5	12.5	7	15
							8			
1270	6.5	8.5	7.5	>7	>12	12.5	>10.5	21.5		
1271	>20	>18	13.5	10.5	9	11.5	>15	>7	17	
1308	5.5	8.5	8.5	7.5	7.5	7	8	7		
1309	3	4	6	6	5	4	4	5.5	5	
1351	12	10.5	12.5	3.5	10	13	7.5	0	10.5	9.5
1359	11	10.5	11	16.5	11	13.5	10	8	12.5	10.5
1373	10	9.5	11	6	8.5	8	10	14		
1404	7	8	7	6	7.5	7	7	6		
1501	9.5	0	8.5	11.5	12	8.5	8.5	7	13.5	12

* Split oestrus.

TABLE 4

ANIMAL	HEAT RECORDS OF UNILATERALLY OVARIECTOMIZED ANIMALS DURING CONSECUTIVE CYCLES BEFORE AND AFTER OPERATION (LENGTH OF HEAT IN HOURS)																
969	9.5	4.5	7.5	6.5	O	0	0	5.5	10.5	0	5	7	I	7.5	8.5	0	
1021	O	11	6	6	>19.5	14	0	13.5	I	>5	10.5	17	16.5				
1023	0	0	O	6.5	4	13.5	6.5	6.5*	10.5	11	I		7.5	12.5	3*		
								14							>4		
1108	9	7.5	10	10.5	10	12.5	O	9.5	13.5	I	>2*		9.5	15.5	7.5*		
											12.5				>4.5		
1236	12.5	12.5	13	>16*	O	20	3	17.5	16.5	31	18.5*						
				>5.5							13						
1269	8.5	8	7.5	4.5	O	5.5	8.5	6	8	9	10	15.5	I	10.5	5.5*	8.5	
															>6.5		
1354	14	10	12	11	O	8.5	10	3.5	13	10.5	10.5	>10.5	I	>3	0	0	
1508	8	5	8	>6.5	O	4.5	11	7.5	16	15	13	12.5	I	12.5	11	11	8.5
1509	10	>6.5	>16	9	O	5	10	6.5	8	>8.5	7.5	6.5	I	>6	6.5	8	8
1521	4.5	6	5.5	9.5	O	4	8	7	>8	10	8	I	10	9.5	5.5	8.5	

* Split oestrus.

O = operation, I = interval of 5½ months.

56

131

The most plausible suggestion is that some fairly stable characteristic is involved, such as the sensitivity to hormonal action or the individuality of the mechanism controlling the manner of ovulation. Both influence the length of individual heat periods and both are factors which may well have an inherent or constitutional basis.

Mounting activity

Early in the work an accurate record of mounting activity was not possible although certain data of value were obtained. These were later supplemented by data collected during observations over a 10-day period when there was scarcely a minute that the 98 animals which came into heat were not being watched. As a result a more adequate discussion of the subject can now be given.

The conclusion that no relationship exists between the frequency of mounting and the length of heat has been confirmed. The length of heat averaged 8.8 hours in 27 animals that mounted from 8 to 47 or more times, and 8.9 hours in 30 animals which were recorded as not having mounted at all (19).

When the early observations were being made it was estimated that mounting is displayed during as many as 85 per cent of the heat periods. The estimate was approximately correct because of the 98 animals observed most carefully 89 mounted. Of these, 13 mounted once each, 21 from 2 to 5 times each, 15 from 6 to 9 times, 12 from 10 to 19 times, 10 from 20 to 29 times, 6 from 30 to 39 times, and 12 from 40 to 96 times each.

Mounting may be displayed when the first heat occurs (table 1). One animal mounted 47 times and two others 20 times each. Altogether 24 mounted at least once and it is probable others also mounted because not all the mounting was recorded. It is thought, therefore, that mounting is as common or nearly as common at the beginning of cyclic sexual activity as later.

A more accurate association of the time of mounting with the time of heat has been made (fig. 1). The earlier observations suggested that mounting is essentially a prooestrous behavior. Actually, however, the peak of mounting activity in a group of animals coincides with the beginning of heat, and more mounting is displayed from then on than before heat. Because of this we feel the descriptive term "prooestrous behavior" should not be used and that "mounting activity" is better.

The factors involved in the production of mounting activity have been puzzling from the beginning and it is only recently that a solution of the problem has seemed near. A widespread opinion is that excessive mounting is caused by the presence of cystic follicles (4, 7, 10, 12, 15). In the guinea pigs we have

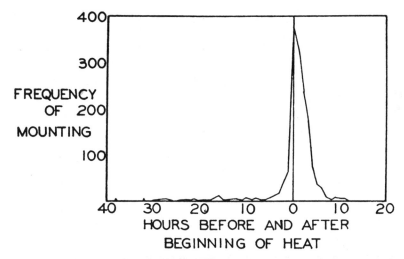

Fig. 1. Total Mounting Activity Displayed by 98 Animals Observed During One Heat Period Each

studied no such indications have been found. The ovaries from some of the most active animals were completely normal and the only follicular cysts encountered, unless follicles in the split oestrus-animals may be considered cystic, were in the ovaries from animals which neither came into heat nor mounted (19).

Early in the work no relationship was thought to exist between the number of maturing follicles and the frequency of mounting. Apparently this conclusion should be corrected, because when the ovaries were examined microscopically, the average number of maturing follicles was found to be greater in active than in quiet animals. It seems, therefore, that extensive mounting activity frequently is associated with the maturation of more than the average number of follicles. If this is true, quantitative factors must be involved. However, as with the length of heat, variations in the sensitivity of individuals may also be important.

As far as the rôle of specific hormones is concerned, only a beginning has been made. Ball (1) states that clitoris enlargement and the appearance of more male sex activity than normally appears in untreated female rats followed the injection of testosterone propionate over periods of 2 to 3 months. In the normal guinea pig injections of 4 mg. of this hormone daily for 17 days caused development of the clitoris into a penis-like structure and partially suppressed follicular growth, but did not cause any noticeable display of mounting activity (Boling and Hamilton, unpublished). In the spayed guinea pig the frequency of mounting was no greater when the administration of theelin was followed by such androgens as androstenediol, androstenedione, or dehydroandrosterone acetate than after theelin alone. Apparently, therfore, to the extent the guinea pig has been studied, androgenic action is not involved as in the rat.

An entirely different possibility is suggested by an analysis of certain experimental data. Of the 98 normal animals observed especially for mounting activity, all but 9 mounted at the time of heat. To be contrasted with these are 34 normal animals in which heat was produced on the 13th to 15th day of the cycle by intravenous injection of luteinizer, and 16 normal animals in which heat was produced on the 13th to 15th day by the subcutaneous injection of progesterone. In four of the latter the progesterone was preceded by 40 I.U. (International Units) oil-soluble theelin and in one by 1000 I.U. Progynon-B. Of the 34 animals injected with the luteinizer, mounting activity was displayed by only three, all of which had been injected on the 15th day and therefore fairly late in the cycle. Of the 16 animals injected with progesterone, mounting activity was displayed by only two, both of which had received the preliminary injections of theelin.

The low frequency of mounting in the experimental groups

has been taken to indicate that mounting activity like heat is produced by the follicular fluid hormones, and that those in mature follicles, by virtue of their quantity, quality or duration of action, are more effective than those in smaller follicles. The results from unpublished experiments support this hypothesis and indicate that both oestrin and progesterone are involved and that the action is synergistic as in the production of heat. But the respective mechanisms are not believed to be the same, partly because no positive correlation exists between the frequency of mounting activity and the length of heat and partly because oestrin and progesterone injections that will produce heat in 100 per cent of the animals are followed by only irregular results as far as mounting is concerned. The investigation is still not sufficiently advanced to permit any further statement. A detailed report will be given later.

A final factor for which allowance must be made is the influence of other animals. Some influence may exist, but if mounting by one animal does stimulate mounting by others, the latter are affected only if they are oestrous or prooestrous individuals and therefore have been subjected to hormonal action.

Cycle-length

The regularity of cycle-length is noted elsewhere (17), but the subject is mentioned again because more can be said about certain irregularities than was possible before.

New data bearing on 556 intervals from the beginning of one heat period to the beginning of the next have been obtained. Of these, 508 were normal cycles from 14 to 18 days in length. Twenty-two intervals are from 29 to 34 days in length. Examination of the ovaries indicates that intervals of this length represent two cycles, ovulation without heat having occurred at the beginning of the second. The regularity with which the vaginal membrane opened and closed when the interval was from 45 to 79 days suggests a similar explanation for intervals of these lengths in three animals, except that more than one heat period was missed.

The short cycles observed in 3 animals following unilateral

ovariectomy were caused by removal of the newly formed corpora lutea (5).

At least one of the six intervals from 23 to 26 days in length was caused by failure of ovulation and corpus luteum formation at the expected time. New follicular development required about nine days. Consequently, the interval between the normal heat, which occurred at the end of this time, and the preceding heat period was 25 days, 18 hours. But this interval should not be regarded as a single prolonged cycle. It represents two cycles, a first of normal length and a second short cycle similar to that which occurs in the absence of a corpus luteum. Supplementary observations suggest that the five other intervals of this length can be accounted for in the same way.

No explanation can be given for the remaining 14 intervals of irregular length except that the 3-hour heat period recorded for animal 1073 (table 2) after an interval of 19 days, 16½ hours, occurred when the animal was sick. Likewise, the observations have not enabled us to determine how irregularities of cycle-length are associated with the age of the animal. In a general way, irregularities are more common in young and old animals than in individuals in the prime of life, but it is not known to what extent the specific irregularities that have been encountered are associated with age.

The split oestrus

This term was applied to the type of heat in which the period of sexual receptivity is interrupted by one or more intervals when the copulatory response is not given. In the earlier work the split oestrus was detected only six times. During the more recent observations, 68 cases were encountered in 48 animals. This number represents a frequency between three and four times that reported originally, but may be high because certain animals in which the split oestrus tended to recur were reserved for observation over several cycles.

Before the newer data were studied, little was known about the abnormality except that the first part of heat usually was shorter than the second, the length of the interval when the heat response could not be obtained varied from 1 to 14 hours, and the vaginal changes proceeded normally despite the interruption of heat (16). Since then considerable new information has been obtained.

The earlier indication that the first part of the split oestrus usually is shorter than the second has been confirmed. In only 10 of the 68 cases was the first part longer than the second. The length of the first part averaged 4.2 hours, range 0.5 to 18.5 hours, and that of the second averaged 9.7 hours, range 2.5 to 19.5 hours.

The split oestrus may be composed of three parts as well as two and presumably of even more. When there were three parts, each of the first two was shorter than the third. For the 8 animals in which this happened the length of the first part averaged 1.5 hours, range 0.5 to 2.5 hours; the length of the second averaged 2.4 hours, range 0.5 to 7.0 hours; and the length of the third averaged 11.1 hours, range >7.0 to 15.5 hours.

The length of the intervals varied greatly. Sixteen were 1 or 1.5 hours long. Forty varied from 2 to $12\frac{1}{2}$ hours, 13 varied from 13 to $23\frac{1}{2}$ hours and 7 varied from 26 to 51 hours. No relationship was found between the length of the interval and the length of either part of heat.

Mounting activity may be displayed during either part of the split oestrus and during the interval between. However, it is of more frequent occurrence shortly before and during the second part of heat than prior to or during the first part; the detection of the abnormality usually depends on this circumstance.

Although the data are not adequate to provide complete information bearing on the relationship between the age of the animal and the split oestrus, this type of behavior is believed to be more common in middle-aged and old than in young animals. Of the 189 heat periods that were among the first four displayed by 54 animals of known age, no instance of split oestrus was shown (table 1). Of the 89 heat periods displayed by 26 of the same animals 5 to 11 months later, 3 were of the split-oestrus type. Of the 304 recorded from 39 animals observed during 1935–36 (tables 2 and 3), 5 were of the split-oestrus type. Of the 143 recorded from the same animals during 1936–37, 10 were of this type.

Two possible explanations for the split oestrus were given when the ovarian condition in such animals was described (19). One is that the first part of heat is produced by the action of oestrin and that the second part is caused by the secondary action of progesterone. The other explanation is that the split oestrus is a type of prolonged heat and a consequence of delayed ovulation. The first part of the latter suggestion is supported by the fact that the split oestrus is more likely to occur in animals in which heat is long. Of the 45 animals that displayed this behavior and were observed throughout four or more heat periods, heat was longer than average in 34, of average length in 9, and shorter than average in only 2. The second part of the suggestion is supported by the fact that ovulation occurred during or after the second part of heat in 10 animals and could have occurred at this time in the remaining 3 of 13 which were killed when the point was being investigated.

In a personal communication (Frederick N. Andrews), a condition similar to the split oestrus is described for the horse. This would indicate that this type of sexual behavior is not peculiar to the guinea pig. It is also possible that the prolonged heat in cattle which is associated with the failure of the mature follicle to rupture (8, 9, 15) is comparable with the split oestrus.

The incidence of heat at each of the 24 hours of the day

When the data from the earlier observations were being summarized, graphs were prepared showing the time of day when animals are most likely to be found in heat. A shift in this time coincided with the seasonal change in the time of sunset and the approach of darkness. After more extensive observations had been made, a similar summary of the newer data seemed advisable (fig. 2). The predominant occurrence of heat at night is seen as before, but no evidence is shown for the seasonal shift that was indicated previously. If it is possible to determine what nocturnal factor stimulates the beginning of heat, the problem of seasonal differences in the time of maximum heat occurrence can be reopened.

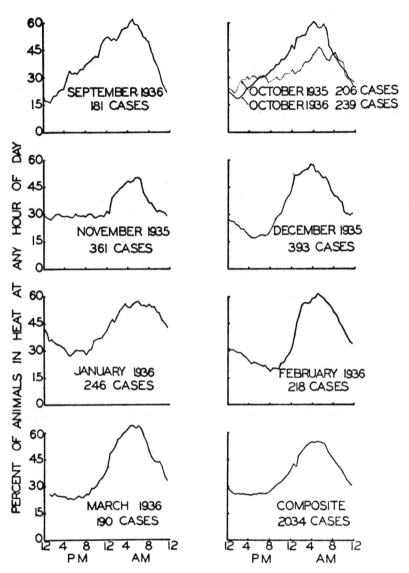

FIG. 2. MONTHLY AND COMPOSITE RECORDS OF HEAT OCCURRENCE DURING THE
24 HOURS OF THE DAY

THE MEASUREMENT OF SEXUAL DRIVE

Although the female guinea pig has not been used in investigations of sexual drive, it would appear to be admirably adapted to such studies. The display of sexual behavior is definite and measureable and probably more data bearing on behavior and the associated morphological changes have been obtained for this species than any other laboratory mammal. However, if the guinea pig is to be used, it is necessary to emphasize that such manifestations of sexual drive as the display of the copulatory response on the one hand, and running[5] and mounting activity on the other, are not to be weighted equally nor may inferences as to the extent of one be made from the other. Mounting activity and sexual receptivity usually are closely associated in point of time, but no correlation exists between the frequency of mounting and the length of heat. Consequently, in any measurement of sexual drive, mounting activity and receptivity should be regarded as separable components of a sexual behavior complex and measured directly by whatever means are considered most appropriate.

In certain respects the relationship between mounting activity and heat corresponds to that between vaginal changes and heat. The temporal relationship is close and often the time of heat can be correctly inferred from the activity of the animal. But mounting, like vaginal changes, may occur in the absence of heat and neither provides a means by which the beginning and end of heat can be determined accurately. The heat response is specific and neither the vaginal condition nor the display of mounting is a dependable indicator of the willingness to mate.

What has been said is known to apply only to the guinea pig. Because the rat has been used so extensively for investigations of sexual drive, it would be interesting to ascertain if the same lack of correlation exists between activity and length of heat in this species. The answer to the question awaits a means of determining the length of heat in the rat. In the meantime,

---

[5] Our impression is that a high correlation exists between running and mounting and that the latter may be used as a measure of the former. The point is being investigated.

what has been found in the guinea pig demonstrates the importance of recognizing the different aspects of sexual activity and reëmphasizes the point made by Stone (14), that in studies of drives more attention must be given to valid and reliable indicators.

SUMMARY AND CONCLUSIONS

Continuous day and night observations of the same animals during 5 to 11 consecutive reproductive cycles one year and 3 to 4 the next, and certain experimental procedures have provided new data bearing on cyclic sexual behavior in the guinea pig.

1. The age when the first heat period occurred varied from 33 to 134 days and averaged 67.8 days. The age when first rupture of the vaginal membrane occurred varied from 33 to 111 days and averaged 58.2 days. The interval between the first opening of the membrane and the first heat period tended to become longer as the latter was delayed.

2. The average length of each of the first two heat periods was shorter than that of the subsequent ones. This is attributed to the lower sensitivity to hormonal action of young animals which previously have not come into heat.

3. In many animals a tendency toward consistency of heat-length is shown which may last during much of the reproductive life.

4. Mounting activity is displayed most frequently at the beginning of heat, but many variations are shown. The factors responsible for this behavior are still not clear, but both oestrin and progesterone acting synergistically are involved. The mechanism probably is not the same as that producing heat.

5. Twenty-five of the 45 dioestrous intervals of abnormal length which occurred in otherwise normal animals can be explained by ovulation without heat. Six intervals from 23 to 26 days in length may have been caused by the failure of ovulation and corpus luteum formation at the expected time followed by the development of a new set of follicles which ruptured 9 or 10 days later. The remaining intervals of abnormal length cannot be accounted for.

6. Supplementary data concerning the character of the split oestrus are given. The abnormality tends to occur in animals in which heat is long and is more common in middle-aged and old than in young animals.

7. No evidence was found for a seasonal shift in the time of day when heat is most likely to occur.

8. The lack of correlation between the frequency of mounting and the length of heat is emphasized and the importance of the fact for investigations of sexual drive in the guinea pig is discussed.

REFERENCES

(1) BALL, J. 1937 The effect of male hormone on the sex behavior of female rats. Psychol. Bull., **34, 725.**

(2) BLANDAU, R. J., AND YOUNG, W. C. 1936 Ovum age and pregnancy in the guinea pig. Data from 462 artificially inseminated females. Anat. Rec., **67,** (Suppl. no. 1), 33.

(3) COLLINS, V. J., BOLING, J. L., DEMPSEY, E. W. AND YOUNG, W. C. 1938 Quantitative studies of experimentally induced sexual receptivity in the spayed guinea pig. Endocrinol., **23,** 188–196.

(4) COURRIER, R. 1925 Nymphomanie et ovaires kystiques. Compt. rend. Soc. de Biol., **93,** 674–675.

(5) DEMPSEY, E. W. 1937 Follicular growth rate and ovulation after various experimental procedures in the guinea pig. Am. Jour. Physiol., **120,** 126–132.

(6) DEMPSEY, E. W., HERTZ, R. AND YOUNG, W. C. 1936 The experimental induction of oestrus (sexual receptivity) in the normal and ovariectomized guinea pig. Am. Jour. Physiol., **116,** 201–209.

(7) DETLEFSEN, J. A. 1914 Genetic studies on a Cavy species cross. Carn. Inst. Wash., Pub. no. 205, pp. 134.

(8) FREI, W., UND LUTZ, E. 1929 Der heutige Stand der Forschungen über das Oestrushormon und die Nymphomanie des Rindes. Virchow's Arch. f. path. Anat., **271,** 572–622.

(9) HAMMOND, J. 1927 The physiology of reproduction in the cow. London: Cambridge University Press.

(10) HARRIS, G. W. 1937 The induction of ovulation in the rabbit, by electrical stimulation of the hypothalamo-hypophyseal mechanism. Proc. Roy. Soc., London, ser. B., **122,** 374–394.

(11) MYERS, H. I., YOUNG, W. C., AND DEMPSEY, E. W. 1936 Graafian follicle development throughout the reproductive cycle in the guinea pig, with especial reference to changes during oestrus (sexual receptivity). Anat. Rec., **65,** 381–401.

(12) PEARL, R., AND SURFACE, F. M. 1915 Sex studies. VII. On the assumption of male secondary characters by a cow with cystic degeneration of the ovaries. Maine Agric. Exp. Station, Bull. no. 237, 65–80.

(13) Stone, C. P. 1926 The family resemblance of female rats with respect to (1) the ages of first oestrus and (2) the body weights. Am. Jour. Physiol., 77, 625-637.

(14) Stone, C. P. In press. Chapter on Sex Drive. Allen, Sex and Internal Secretions, Second Edition. Williams and Wilkins, Baltimore.

(15) Williams, W. L., and Williams, W. W. 1921 The diseases of the genital organs of domestic animals. Ithaca, N. Y.

(16) Young, W. C. 1937 The vaginal smear picture, sexual receptivity and the time of ovulation in the guinea pig. Anat. Rec., 67, 305-325.

(17) Young, W. C., Dempsey, E. W. and Myers, H. I. 1935 Cyclic reproductive behavior in the female guinea pig. Jour. Comp. Psychol., 19, 313-335.

(18) Young, W. C., Dempsey, E. W., Hagquist, C. W. and Boling, J. L. 1937 The determination of heat in the guinea pig. Jour. Lab. and Clin. Med., 23, 300-302.

(19) Young, W. C., Dempsey, E. W., Myers, H. I. and Hagquist, C. W. 1938 The ovarian condition and sexual behavior in the female guinea pig. Am. Jour. Anat., 63, 457-487.

Reprinted from *Amer. J. Physiol.*, **116**, 201–209 (1936)

15

THE EXPERIMENTAL INDUCTION OF OESTRUS (SEXUAL RECEPTIVITY) IN THE NORMAL AND OVARIECTOMIZED GUINEA PIG[1]

EDWARD W. DEMPSEY, ROY HERTZ AND WILLIAM C. YOUNG

From the Arnold Laboratory, Brown University

Received for publication March 7, 1936

Notwithstanding the reports that spayed mice (Allen et al., 1924; Marrian and Parkes, 1930; and Wiesner and Mirskaia, 1930), rats (Allen et al., 1924; Hemmingsen, 1933), rabbits (Lacassagne and Gricouroff, 1925), and dogs (Kunde, D'Amour, Carlson and Gustavson, 1930) copulate following the injection of either follicular fluid or oestrin, an analysis of the mating behavior described for the guinea pig (Young, Dempsey and Myers, 1935a, b) has yielded several lines of evidence indicating that oestrin alone is not sufficient to produce mating behavior.

When it was learned that the one ovarian change which is associated with the beginning of heat is the beginning of the preovulatory swelling (Myers, Young and Dempsey, 1935), and that injection of the luteinizing hormone causes the preovulatory swelling (Foster and Hisaw, 1935), it seemed the luteinizing hormone might be the extra-ovarian factor we had postulated to be associated with oestrin in causing heat. Evidence will be presented below which partially supports this hypothesis. It now appears, though, that the luteinizing hormone does not directly produce heat, but stimulates the preovulatory swelling, ovulation and the production of progesterone, and it is this latter substance acting synergistically with oestrin that causes heat. The experiments upon which this conclusion is based are described in the following sections.

The injection of oestrin preparations into spayed and normal females. When water soluble theelin was injected subcutaneously into spayed females, heat followed only three of fifty-seven injections after latent periods of 33 to 42 hours (table 1). This low percentage of positive results is not attributable to the injection of insufficient quantities, since a four-fold increase in the minimal effective dose did not produce heat.

Theelin-in-oil was more efficacious in producing heat than water soluble theelin, seventeen of fifty-four injections being followed by oestrous periods of one to twenty-one hours. Still better results were obtained

[1] This investigation was supported by a grant from the Committee for Research in Problems of Sex, National Research Council.

with dihydroxyoestrin benzoate (Progynon-B, Schering). Eleven of seventeen injections were followed by heat periods of six to seventy-seven hours. Two unusual features of the heat induced by this latter substance are its excessive length (sixty-six and seventy-seven hours) in two instances and the occurrence of a "split oestrus" (Young, Dempsey and Myers, 1935a, p. 323) in four instances.

It was thought the administration of several small injections followed by a large injection might more nearly simulate the normal changes in oestrin level in the animal. Accordingly, five animals were injected twice daily with 12.5 I.U. theelin-in-oil, followed by 1000 I.U. four days after the first injection, and eight animals were injected four times daily with 10 I.U. followed by 1000 I.U. five days after the first injection. The results obtained were not different. Altogether, five animals came into heat. Of these, two came into heat while the small injections were being made, but not later after the large quantity was injected; two came into heat once while the small quantities were being injected and again after the large injection, and one came into heat only after the large injection. Since only five of thirteen animals came into heat, the administration of several small injections followed by a large injection is no more effective than a single large injection.

The injection of theelin into normal animals was undertaken with the thought that the normal oestrus might be lengthened by supplying additional theelin. The unexpected result was obtained that animals with intact ovaries respond to such injections less frequently than spayed individuals. Only two of ten animals injected with theelin-in-oil at the beginning or the end of heat came into heat again, and in no case was the normal oestrus longer than the previous normal periods.

Since it was thought the absorption time of the oil soluble theelin might be so long that it would defeat our purpose of augmenting the animal's own supply of theelin, another series of injections was made with water soluble theelin (table 2). Nine animals were injected with 25 R.U. every hour for four hours, beginning at the onset of oestrus and continuing until at least two hours after the end of oestrus, and three animals were injected with 25 R.U. every hour for four hours, beginning with the end of oestrus. Again, the heat period was not lengthened, the average for the nine animals injected from the beginning of oestrus being 6.5 hours which is 1.5 hour shorter than the average for the colony.

A last group of normal animals was composed of six injected with 1000 I.U. theelin-in-oil on the eighth day of the dioestrum and seven injected on the twelfth, thirteenth or fourteenth day. Only one animal, injected on the eighth day, came into heat after a latent period of seventeen hours. This was the only time a latent period of less than twenty-eight hours followed the injection of oestrin. Examination of the ovaries of the twelve day animals forty-eight hours after the injection revealed normal large follicles but no ovulation points. The next oestrus occurred either as though nothing had happened (2 cases), or following a delay of three (4 cases) or four (1 case) days. Apparently, the action of theelin late in the cycle slightly retards, rather than hastens, the appearance of the next oestrus.

When the animals injected with oestrin preparations are considered as

TABLE 1

Effect of oestrin preparations in causing heat in spayed animals

PREPARATION	AMOUNT	NUMBER OF ANIMALS	NUMBER OF INJECTIONS	INJECTIONS FOLLOWED BY HEAT		INJECTIONS FOLLOWED BY MOUNTING	LATENT PERIOD	LENGTH OF HEAT
				Number	Percent			
							hours	hours
Theelin (water sol).....	100 R.U. or more	24	57	3	5.3	15	33–42	1–12
Theelin (oil sol)........	1000 I.U.	30	54	17	31.0	12	35.3	1–21
Progynon-B............	1000 I.U.	17	17	11	65.0	1	33.4	6–77

TABLE 2

Effect of oestrin preparations in causing heat in normal animals

PREPARATION	AMOUNT	STAGE IN CYCLE	NUMBER OF ANIMALS	INJECTIONS FOLLOWED BY HEAT	INJECTIONS FOLLOWED BY MOUNTING	LATENT PERIOD
						hours
Theelin (water sol)....	100 R.U. or more	Onset of oestrus	9	0	1	
	100 R.U. or more	End of oestrus	3	0	1	
Theelin (oil sol).......	1000 I.U.	Onset of oestrus	7	1	0	40.5
	1000 I.U.	End of oestrus	3	1	0	31.5
	1000 I.U.	8th day	6	1	0	17.0
	1000 I.U.	12th day	7	0	0	

a whole, the following features are noted. 1. An injection effective at one time may be ineffective at another time. 2. Divided injections of small quantities of oestrin frequently are quite as effective as large quantities. 3. Normal animals respond to oestrin less frequently than spayed animals. 4. When heat occurs following a single injection of water-soluble theelin, theelin-in-oil, or dihydroxyoestrin benzoate, the length of the latent period is the only reasonably constant feature. In no instance following thirty-one effective injections into spayed animals was the latent period less than twenty-eight or more than forty-five hours.

Altogether, the data given above emphasize that the results following oestrin injections are characterized by irregularity and are difficult to reconcile with the view that oestrin alone is responsible for the production of heat.

The injection of luteinizing hormone into normal and spayed females. When 0.2 gram equivalent of sheep luteinizing hormone extracted according to the method of Fevold, Hisaw, Hellbaum and Hertz (1933) was injected intravenously into the marginal ear vein of twenty-four animals on the twelfth to fourteenth day of the cycle, nineteen came into heat after an average latent period of 5.8 hours (table 3), a result which shortened the dioestrum an average of 2.6 days. The heat period produced averaged only 5.6 hours in length, and is therefore shorter than the normal oestrus

TABLE 3

Effect of intravenous injections of luteinizing hormone preparations in causing heat in normal animals

PREPARATION	AMOUNT	STAGE IN CYCLE	NUMBER OF ANIMALS	INJEC- TIONS FOLLOWED BY HEAT	LATENT PERIOD
					hours
Sheep L.H................	0.2 g. eq.	12th–15th day	24	19	5.8
Mare serum L.H..........	0.4 g. eq.	12th–15th day	6	4	6.1
Antophysin..............	25–100 R.U.	12th–15th day	7	5	6.4
Sheep L.H................	0.2 g. eq.	8th day	4	0	
Sheep L.H................	0.2 g. eq.	4th day	4	0	
Sheep L.H................	0.2 g. eq.	End of oestrus	3	0	

shown previously by these animals which averaged 7.9 hours. Laparotomy in fifteen animals and histological examination from seven cases has shown that ovulation and luteinization occur following this treatment.

Both heat and ovulation were also produced at this time in four of six cases by a luteinizing hormone fraction of pregnant mare serum extracted according to Fevold's method, and in five of seven cases by the pregnancy urine concentrate, Antophysin. Since these substances contain no lactogenic hormone (Riddle, Bates and Dykshorn, 1932), growth hormone, adrenotropic factor or thyreotropic factor (Evans, Meyer and Simpson, 1933), the above results probably can be attributed to the luteinizing hormone alone.

When the luteinizing hormone was injected into four animals on the fourth day and four animals on the eighth day of the dioestrum, and into three animals at the end of heat, oestrus did not follow. Sexual receptivity, then, can be produced by administration of the luteinizing hormone only when there are follicles in the ovary mature enough to be ovulated.

147

The next normal oestrus following the injection of the luteinizer occurred nine to fifteen days later, the average being 13.2 days. This suggests that the lutein tissue formed following the injections was hypofunctional, since animals from which the corpora lutea have been removed soon after the end of oestrus return to heat and ovulate in about this time (Loeb, 1911; and unpublished data from this laboratory).

The failure of luteinizing hormone to produce oestrus early in the cycle suggested that possibly sufficient oestrin to condition the animal had not been produced before the twelfth day. However, when two six-day and two eight-day animals were given 1000 I.U. theelin each and forty-eight hours later were injected with luteinizing hormone, heat did not occur. Since the presence of a functional corpus luteum at this time might counteract the effect of the theelin, spayed animals were injected with 1000 I.U. theelin, and forty-eight hours (5 animals) and twenty-four hours (5 animals) later were injected intravenously with luteinizing hormone. Again, heat

TABLE 4

Effect of 40 I.U. oestrin followed by 0.2 I.U. progesterone in causing heat in spayed animals

TIME BETWEEN INJECTIONS	NUMBER OF ANIMALS	INJECTIONS FOLLOWED BY HEAT	INJECTIONS FOLLOWED BY MOUNTING	LATENT PERIOD
hours				*hours*
0	4	0	0	
12	8	2	0	9.5
24	8	6	1	6.6
36	3	3	1	4.3
48	24	24	5	4.8

did not occur. Likewise, when four animals, all of which previously had responded to injections of oestrin alone were given 1000 I.U. dihydroxyoestrin benzoate and twelve hours later were injected with luteinizer, heat occurred in only two animals after latent periods of 30 and 45 hours, the usual latent period for oestrin. Apparently, therefore, the combination of oestrin and luteinizing hormone can neither produce oestrus regularly nor shorten the latent period of oestrin. This indicates that the action of the luteinizer in producing oestrus is mediated through the ovary, and since the luteinizing hormone causes the formation of the corpus luteum, the next attempts to produce oestrus involved injections of the corpus luteum hormone.

The injection of progesterone into spayed and normal females. Because all the known progesterone reactions occur only after an initial treatment with oestrin, a similar procedure was employed. In a first series twenty-eight spayed females were injected subcutaneously with 40 I.U. theelin over a period of forty-eight hours, followed by 0.05 to 0.2 I.U. of progesterone (Proluton, Schering). The twenty-four animals which received the larger amount came into heat following latent periods of three to nine hours, the average being 4.8 hours. One of the animals receiving 0.1 I.U. came into heat and 0.05 I.U. was not effective in either of two cases.

To determine whether an initial treatment with oestrin is necessary, four spayed animals were injected with 0.2 I.U. progesterone alone. Not one of the four came into heat. Theelin injections were then followed 0, 12, 24 and 36 hours later by progesterone (table 4). No one of four animals injected with the two preparations simultaneously came into heat. Of the others, two of eight in which the injections were twelve hours apart, six out of eight in which the injections were twenty-four hours apart, and all in which the injections were thirty-six or more hours apart came into heat. Therefore, an initial treatment with oestrin at least twelve hours prior to the injection of progesterone is necessary.

A corollary to the above developed when progesterone was injected at the end of oestrus. Whether the oestrus was normal (3 cases), induced by oestrin alone in spayed animals (2 cases) or induced by oestrin and progesterone in spayed animals (2 cases), there was no return of heat. Apparently, once oestrus has been evoked, the animal must again be conditioned before heat can occur.

In the case of some of the oestrin-progesterone reactions, a rather definite quantitative relationship has been shown to be necessary (Hisaw and Leonard, 1930). However, when four spayed animals were injected with 1000 I.U. theelin and forty-eight hours later with 0.2 I.U. progesterone, all four came into heat. Likewise, when four injections of 250 I.U. each were made over the forty-eight hour period, followed by 0.2 I.U. progesterone, oestrus occurred. Thus, the result was no different from that obtained following the smaller amounts of oestrin administered in the earlier experiments, indicating that there is considerable latitude in the quantitative relationships of the reaction.

Fourteen normal animals were injected with progesterone alone on the twelfth to fifteenth days of the dioestrum. Of those injected on the fourteenth to fifteenth days, seven of ten came into heat. Of those injected on the twelfth and thirteenth day, unlike similar animals injected with luteinizing hormone, heat occurred in only one of four animals. When, however, such animals were injected with 40 I.U. theelin and thirty-six hours later with progesterone, oestrus followed in four of six animals. Laparotomies following these induced heat periods showed that ovulation had not occurred in any of the nine animals examined. Despite the operation, six of the nine animals came into heat again spontaneously four or five days after the experimental oestrus and second laparotomies in three animals showed that ovulation had occurred then. In these cases, then, it would seem that the substance which normally is produced at ovulation and which causes oestrus was introduced into suitably conditioned animals and produced sexual receptivity without ovulation.

On the basis of the results obtained from these experiments, it seems likely that progesterone is involved very directly in the production of sexual receptivity in the guinea pig.   The chain of events is thought to be approximately as follows: The rise in the oestrin level which occurs toward the end of the cycle conditions the animal for oestrus.   The appearance of oestrus itself, however, depends upon the subsequent action of progesterone which is produced in the still unruptured follicle after it has been acted upon by the luteinizing hormone of the hypophysis.

DISCUSSION.   While this investigation was in progress, Witschi and Pfeiffer (1935) reported that mating reactions can be induced in rats by injections of luteinizing hormone, provided mature follicles are present. Since ovulation also occurred, it was not clear to them whether the mating instincts were aroused directly by the luteinizer or by concomitant changes occurring in the ovary.   The production of sexual receptivity in spayed animals by injections of oestrin and progesterone and our failure to evoke mating reactions by injections of luteinizing hormone and oestrin would seem to answer this question as far as the guinea pig is concerned. The rôle of the luteinizer is not direct, but acting upon the ovary, it causes the production of progesterone which brings the animal into heat.

Our failure to hasten ovulation in normal animals injected with oestrin or progesterone, or both, on the twelfth to fifteenth days of the cycle has caused us to question the hypothesis (Lane, 1935) that the release of the luteinizer is effected when an increased level of oestrin acts upon the pituitary.   This failure to hasten ovulation can not be attributed to the immaturity of the follicle, since ovulation can be produced at this time by injections of luteinizer.   In addition, the approach of darkness, an environmental factor, influences the time of oestrus (Dempsey, Myers, Young and Jennison, 1934) and therefore, it would seem, the time of the release of the luteinizer.

The current theories regarding the relationships between the pituitary and the rest of the endocrine system render caution advisable in assigning the cause of oestrus to the hormonal mechanism we have described.   For example, we do not know why oestrin alone brings some animals into heat. A possibility is that there is a conversion of oestrin to progesterone in the animal.   Nevertheless, the data which indicate that oestrus is caused by the action of progesterone after the animal has been conditioned are felt to clarify the ultimate cause more than the hypothesis that oestrin alone is responsible.   In addition, a procedure that can be depended upon to bring females of at least one species into heat should be useful as a point of departure from which studies of the complete regulatory mechanism can be continued.

Indispensable assistance in observing the animals has been given us by Mr. R. J. Blandau, Mr. J. L. Boling and Mr. C. W. Hagquist who are working on other aspects of the problem. Appreciation is also expressed for the generous contribution of theelin provided through the courtesy of Dr. Oliver Kamm by Parke, Davis and Co., of Antophysin, supplied by the Winthrop Chemical Co., of mare serum, supplied by Chappel Bros., and of sheep luteinizing hormone given us by Dr. H. L. Fevold of Harvard University.

SUMMARY

1. Only 14 per cent of injections of 1000 I.U. theelin in 21 normal females and 31 per cent of 54 injections in 30 spayed females were followed by sexual receptivity. The latent period averaged 35.3 hours.

2. Intravenous injections of 0.2 gram equivalent of sheep luteinizing hormone produced heat and ovulation in 19 of 24 cases when injected into normal animals on the 12th to 15th days of the dioestrum, but were ineffective earlier in the cycle in 15 cases. The latent period averaged 5.8 hours. Similar injections of luteinizing hormone into spayed females were ineffective and the combination of oestrin and luteinizing hormone did not enhance the efficiency of the oestrin alone in 15 cases.

3. Injections of 40 I.U. or more of oestrin, followed 36 to 48 hours later by 0.2 I.U. progesterone invariably produced heat in 24 spayed females. The latent period averaged 4.8 hours after the injection of the progesterone. Seven of nine normal animals injected with progesterone alone on the 14th or 15th day of the cycle came into heat 5 to 10 hours after the injection, but did not ovulate.

4. On the basis of these data, it is postulated for the guinea pig that after conditioning with oestrin, oestrus is caused by the action of progesterone, produced under the influence of the luteinizing hormone during the preovulatory enlargement of the follicle.

REFERENCES

ALLEN, E. ET AL. Am. J. Anat. **34**: 133, 1924.

DEMPSEY, E. W., H. I. MYERS, W. C. YOUNG AND D. B. JENNISON. This Journal **109**: 307, 1934.

EVANS, H. M., K. MEYER AND M. E. SIMPSON. Mem. Univ. Calif. **11**, 1933.

FEVOLD, H. L., F. L. HISAW, A. HELLBAUM AND R. HERTZ. This Journal **104**: 710, 1933.

FOSTER, M. A. AND F. L. HISAW. Anat. Rec. **62**: 75, 1935.

HEMMINGSEN, A. M. Skand. Arch. f. Physiol. **65**: 97, 1933.

HISAW, F. L. AND S. L. LEONARD. This Journal **92**: 574, 1930.

KUNDE, M. M., F. E. D'AMOUR, A. J. CARLSON AND R. G. GUSTAVSON. This Journal **95**: 630, 1930.

LACASSAGNE, A. AND G. GRICOUROFF. Compt. Rend. Soc. Biol. **93**: 928, 1925.

LANE, C. E. This Journal **110**: 681, 1935.

LOEB, L. Deutsch. med. Wchnschr. **37**: 17, 1911.

MARRIAN, G. F. AND A. S. PARKES. J. Physiol. **69**: 372, 1930.

MYERS, H. I., W. C. YOUNG AND E. W. DEMPSEY.   Proc. Am. Soc. Zool., Anat. Rec. **64:** 53 (Suppl.), 1935.

RIDDLE, O., R. W. BATES AND S. E. DYKSHORN.   Proc. Soc. Exper. Biol. and Med. **29:** 1211, 1932.

WIESNER, B. P. AND L. MIRSKAIA.   Quart. J. Exper. Physiol. **20:** 273, 1930.

WITSCHI, E. AND C. A. PFEIFFER.   Anat. Rec. **64:** 85, 1935.

YOUNG, W. C., E. W. DEMPSEY AND H. I. MYERS.   J. Comp. Psychol. **19:** 313, 1935a.
Proc. Am. Assoc. Anat., Anat. Rec. **61:** 65 (Suppl.), 1935b.

Reprinted from *Endocrinology*, **25**, 359–364 (1939)

16

THE ESTROGEN-PROGESTERONE INDUCTION OF MATING RESPONSES IN THE SPAYED FEMALE RAT[1]

JOHN L. BOLING AND RICHARD J. BLANDAU

From the Arnold Biological Laboratory, Brown University

PROVIDENCE, RHODE ISLAND

ESTROGEN and progesterone administered in the proper sequence have been found to be more effective than estrogen alone in inducing sexual receptivity in the guinea pig (1). Since this was the only species in which this action of estrogen and progesterone had been observed, it seemed desirable to ascertain if these hormones acting synergistically are not also effective in inducing heat in another species such as the rat. The rat was chosen because the occurrence of mating in 'constant estrous animals' which ordinarily do not mate, ovulate, or form corpora lutea has been reported following the injection of luteinizing hormone (2, 3). The induced mating was associated with ovulation and corpus luteum formation and suggested a relationship between the appearance of the corpora lutea and mating, similar to that postulated for the guinea pig. The possibility has been investigated and data, which indicate that in this respect the rat is similar to the guinea pig, are given below.

EXPERIMENTAL

Thirty 3 to 7-month-old spayed rats of the Wistar strain were used. Estrogen (estradiol benzoate) and progesterone[2] in corn oil were injected subcutaneously. The animals were given a rest period of 10 or more days between each series of injections.

The procedure for all experiments was essentially as follows. The estrogen was administered in a single injection, with the exception of 10 animals which received 100 R.U. in 3 doses of 50, 25 and 25 R.U. over a period of 48 hours. Hourly observations for evidences of heat were begun 24 to 40 hours after injection of the estrogen and continued until the 48th, 72nd or 96th hour. Except as noted below the animals which did not come into heat on the estrogen received a single injection of 0.2, 0.3 or 0.4 I.U. progesterone. Hourly observations were then continued until the end of heat or, when heat did not occur, for 20 to 24 hours.

Two carefully graded systems have been proposed for classifying the widely varying intensities of heat responses observed during the normal cycle of the rat (4, 5). In the present work only two types of behavior were em-

---

[1] The investigation was supported by a grant from the Committee for Research in Problems of Sex, National Research Council; grant administered by Prof. William C. Young.

[2] Progynon-B and proluton for which we are indebted to Dr. Erwin Schwenk of the Schering Corporation.

ployed in recording the response to hormonal action: first, that displayed by animals which did not respond immediately to the attention of the male, and second, that displayed by animals which responded immediately, even to less vigorous males, and gave a marked lordosis when stroked on the dorso-posterior region. The latter type of behavior is characteristic of that displayed by normal animals at the height of heat. A negative response was considered to have been obtained only when the female would not mate after repeated attempts of vigorous males.

TABLE 1. THE INDUCTION OF HEAT BY INJECTION OF ESTROGEN (ESTRADIOL BENZOATE)

No. of injections	Quantity of injected estrogen	Length of observation after estrogen injection	No. of injections followed by heat	Latent period	
				Average	Range
	R.U.	hours		hours	hours
44	5	48	0.0	—	—
6	5	72	0.0	—	—
38	10	72	6	52	48 to 54
19	10	96	7	56	<39 to 74
10	20	72	2	49.5	49 to 50
20	100	96	18	50	38 to 67

Induction of Heat by Estrogen Alone

Five R.U. of estradiol benzoate given in a single injection did not induce mating in any one of 44 animals observed for 48 hours, or in any one of 6 animals observed for 72 hours after the injection (table 1). The only indication of sexual activity was a quivering of the ears (4, 5) without mating which was observed in one animal beginning the 45th hour after injection.

The injection of 10 R.U. of estradiol benzoate was followed by heat in 6 of 38 animals observed 72 hours, and in 7 of 19 observed 96 hours. The latent period between injection and the beginning of heat averaged 52 hours in animals of the first group, with a range of 48 to 54 hours. In animals of the second group, the latent period between injection and the beginning of heat averaged 56 hours, and ranged from <39 to 74 hours. This group was established when it was found that in 2 animals the latent periods for estrogen-induced heat were 72 and 74 hours. The possibility existed, therefore, that had the first group of animals been observed longer than 72 hours, additional cases of estrogen-induced heat would have been found. Data which were obtained subsequently make this possibility seem unlikely. In the first place, in no other case among the animals in the second group did a latent period exceed 63 hours. Secondly, 20 animals were injected with 100 R.U. of estrogen and observed for 96 hours. In no one of the 18 which came into heat did the latent period exceed 67 hours.

In the case of the 2 groups of animals injected with 10 R.U. of estrogen, the length of heat varied from 1 to 7 hours. The intensity of the response was usually of a low grade; vigorous males were required to obtain a mating from most animals. Quivering of the ears was observed in the animals which came into heat and in some in which heat did not occur.

Two of 10 animals given single injections of 20 R.U. of estradiol benzoate mated after latent periods of 49 and 50 hours. One animal remained in heat 4.5 hours while the other mated intermittently for 21 hours. This intermittent mating was characteristic of heat induced by large injections of estrogens alone. The intensity of the mating responses was similar to that of responses observed after the injection of 10 R.U. of estrogen.

In order to determine what percentage of animals could be brought into heat by large injections of estrogen, 20 were injected with 100 R.U. of estradiol benzoate. Ten received the entire amount in one injection. The remaining 10 were given an initial injection of 50 R.U. followed by injections of 25 R.U., 24 and 48 hours later. Nine of the 10 animals receiving a single injection came into heat after latent periods which averaged 50 hours, range 46 to 67 hours. Similarly, 9 of 10 animals receiving 3 injections came into heat. The latent periods averaged 49 hours and ranged from 38 to 67 hours. In the first group of animals heat was displayed for an average of 19 hours (range 2 to 29 hours) although frequently it was intermittent. In the second group the period during which heat was displayed averaged 112 hours with a range of 3 hours to 10 days. In this group, also, especially during the daylight hours, long periods occurred when the females would not mate with even the most vigorous males. Furthermore, in both groups, when the animals would mate, the intensity of the responses varied greatly. Occasional animals showed a very intense heat, but generally the responses were not as intense as those displayed during a normal or an estrogen-progesterone induced heat.

Induction of Heat by Estrogen-Progesterone Injections

Twenty-two of the 44 animals injected with 5 R.U. estradiol benzoate and which did not come into heat following this treatment, received 0.4 I.U. of progesterone 48 hours later. In this group progesterone was injected on the 48th hour because other experiments had demonstrated that 5 R.U. of estrogen is not sufficient to keep all animals conditioned 72 hours. Twenty of the animals came into heat after a latent period which averaged 4.5 hours and ranged from 3 to 8 hours (table 2). The length of heat averaged 10.5 hours (range 2 to 14 hours). In general the response of these animals was not as intense as that observed in normal animals or in animals conditioned with larger quantities of estrogen. Vigorous males were needed for most matings.

Fifteen of the 44 animals injected with 10 R.U. estradiol benzoate which did not come into heat received 0.4 I.U. progesterone on the 72nd hour. All came into heat. The latent period between the progesterone injection and heat averaged 2.7 hours (range 1.5 to 4 hours). The length of heat averaged 11.5 hours (range 7.5 to 15.5 hours). The intensity of the response was comparable to that of normal animals. Lordosis could easily be elicited by non-aggressive males and in the absence of males the copulatory response was readily obtained by stroking the animals on the back.

When the animals which had failed to come into heat following the larger dosages of estrogen were injected with progesterone, the results were the same. Eight injected 72 hours previously with 20 R.U. of estrogen and 2 in-

155

jected 96 hours previously with 100 R.U. came into heat after receiving 0.4 I.U. progesterone. The latent period after progesterone injection and the length of heat were about the same as those recorded when animals were injected with 10 R.U. estrogen followed by progesterone (table 2).

TABLE 2. THE INDUCTION OF HEAT BY ESTROGEN-PROGESTERONE ACTION

No. of injections	Quantity of injected estrogen	Interval between injections	Quantity of injected progesterone	No. of injections followed by heat	Latent period		Length of heat	
					Average	Range	Average	Range
	R.U.	hours	I.U.		hours	hours	hours	hours
22	5	48	0.4	20	4.5	3–8	10.5	2–14
22	5	48	0.3	15	5.2	3–11	9.6	2.5–16
15	10	72	0.4	15	2.7	1.5–4	11.5	7.5–15
15	10	72	0.3	15	2.4	2–3	9.6	9–11
8	20	72	0.4	8	2	2–2	unknown	11 to 23
2	100	96	0.4	2	3	3–3	9.5	9 & 10
6	10	96	0.4	6	3	3–3	unknown	unknown

When the quantity of injected progesterone was reduced from 0.4 I.U. to 0.3 I.U. a slight decrease in effectiveness was noted. Fifteen of 22 animals previously injected with 5 R.U. of estrogen came into heat when 0.3 I.U. of progesterone was injected 48 hours later. All of 15 animals receiving 0.3 I.U. of progesterone after an initial injection of 10 R.U. of estrogen came into heat, but the response was not as intense as that of animals receiving 0.4 I.U. of progesterone after the same amount of estrogen. When the quantity of injected progesterone was reduced to 0.2 I.U., only 10 of 25 adequately conditioned animals came into heat and the possibility cannot be excluded that several of these were brought into heat by the action of the estrogen alone. It is apparent, therefore, that under the conditions of this experiment, 0.3 I.U. to 0.4 I.U. of progesterone is the optimal dosage of this hormone.

Effect of Additional Estrogen Following an Initial Injection of Estrogen Adequate to Condition Animals for an Estrogen-Progesterone Induced Heat

Seventeen animals were injected with 10 R.U. of estradiol benzoate. Five came into heat on the estrogen alone. Ninety-six hours after the initial injection 6 of the 12 animals which did not come into heat received 20 R.U. of estrogen. At the same time the 6 remaining animals were injected with 0.4 I.U. of progesterone. The 6 which received the progesterone came into heat 3 hours after injection, while none of the 6 receiving the additional estrogen had come into heat within 9 hours, when observations were stopped.

When the data presented above are considered as a whole, the importance of progesterone for the induction of heat will be apparent. To be sure, the injection of estrogen is frequently followed by heat or by the other signs of a low grade 'estrous behavior,' but estrogen and progesterone acting synergistically were nearly always effective, even when small quantities were used. In the experiments described above, 137 animals were injected with from 5 to 100 R.U. of estrogen and observed well beyond the interval during which the estrogen-induced heat is likely to occur. Of these, only 33 came into heat. Fifty-three of the remaining 104 were subsequently injected with 0.4 I.U. of progesterone. Of these, 51 displayed a normal heat response, and since the 2 in which heat did not occur were among those injected with the smallest quantity of estrogen, it is probable they were not properly conditioned. In the 31 animals in which 10 R.U. or more of estrogen was injected and did not cause heat, 0.4 I.U. of progesterone was effective in all.

The numerous observations of those who have studied morphological and behavior changes associated with reproduction, indicating that some animals are more sensitive to estrogens than others, are confirmed. Very early in the study animals were noticed which were more responsive to estrogen than others.

DISCUSSION

When the history of attempts to reproduce sexual receptivity in spayed rats is reviewed, the hypothesis that heat in this species is caused by the synergistic action of estrogen and progesterone is not irreconcilable with earlier observations. Allen, et al. (6), Hemmingsen (4), Ball (5), and Hemmingsen and Krarup (7) induced heat in spayed rats by injecting estrogens. The results, however, were never uniform because very large quantities were not effective in some animals while comparatively small quantities were effective in others. Other important observations were those of Witschi and Pfeiffer (2) and Ball (3) to the effect that mating in 'constant estrous rats' can be induced by the injections of luteinizing hormone and is associated with ovulation and corpus luteum formation. Cole (8) found that mating will occur in immature rats at the earliest ages at which ovulation can be induced regularly. The supposition which followed was that a factor supplementary to estrogen is involved, and as in the guinea pig (1), the action of progesterone seemed possible. The best evidence for the likelihood of this supposition has been the success with which heat was induced when progesterone was administered following a conditioning injection of estrogen. In a recent personal communication Doctor Ball has told us that she too has been able to induce heat in spayed rats by estrogen-progesterone treatment when the quantity of injected progesterone is larger than she used previously (3).

The possibility that heat in the rat is stimulated by the synergistic action of estrogen and progesterone suggests that the corpus luteum must be actively producing hormone at the time of its formation or even earlier in the non-pregnant animal. Morphological evidence for such activity is reported by Long and Evans (9). Physiological evidence, however, was lacking prior to the present work and Astwood's observation (10) that increase in water content of the uterus which is attributed to estrogenic action is inhibited prior to ovulation by the action of "an ovarian secretion having the properties of a corpus luteum hormone."

The data which have been presented suggest that the hormonal factors responsible for the occurrence of heat in the rat and guinea pig are essentially similar. However, certain quantitative differences exist. On the whole,

157

estrogen is more effective in causing heat in the rat than in the guinea pig. Injections of 1000 I.U. (200 R.U.) of estradiol benzoate induced heat in only 60% of the injected guinea pigs (1), whereas in the rat 100 R.U. of the same hormone caused heat in 90% of the animals. On the other hand, the rat appears to be less sensitive to progesterone. One-tenth of an international unit is an effective quantity of progesterone in the guinea pig, whereas the rats we employed did not respond regularly until 0.3 I.U. to 0.4 I.U. were injected. This species difference in response to corpus luteum hormone is also indicated by the results from studies of the effect of progesterone on ovulation and heat in normal animals. Dempsey (11) found that 0.05 I.U. of progesterone daily would inhibit ovulation and heat in the guinea pig, while Phillips (12) found that 1.5 I.U. per day was necessary to inhibit ovulation and heat in the rat.

SUMMARY

A small injection of estrogen followed after a suitable interval by an injection of progesterone is more effective than a larger quantity of estrogen alone in inducing sexual receptivity in the spayed female rat. With respect to length and intensity, the estrogen-progesterone induced heat resembles that of normal animals more nearly than the estrogen induced heat. It seems likely, therefore, that in the normal rat as in the guinea pig sexual receptivity is caused by the synergistic action of estrogen and progesterone produced in the still unruptured follicles. The rat appears to be more sensitive than the guinea pig to estrogen-action and less sensitive to progesterone-action.

We wish to express our appreciation of the very helpful advice given by Dr. William C. Young during the progress of this investigation.

REFERENCES

1. DEMPSEY, E. W., R. HERTZ AND W. C. YOUNG: Am. J. Physiol. 116: 201. 1936.
2. WITSCHI, E., AND C. A. PFEIFFER: Anat. Rec. 64: 85. 1935.
3. BALL, J.: Proc. Soc. Exper. Biol. & Med. 35: 416. 1936.
4. HEMMINGSEN, A. M.: Skandinav. Arch. f. Physiol. 65: 97. 1933.
5. BALL, J.: J. Comp. Psychol. 14: 1. 1937.
6. ALLEN, E., et al.: Am. J. Anat. 34: 133. 1924.
7. HEMMINGSEN, A. M., AND N. B. KRARUP: Biologiske Meddelelser 13: 3. 1937.
8. COLE, H. H.: Am. J. Physiol. 119: 704. 1937.
9. LONG, J. A., AND H. M. EVANS: Mem. Univ. Calif. 6: 1. 1922.
10. ASTWOOD, E. B.: Am. J. Physiol. 126: 162. 1939.
11. DEMPSEY, E. W.: Am. J. Physiol. 120: 126. 1937.
12. PHILLIPS, W. A.: Am. J. Physiol. 119: 623. 1937.

Reprinted from *Behaviour*, **10**, 340–354 (1957)

17

STRAIN DIFFERENCES IN THE BEHAVIORAL RESPONSES OF FEMALE GUINEA PIGS TO ALPHA-ESTRADIOL BENZOATE AND PROGESTERONE [1])

by

ROBERT W. GOY [2]) and **WILLIAM C. YOUNG**

(Department of Anatomy, University of Kansas, Lawrence, Kansas, U.S.A.)

(With 2 Figs)

(Rec. 16-II-1956)

THE PROBLEM

Genetic, experiential and hormonal factors have been shown to be important for the determination of sexual behavior patterns in the male guinea pig (VALENSTEIN, RISS and YOUNG, 1954; VALENSTEIN and YOUNG, 1955; VALENSTEIN, RISS and YOUNG, 1955). The present study extends the work to the role of genetic and hormonal factors and their interactions in the determination and display of estrous behavior by the female.

SUBJECTS, TREATMENT AND TESTING PROCEDURES

In studies of the male the pattern of behavior of the intact animal is easily determined (YOUNG and GRUNT, 1951; VALENSTEIN, RISS and YOUNG, 1954). Investigation of the intact female, on the other hand, presents a more difficult problem. Sexual behavior is displayed only at cyclic intervals, heat is of several hours duration, it shows a varying intensity and, within any 24-hour period, the exact time when animals will come into heat is unpredictable (YOUNG, DEMPSEY and MYERS, 1935; YOUNG, DEMPSEY, HAGQUIST and BOLING, 1939). For these reasons the patterns of behavior which characterize individual animals cannot be determined accurately from occasional examinations and continuous day and night observation is necessary. It has been learned, however, that the administration of suitable quan-

---

1) This investigation was aided in part by research grant M-504(C2) from the National Institute of Mental Health, National Institutes of Health, Public Health Service and in part by a grant from the Committee for Research in Problems of Sex, National Academy of Sciences-National Research Council.

2) Public Health Service Research Fellow of the National Institute of Mental Health.

tities of an estrogen and progesterone to spayed females is followed after a predictable interval by a display of behavior which is similar to that exhibited prior to ovariectomy (BOLING, YOUNG and DEMPSEY, 1938; YOUNG and RUNDLETT, 1939). In the present experiment, therefore, spayed, injected females were used.

Data were collected for each of the two components composing the pattern of behavior in the female; 1) the copulatory response, and 2) a male-like mounting behavior (AVERY, 1925; YOUNG, DEMPSEY and MYERS, 1935). Although a close temporal relationship exists, the two components are not related quantitatively and there is evidence that they are mediated by different mechanisms (YOUNG, DEMPSEY, HAGQUIST and BOLING, 1939).

Fifty-three females were randomly selected from the general colony maintained in the laboratory. Fifteen were from the highly inbred strain 2, 17 from the highly inbred strain 13, and 21 from a genetically heterogeneous group (for convenience referred to as strain T). The animals were 3 to 6 months of age at the time of selection and all were spayed within 10 days after selection. The operation included the bilateral removal of the ovarian ligament, ovarian fat pad, ovary, follicular tube, and the proximal five millimeters of the uterine horn. Two months later tests for responsiveness to α-estradiol benzoate and progesterone were begun.

In a first phase of the work the influence of the genetic background on estrous behavior was studied. The females were divided into five observational groups with approximately the same proportions of the different strain in each. Three tests were given to each group. An injection of 100 I.U. of α-estradiol benzoate was followed 36 hours later by 0.2 I.U. of progesterone [1]).

A second phase of the work was an effort to answer two questions: Do the strains differ in their responsiveness to α-estradiol benzoate? Once a threshold is reached, is the intensity of the response related to the amount of α-estradiol benzoate, or, as in the male (GRUNT and YOUNG, 1952, 1953; RISS and YOUNG, 1954), are supra-liminal quantities without visible effect on the behavior?

These problems were studied after the first phase of the work had been completed. Forty-four of the 53 females used in the first phase were divided into four observational groups, each confined to a single cage. The strains were proportionally represented in each group. A series of tests was begun to determine the effect of 25, 50, 200, 400, or 800 I.U. of α-estradiol benzoate

---

[1]) Alpha-estradiol benzoate and progesterone were supplied by the Schering Corporation, Bloomfield, N.J.

23

followed 36 hours later with 0.2 I.U. of progesterone. When the estrogen was administered each observational group received a different quantity. The order in which each group received the different amounts was randomized. The effects of 100 I.U. of the estrogen were taken from the data obtained during the first phase of the work, but only for the animals which were also used in the second phase.

All injections were subcutaneous in the right or left axilla. Mazola corn oil was used as the medium and each injection was 0.5 cc. in volume. The progesterone was given in the early morning and the animals were observed continuously for 15 consecutive hours.

Each was tested at intervals of at least 3 and not more than 5 weeks. On test days animals were removed from their home cages when the progesterone was injected, placed in cages without tops and fingered hourly. When necessary to facilitate identification of individuals, geometric designs were painted on the sides and back.

MEASURES OF ESTRUS

1) Latency of heat is the length of the interval between the injection of progest rone and the elicitation of the first lordosis.

2) Duration of heat is the number of hours lordosis can be elicited. Toward the end of a heat period, lordoses become feeble and difficult to elicit, and an operational criterion is necessary to determine when an animal shall be classified as unresponsive. For this purpose, each animal is stroked or fingered five times and if no lordosis is displayed the animal is considered to be out of heat. Occasionally an animal is out of heat at the time of one fingering test, but responds an hour later. In these cases the animal is recorded as out of heat at the time of the first failure to respond unless lordoses are obtained for two or more consecutive hours. This occurred only 3 times in 377 tests.

3) The duration of the maximum lordosis in seconds. This measure has not been reported in the literature on estrous behavior, therefore a detailed description is given:

The lordosis reflex includes several components, an arching or straightening of the back, elevation of the pudendum, displacement of the rear feet laterally and caudally so that a wide stance is taken, and emission of a low gutteral growl. When an estrous female is stroked lightly in a caudo-cephalad direction all components of the lordosis are displayed nearly simultaneously. If the stroking is continued (prolonged stimulation), the full reflex will be maintained for a time which varies with the genetic background and the

161

phase of estrus. If stimulation is continued until voluntary termination is produced, the duration of the reflex can be measured with a stop-watch. The response may be considered terminated when any one of the following signs is evident: a) a sudden or gradual return of the back and pudendum to a normal position; b) a sudden return of the feet to the normal position and a loss of the wide stance characteristic of the reflex; c) kicking with the hind feet; d) dashing forward; e) squatting; f) urinating; and g) an abrupt termination of the growl accompanied by a soft squeal.

A stop-watch is started immediately on display of lordosis and stopped as soon as the complete response is no longer apparent. The duration of the response in seconds is recorded as a measure of the intensity of the reflex. The duration of the longest lordosis elicited from each animal is used to estimate the mean maximum lordosis characteristic of the strain. The proximity of the maximum lordosis to the beginning of heat is also found useful in characterizing the strains.

4) Male-like mounting behavior. Mounts accomplished by an individual are classified as a) c o m p l e t e   m o u n t s at the posterior end including clasping and pelvic thrusts, b) p o s t e r i o r   m o u n t s without pelvic thrusts, and c) a b o r t i v e   m o u n t s which are not posteriorly oriented, do not involve clasping, and usually are not accompanied by pelvic thrusts. Recorded mounting activity is usually preceded by locomotor activity best described as prowling or standing in one place and treading the floor of the cage with the hind feet. Both treading and prowling, when they precede mounting, are accompanied by the typical low gutteral growl or chatter.

5) Per cent of females brought into heat by the hormonal treatment.

HOUSING AND MAINTENANCE

Each observational group was housed in one of a battery of four cages. Each cage was 4' wide, 2' deep, and 1' high. The floor of each cage was covered with shavings. Water and food were continuously available. The diet consisted of rabbit pellets and oats supplemented daily with green vegetables and alfalfa. The temperature of the laboratory is maintained at 70 ° to 75° F.

RESULTS

A. Estrous behavior following injection of each strain with 100 I.U. of α-estradiol benzoate and 0.2 I.U. of progesterone.

The data are summarised in Table I. From strain to strain, consistent differences in the patterns of behavior exist. For example, strain 2 displayed on every test the shortest latency and longest duration of estrus, a lordosis intermediate in its maximum duration, and the lowest frequency of mounting. Except for the well known inverse relationship between latency and duration of heat (COLLINS, BOLING, DEMPSEY and YOUNG, 1938), the independence of the other measures in their responsiveness to the hormones is conspicuous.

TABLE I

Responses of spayed female guinea pigs injected with 100 I.U. α-estradiol benzoate and 0.2 I.U. of progesterone

Test 1

	N (animals)	Per cent in heat	Latency of heat in hours	Duration of heat in hours	Duration of maximum lordosis in seconds	Mean number of Mounts
2	15	93.3	3.1	7.0	16.5	0.4
T	21	95.2	5.6	4.4	14.5	7.8
13	17	88.2	5.7	4.7	26.5	32.6

Test 2

2	15	100	3.4	8.0	16.6	0.6
T	21	95.2	6.0	4.8	11.4	5.6
13	17	94.1	5.8	5.0	24.6	31.2

Test 3

2	15	100.	3.9	6.2	14.7	1.8
T	21	100	6.0	4.0	10.0	11.5
13	17	88.2	6.1	5.2	25.5	45.3

Means (estimated from individual averages)

2	15	97.8	3.4	7.2	15.9	1.0
T	21	96.8	5.9	4.3	11.7	7.3
13	17	90.2	5.7	5.0	25.2	37.7

When the H statistic (WALKER and LEV, 1953) was used, analysis of the combined results of the three tests indicated that most of the differences among the strains are significant. Large values of H corresponding to probabilities <.001 were obtained for every measure except the per cent of animals in heat.

The differences between strains were evaluated with the MANN-WHITNEY U test (AUBLE, 1953). All individual comparisons were significant at the

.001 level of confidence except for the per cent in heat and the difference between strain 13 and strain T females with respect to latency of heat.

A strain difference not revealed in Table I concerns the duration of lordosis on consecutive hours of heat. When the data were plotted for all the females regardless of the amount of injected estrogen curves were obtained which were markedly different (Fig. 1). The maximum lordosis

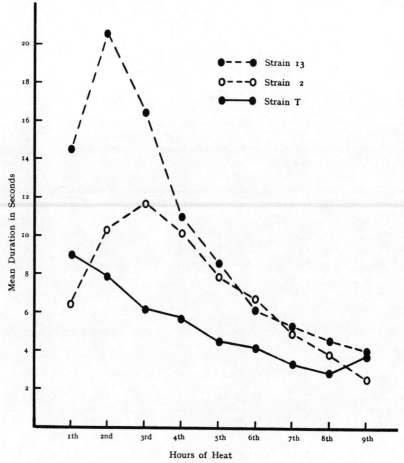

Fig. 1. Intensity of lordosis on successive hours of heat regardless of the amount of α-estradiol benzoate.

was elicited during the third hour of heat for most strain 2 females, but for strain T and strain 13 females the maxima were most often elicited during the first and second hours, respectively. The data confirm the suggestion based on qualitative observations (YOUNG, DEMPSEY and MYERS, 1935) that lordosis is stronger at the beginning than at the end of heat.

B. Estrous behavior following injections of each strain with different amounts of α-estradiol benzoate and 0.2 I.U. of progesterone.

The data are summarized in Table II. A conspicuous result is that the genetically determined patterns are discrete and remain constant despite increments in the hormone. When the single measures were followed the duration of estrus increased and latency of heat decreased, but the other measures remained approximately the same.

TABLE II

Responses of spayed female guinea pigs injected with different amounts of α-estradiol benzoate and 0.2 I.U. of progesterone

I.U. of α-estradiol benzoate	Strain	N (tests)	Per cent in heat	Latency of heat in hours	Duration of heat in hours	Duration of maximum lordosis in seconds	Mean number of Mounts
	2	14	41	6.0	6.0	10.3	0
25	T	13	15	6.5	5.0	10.5	15.5
	13	17	0	—	—	—	—
	2	14	100	4.5	6.2	13.9	2.5
50	T	13	85	6.4	3.3	10.4	19.4
	13	17	53	6.1	4.2	23.9	46.2
	2	42	98	3.4	7.2	15.6	0.9
100	T	39	97	6.0	4.3	11.0	8.7
	13	51	90	5.8	5.0	25.2	37.7
	2	14	100	3.4	7.5	14.2	1.0
200	T	13	100	5.5	4.9	10.7	13.7
	13	16	94	4.9	4.9	24.6	45.1
	2	14	100	3.0	8.4	14.2	0.3
400	T	13	92	4.7	6.0	10.9	14.4
	13	17	100	4.4	6.1	23.0	55.9
	2	14	100	2.6	9.6	13.9	0.5
800	T	13	92	5.1	6.3	9.6	21.1
	13	16	100	4.6	6.8	21.6	39.2

Analysis of the data is most accurate if the animals responding at every level of hormone are considered regardless of strain (Table III). When the data for duration of heat are analyzed by a non-parametric analysis of variance, a value of 564.6 is obtained for X^2r (P < .001). Latency of heat, varying inversely with amount of hormone, yields a X^2r equal to 343.7 (P < .001). Values for latency of 3.7 hours at 400 I.U. of α-estradiol benzoate and 3.8 hours at 800 I.U. suggest that an asymptote exists for this measure. Within the limits of the values of α-estradiol benzoate used, no such asymptote is suggested for the duration of heat. Neither the duration of the maximum lordosis in seconds nor the frequency of mounting varied in any systematic manner with the quantity of hormone administered. There was an apparent increase in frequency of mounting behavior when 800 I.U. were given, but it is not statistically significant. When WILCOXON's test (1949) is applied to the data, T equals -91, P >.10. For this analysis, 8 strain 13 and 11 strain T females which responded on the 50 I.U. and 800 I.U. tests were used.

TABLE III

Responses of 30 spayed female guinea pigs regardless of strain to different amounts of α-estradiol benzoate and 0.2 I.U. of progesterone

	Latency in hours	Duration in hours	Maximum Lordosis in seconds	Frequency of Mounting [1]
50 I.U.	5.6	4.8	14.2	25
100 I.U.	4.6	5.9	15.2	19
200 I.U	4.3	6.2	15.3	27
400 I.U.	3.7	7.3	14.7	25
800 I.U.	3.8	8.0	14.4	32

1) Data exclude values obtained from 14 strain 2 females inasmuch as these animals typically show little mounting behavior.

Of particular significance in connection with the fact that the duration of maximum lordosis did not vary with the amount of α-estradiol benzoate was the finding that the shape of the intensity curve also did not change. This result was obtained with all three strains, but the data are presented only for the strain 2 females (Fig. 2). It will be recalled that the duration of estrus increased with the larger amounts of hormone. The question arose as to what alterations in the intensity curve resulted from the prolongation of estrus. Inspection shows that estrus was prolonged by the addition of feeble lordoses at the tail end of the curves rather than at the beginning or during the peak. It is conspicuous that the maximum lordosis occurred during the third hour of heat regardless of the increase in duration of heat with the larger amounts of hormone.

Additional evidence of strain difference was revealed in this phase of the work which could not appear when a constant amount of hormone was used. The strains differ markedly with respect to the per cent in heat at the various levels of α-estradiol benzoate (Table II). For strain 2 females 100 per cent were in heat when only 50 I.U. of hormone were injected. The strain T females reached the same level when 100 I.U. were given, but strain 13 females did not show 100 per cent response until 400 I.U. had been administered.

The marked sensitivity of strain 2 can be demonstrated in another way. In the entire series of 115 tests given to strain 2 females, estrous reactions prior to the injection of progesterone were recorded 13% of the time. Estrogen-induced heats were observed in only 2% of tests on animals from the

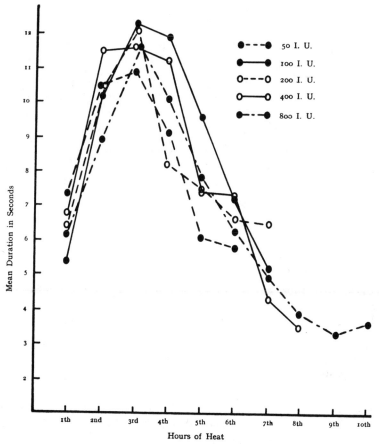

Fig. 2. Intensity of lordosis on successive hours of heat in strain 2 females given different amounts of α-estradiol benzoate and 0.2 I.U. of progesterone. Curves end at point where 50% of the females were out of heat.

strain T, and never in strain 13 females. No problem arises as the result of the incidence of estrogen-induced heats within strain 2 inasmuch as qualitative difference between such estrous reactions and those induced by estrogen-progesterone treatment are easily distinguished. Chiefly, the lordosis elicited during an estrogen-induced heat is nearly always feeble and lacks one or more of the behavioral components of growling, quivering and extension of the hind legs. Further, the estrogen-induced heat is intermittent and never shows the character'stic changes in intensity found in the estrogen-progesterone-induced heat and in that displayed by the intact female.

DISCUSSION

Completion of the present study enables us to bring together for the first time information bearing on the sexual behavior of males and females from the same genetic strains. For both sexes the genetically determined differences between the strains are conspicuous. In males the vigor of sexual behavior in the genetically heterogeneous stock (strain T) is greater than in the inbred strains (VALENSTEIN, RISS and YOUNG, 1954). Within the two inbred strains, total scored behavior is nearly the same, as is the average interval between the beginning of the test and ejaculation. There are slight differences in the other measures, but they are generally not significant. Beyond the impression of a greater sluggishness of behavior on the part of the strain 13 animals, a sharp distinction between the males of the two strains cannot be made.

During the investigation of the female, strain differences were found to be greater. A comparison is facilitated by an examination of the rank-order of the measures (Table I). The measures are not equivalent and any criterion of vigor is arbitrary, but if the length of time vigorous lordoses can be elicited (Fig. 1), the duration of the maximum lordosis, and the amount of male-like mounting are assumed to be the most valid expressions of vigor, it is clear that the sexual vigor of the inbred strains is greater than that of strain T. This order of sexual vigor is opposite that characteristic of the males from these strains.

The value of such a relationship for the perpetuation of the strains will be apparent. If a male is sluggish in the sense that more time is required for the organization of the adult pattern with experience, or in the sense that a longer time is required to achieve ejaculation, then females which display vigorous reactions for relatively long intervals will increase the probability of both the organization of sexual behavior and the achievement of ejaculation by the male.

A comparison of the responsiveness of males and females to supra-liminal amounts of hormone is of interest. When the male was being studied, above-threshold quantities of testosterone propionate did not alter the characteristic pattern or the vigor of the behavior (GRUNT and YOUNG, 1952, 1953; RISS and YOUNG, 1954). On the other hand, the administration of larger than threshold amounts of α-estradiol benzoate to the female lengthened the duration of heat, but did not affect either mounting behavior or the duration of the lordosis reflex. This increase in the duration of heat requires close examination because it represents a divergence from a principle of gonadal hormone action for the male, and because it is in apparent disagreement with results reported earlier for the female (COLLINS, BOLING, DEMPSEY and YOUNG, 1938). These investigators reported that although the duration of heat increased somewhat with supra-liminal doses of estrone, the increase in duration was not proportional to the increase in hormone. The consideration of the data which led to this conclusion may have been influenced by the opinion commonly held in the thirties that a direct relationship exists between hormonal dosage and response. At the same time, the conclusion suggested by the present study, that estrus is prolonged by supra-liminal dosages, must be evaluated carefully before it can be accepted as a true divergence from the principle of gonadal hormone action in the male.

The prolongation of estrus is best evaluated by examining the intensity curves obtained with different amounts of α-estradiol benzoate (Fig. 2). When this is done it is clear that estrus was extended by the addition of weak or feeble responses. One may question then whether the increase in duration of estrus represents a real increase in the period of receptivity to male copulatory attempts. It is known, from many mating tests in this laboratory, that females may respond to fingering with a feeble lordosis both before and after the period when they will accept a male. The increase obtained from supra-liminal amounts of α-estradiol benzoate does not appear from an examination of the intensity curves for the strain 2 females (Fig. 2) to result in any increase in the length of time vigorous lordoses (arbitrarily defined as 9 seconds or longer) can be elicited. This is also true of the lordoses displayed by the strain 13 and strain T females. For this measure, therefore, as well as for the duration of the maximum lordosis reflex and male-like mounting, the principle of gonadal hormone action appears to be the same as for male behavior.

At this point the distinction between "responsiveness" and "vigor" should be explained. As we are using the terms in our discussion of the female, responsiveness is a measure of the effectiveness of a hormone in inducing estrus, regardless of the character of the induced estrus. Vigor refers to the character of the pattern of behavior, particularly the length of time strong lordoses can be elicited, the duration of the maximum lordosis, and the

169

amount of male-like mounting. In the case of the male, responsiveness would be defined as the effectiveness of a hormone in maintaining or restoring the capacity for ejaculation. Vigor would refer to the character of the total pattern including such measures as length of the interval between the beginning of a test and ejaculation, the number of intromissions preceding ejaculation, and the length of the recovery period following ejaculation.

It is clear from the present study of the female that both responsiveness and vigor are genetically determined and that they are separable in the sense that a responsive animal may be high or low in vigor and vice versa. Of the animals we have studied, strain 13 females appear to be highest in vigor despite the fact that they are lowest in responsiveness, strain 2 females are highest in responsiveness but intermediate in vigor, and strain T females are relatively low in vigor but intermediate in responsiveness.

Tacit recognition of the distinction is contained in statements that gonadal hormones (YOUNG, DEMPSEY, MYERS and HAGQUIST, 1938; YOUNG, 1941; YOUNG and FISH, 1945; GRUNT and YOUNG, 1952, 1953) and certain adrenal cortical hormones (ROME and BRACELAND, 1950, 1951, 1952; TRETHOWAN and COBB, 1952) bring out just what is there. Until now however, conclusive evidence for the distinction between responsiveness and vigor has not been presented. When the male was being studied it was known that the administration of a given amount of testosterone propionate would be followed by the display of one pattern of behavior or another, depending on the behavior prior to castration (GRUNT and YOUNG, 1952, 1953), but the responsiveness of males from the three genetic strains to different levels of hormone was not determined.

Difficult of reconciliation is the lack of relationship between the amount of male-like mounting and the quantity of injected α-estradiol benzoate found in the present study and the direct relationship between mounting behavior and the number of rupturing follicles reported earlier (YOUNG, DEMPSEY, MYERS and HAGQUIST, 1938). The possibility of a chance relationship in the older study would seem to be minimized by the high critical ratio. The possibility that some hormone produced by the ovarian follicles such as estrone or progesterone rather than estradiol might be the effective substance is unlikely. When these substances were given in above-threshold quantities as a part of a different study (unpublished), there was no increase in mounting beyond that displayed when the smaller amounts were administered. Disturbing as the inconsistency is, its resolution at this time is not vitally important. The independance of the mounting behavior and lordosis, except that they are stimulated by the same hormones, is amply demonstrated in the older studies and in this investigation. Explanation of the paradox that in intact females mounting seems to be proportional to the number of rupturing follicles, whereas in spayed females the amount of this behavior is not related to the quantity of exogeneous hormone, will have to come from experiments which we cannot now design.

SUMMARY AND CONCLUSIONS

Using females from the highly inbred strains 2 and 13 and from a genetically hetero-geneous stock (strain T), five measures of sexual behavior were investigated tfor the consistency with which they are displayed by each strain when equal and different quantities of α-estradiol benzoate are followed by a constant amount of progesterone. The five measures were 1) latency of estrus, 2) duration of estrus, 3) duration of the maximum lordosis, 4) frequency of male-like mounting, and 5) per cent of animals brought into heat by the treatment.

From strain to strain there are consistent differences in the five measures. Strain 2 females show the shortest latencies, the longest duration of estrus, the least male-like mounting behavior, and the highest responsiveness to the estrogen. Their maximum lordosis which is intermediate in length is exhibited during the third hour of heat. Strain T females are characterized by long latencies, relatively short heat periods, and are intermediate with respect to mounting behavior and their responsiveness to α-estradiol benzoate. Their mean maximum lordosis is shortest and is exhibited during the first hour of heat. Strain 13 females display a latency and duration of heat similar to those which characterize the strain T females, but more male-like mounting than the other animals. They are least responsive to the estrogen. The mean maximum lor-dosis which is longest in this strain appears during the second hour of heat.

The sexual vigor of strains 2 and 13 is greater than that of the strain T females. This order is opposite that in males from the same strains. The relatively vigorous sexual behavior of the inbred females is postulated to be of value for the perpetuation of these strains.

Latency varied inversely and duration of heat directly with increases in the amount of α-estradiol benzoate, although with respect to the length of time that vigorous lordoses were elicited, above-threshold quantities were without effect on the response. Duration of lordosis and the frequency of mounting were not effected by changes in the amount of hormone. In general, therefore, the principle established for the male, that supra-liminal quantities of gonadal hormone do not alter the characteristic pattern of the behavior, can be extended to the female.

The distinction between responsiveness to the hormone and vigor of the reaction is discussed.

LITERATURE

AUBLE, D. (1953). Extènded tables for the MANN-WHITNEY statistic. — Bull. of the Institute of Educ. Research 1, No. 2, Indiana Univ.

AVERY, G. T. (1925). Notes on reproduction in guinea pigs. — J. comp. Psychol. 5, p. 373-396.

BOLING, J. L., YOUNG, W. C. and DEMPSEY, E. W. (1938). Miscellaneous experiments on the estrogen-progesterone induction of heat in the spayed guinea pig. — Endo-crinology 23, p. 182-187.

COLLINS, V. J., BOLING, J. L., DEMPSEY, E. W. and YOUNG, W. C. (1938). Quantitative studies of experimentally induced sexual receptivity in the spayed guinea-pig. — Endocrinology 23, p. 188-196.

GRUNT, J. A. and YOUNG, W. C. (1952). Differential reactivity of individuals and the response of the male guinea pig to testosterone propionate. — Endocrinology 51, p. 237-248.
—— and —— (1953). Consistency of sexual behavior patterns in individual male guinea pigs following castration and androgen therapy. — J. comp. physiol. Psychol. 46, p. 138-144.
RISS, W. and YOUNG, W. C. (1954). The failure of large quantities of testosterone propionate to activate low drive male guinea pigs. — Endocrinology 54, p. 232-235.
ROME, H. P. and BRACELAND, F. J. (1950). Use of cortisone and ACTH in certain diseases: Psychiatric aspects. — Proc. Mayo Clin. 25, p. 495-498.
—— and —— (1951). The effect of ACTH, cortisone, hydrocortisone and related steroids on mood. — J. clin. exper. Psychopath. 12, p. 184-191.
—— and —— (1952). Psychological response to corticotropin, cortisone and related steroid substances (psychotic reaction types). — J. Amer. Med. Ass. 148, p. 27-30.
TRETHOWAN, W. H. and COBB, S. (1952). Neuropsychiatric aspects of Cushing's syndrome. — Arch. neurol. Psychiat. 67, p. 283-309.
VALENSTEIN, E. S., RISS, W. and YOUNG, W. C. (1954). Sex drive in genetically heterogeneous and highly inbred strains of male guinea pigs. — J. comp. physiol. Psychol. 47, p. 162-165.
——, —— and —— (1955). Experiential and genetic factors in the organization of sexual behavior in male guinea pigs. — J. comp. physiol. Psychol. in press.
—— and YOUNG, W. C. (1955). An experiential factor influencing the effectiveness of testoterone propionate in eliciting sexual behavior in male guinea pigs. — Endocrinology 56, p. 173-177.
WALKER, H. M. and LEV, J. (1953). Statistical Inference. — New York.
WILCOXON, F. (1949). Some rapid approximate statistical procedures. — American Cyanamid Co.
YOUNG, W. C. (1941). Observations and experiments on mating behavior in female mammals. — Quart. Rev. Biol. 16, p. 135-156, 311-335.
——, DEMPSEY, E. W. and MYERS, H. I. (1935). Cyclic reproductive behavior in the female guinea pig. — J. comp. Psychol. 19, p. 313-335.
——, HAGQUIST, C. W., and BOLING, J. L. (1939). Sexual behavior and sexual receptivity in the female guinea pig. — J. comp. Psychol. 27, p. 49-68.
—— and FISH, W. R. (1945). The ovarian hormones and spontaneous running activity in the female rat. — Endocrinology 36, p. 181-189.
—— and GRUNT, J. A. (1951). The pattern and measurement of sexual behavior in the male guinea pig. — J. comp. physiol. Psychol. 44, p. 492-500.
—— and RUNDLETT, B. (1939). The hormonal induction of homosexual behavior in the spayed female guinea pig. — Psychosom. Med. 1, p. 449-460.

ZUSAMMENFASSUNG

Weibchen der extrem ingezüchteten Meerschweinchenstämme 2 und 13 sowie von einer genetisch unreinen Zucht T wurden mit gleich oder verschieden viel α-Oestradiol Benzoat und danach mit gleich viel Progesteron behandelt. Danach wurden die Weibchen vom Stamm 2 in grösstem Prozentsatz und am frühesten brünstig; ihre Brunst hielt am längsten an; sie stiegen am seltensten auf, wie Männchen es tun; ihr Rücken war in der dritten Stunde ihrer Brunst am hohlsten. Stamm T hat die längsten Latenzzeiten, verhältnismässig kurze Brunstperioden, mittlere Prozentsätze des Ansprechens und auch des Aufsteigens nach Männchenart. Die Lordose währt am kürzesten und hat ihren Höchstwert in der ersten Stunde. Die Weibchen von Stamm 13 werden ungefähr ebenso spät und lange heiss wie die von T, steigen aber am häufigsten auf, und die wenigsten Weibchen werden brünstig. Sie krümmen den Rücken am längsten, und zwar in der zweiten Stunde am stärksten. Die Weibchen beider Inzuchtstämme sind geschlechtshungriger als die von T, und umgekehrt die Männchen.

Die Dauer der Brunst war direkt und die Latenz umgekehrt proportional der Oestradiol-Dosis; starke Lordosen waren nicht länger auslösbar und hielten nicht länger an, wenn man unterschwellige Dosen steigerte; ebensowenig die Häufigkeit der Aufsprünge. So gilt der Satz, dass Überdosen von Gonadenhormon das männliche Geschlechtsverhalten nicht ändern, im allgemeinen auch für die Weibchen. Endlich wird der Unterschied zwischen Ansprechbarkeit auf das Hormon und Reaktionsstärke besprochen.

Copyright © 1968 by the Journal of Endocrinology and Dr. H. H. Feder

Reprinted with permission from *J. Endocrinol.*, **40**, 505–513 (1968)

18

PROGESTERONE CONCENTRATIONS IN THE ARTERIAL PLASMA OF GUINEA-PIGS DURING THE OESTROUS CYCLE

H. H. FEDER,* J. A. RESKO AND R. W. GOY

Department of Reproductive Physiology and Behaviour, Oregon Regional Primate Research Center, 505 N.W. 185th Avenue, Beaverton, Oregon 97005, U.S.A.

(*Received 7 August* 1967)

SUMMARY

The quantities of circulating progesterone and its 20-dihydro metabolites in cyclic female guinea-pigs were estimated by gas chromatography with electron capture detection. Forty-four animals were divided into 13 groups of approximately equal size according to stage of the oestrous cycle. The day on which a lordosis response could be elicited was designated Day 0 of the cycle. In the first few hours of sexual receptivity on Day 0 progesterone levels were 0.36 ± 0.07 μg./100 ml. arterial plasma. The ovaries were pre-ovulatory at this stage. After 5 to 12 hr. of behavioural receptivity progesterone levels dropped to 0.04 ± 0.01 μg./100 ml. plasma, and at the termination of receptivity (just after ovulation) progesterone was almost completely undetectable.

The concentration of progesterone increased gradually after ovulation reaching a peak about Day 9 (0.30 ± 0.05 μg./100 ml. plasma) and then fell to undetectable levels by Day 15. The 20-dihydro metabolites of progesterone were not present in significant quantities at any stage of the cycle.

It is concluded that in guinea-pigs a surge of progesterone occurs approximately 8 to 12 hr. before ovulation and coincides with the onset of oestrous behaviour. The plasma concentration of progesterone at this pre-ovulatory stage is at least as high as that measured at any point during the luteal phase.

INTRODUCTION

The experimental induction of oestrous behaviour in some spayed mammals requires, or is facilitated by, the administration of progesterone following a period of oestrogen priming (guinea-pig: Dempsey, Hertz & Young, 1936; rat: Boling & Blandau, 1939; mouse: Ring, 1944; hamster: Frank & Fraps, 1945; cow: Melampy, Emmerson, Rakes, Hanka & Eness, 1957). Direct evidence for the presence of elevated levels of progesterone in peripheral plasma during naturally occurring periods of sexual receptivity has been obtained only in the rat (Feder, Goy & Resko,

---

* Present address: Department of Human Anatomy, South Parks Road, University of Oxford, Oxford.

1967). The hypothesis that endogenous progesterone participates in the induction of spontaneous oestrous behaviour has been regarded as uncertain because the onset of cyclical sexual activity precedes ovulation and the pre-ovulatory ovary of several mammals has been shown to produce little progesterone (see Short, 1967). Moreover, progesterone may not be essential to the induction of oestrus (dog: Leathem, 1938; cat: Harris & Michael, 1964; rat: Lisk, 1965), and it may, in fact, have effects antagonistic to the display of sexual receptivity (Marshall & Hammond, 1945; Baker, Ulberg, Grummer & Casida, 1954; Goy, Phoenix & Young, 1966).

Supporting the hypothesis of a participation of progesterone are reports in which bioassays or chemical techniques demonstrated increased progestational activity during the pre-ovulatory stage (see Rothchild, 1965; Hashimoto & Melampy, 1967; Runnebaum & Zander, 1967). In only one of these studies on spontaneous ovulators (Feder et al. 1967) was it attempted to link progesterone production with the manifestations of sexual behaviour.

In addition to its possible facilitatory effects on sexual behaviour progesterone has been demonstrated to inhibit the display of lordosis in the female guinea-pig (Dempsey et al. 1936; Goy & Phoenix, 1965; Goy et al. 1966; Zucker, 1966). To account for these paradoxical findings, the suggestion was made that small amounts of progesterone might be facilitatory and that larger quantities (presumed to be characteristic of the luteal phase) were inhibitory (Goy et al. 1966). The present study was undertaken to measure the concentration of progesterone in the arterial plasma of guinea-pigs throughout the oestrous cycle in order to test this suggestion. Particular attention was given to the relationship between the pre-ovulatory progesterone level and oestrous behaviour.

MATERIAL AND METHODS

Animals. Forty-four adult (500–900 g.) virgin female guinea-pigs of an albino strain were used. They were kept in groups of six to eight per cage and were handled daily to determine the state of the vaginal membrane. Lights were on from 24.00–12.00 hr. only. The day on which an animal showed the lordosis reflex (arching of the back characteristic of receptive animals) after manual stimulation (Young, Dempsey, Hagquist & Boling, 1937) was designated Day 0 of the cycle. About 15 ml. blood was collected from the left ventricle under pentobarbitone (Nembutal) anaesthesia (approximately 35 mg./kg., intraperitoneally). Heparinized 20 ml. glass syringes were used for blood withdrawals. Animals killed 1, 2, 3, 5, 7, 9, 11, 13 and 15 days after displaying lordosis were bled between 15.00 and 17.00 hr. The progesterone levels during the period of behavioural oestrus were also studied. Blood collections were made at 11.00 hr. on the morning of expected oestrus (Day 16 of a 16-day cycle) at 0–1 hr. after the first lordosis could be elicited, at 3–5 hr., at 5–6 hr., and 7–12 hr., and within 1 hr. after receptivity had terminated. Because oestrous behaviour was not limited to the hours of darkness, receptive animals were killed at various times of day in an attempt to minimize the contribution that possible diurnal variations of progesterone might make to the results. Thus, animals at early stages of heat were killed at 09.30, 10.00, 15.30 and 19.30 hr. Animals at later stages of heat were killed at 07.00, 22.00 and 22.30. For the termination of heat group, samples were taken at 11.30, 15.15, 15.30 and 23.45. Each animal was bled only once, and no samples were pooled. After collection the blood was centrifuged and the plasma frozen at $-20°$ until ether extraction.

The ovaries of animals which were in heat or whose heat period had just ended were removed for histological examination and serially sectioned.

Five additional guinea-pigs were used for an ovariectomy experiment. One was an intact animal bled at Day 8. Four others were ovariectomized on Day 8 (under ether anaesthesia) and blood was withdrawn (under pentobarbitone anaesthesia) 1, 4, 8 and 24 hr., respectively, after gonadectomy.

An animal which showed persistent vaginal opening for over 90 days, almost daily lordosis responses (which were weak and atypical) and excessive male-like mounting behaviour was bled 1 hr. after giving a lordosis response. The ovaries were removed for microscopic examination.

Chemical procedures. For chemical determinations a procedure consisting essentially of ether extraction, thin-layer chromatography, enzymic reduction of progesterone to 20β-hydroxypregn-4-en-3-one, monochloroacetylation, thin-layer chromatography and gas chromatography was used. The details of this method have been described by Van der Molen & Groen (1965). The process for estimating the 20-dihydro metabolites of progesterone is identical with that for progesterone except that the reduction step is omitted. The conditions for gas chromatography used in the present study are given below.

A Barber–Colman gas chromatograph (series 5000) equipped with tritium electron capture detector was used. A pulsing voltage of 40 v with a pulse repetition frequency of 10 kcyc., a pulse duration of 1·3 μsec. and an output impedance of 800 ohm was applied to the detector from a General Radio Company pulse generator (type-1217-B). Washed, silanized U-shaped glass columns (2 ft. long, 5 mm. inner diameter) were packed with Gas Chrom P (80–100 mesh) previously coated with 1 % XE-60. The column temperature was maintained at 220°, the detector at 200° and the flash heater at 250°. Purified nitrogen, used as the carrier gas (25 psi regulator pressure), was filtered through a 13 × molecular sieve (to remove moisture) placed between the cylinder and the gas inlet to the column. Transfer of the sample from a non-silanized 2 ml. test-tube to the gas chromatograph was made by dissolving the sample in 15 μl. toluene and injecting 10 μl. with a Hamilton syringe.

Solvents were analytical grade (Mallinckrodt) and were redistilled before use. Nanograde benzene (Mallinckrodt) was tested on the gas chromatograph for purity and then used without further distillation. Silica gel-G was purified according to Bush (1961).

[7α-^{3}H]Progesterone was obtained from the New England Nuclear Corporation (sp.act. 0·0318 mc/μg.). Isotopic purity was assured by paper chromatography in hexane:benzene (1:1):formamide. Approximately 5000 counts/min. of this material dissolved in methanol was added to each sample (including a water blank) before extraction and used as a check on recoveries up to gas chromatography. Mean recovery of radioactivity was 48·4 % ± 1·2 (s.e.) by the method. The mass of the added isotope was small (about 0·0002 μg.) and not detectable in the gas chromatograph. Plasma samples were assayed singly, not in duplicate.

For the 20-dihydro metabolites, [7α-^{3}H]progesterone was incubated with the supernatant from homogenates of rat pro-oestrous ovaries centrifuged at 23,000 × *g* to prepare [7α-³H]20α-hydroxypregn-4-en-3-one (Wiest, 1959). After incubation, the mixture was extracted with ethylacetate and the extract chromatographed on a thin-layer (cyclohexane:ethylacetate, 1:1). [7α-³H]20β-hydroxypregn-4-en-3-one was prepared by incubating [7α-³H]progesterone with 20β-hydroxysteroid dehydrogenase.

The accuracy and specificity of the technique were assessed. To check accuracy, known amounts of authentic progesterone were added to water and processed. Three samples each of 0·025 μg. and 0·050 μg. and two samples of 0·100 μg. were used. The mean values obtained were 0·023, 0·049 and 0·094 μg., respectively. To control specificity it was thought desirable to demonstrate that chromatographic peaks obtained from plasma with the mobility of progesterone decreased after ovariectomy.

RESULTS

The concentration of circulating progesterone varied during the oestrous cycle (analysis of variance: $P < 0·01$). Between Day 15 (when progesterone was not detected) and the first few hours of sexual receptivity there was a dramatic increase in progesterone. To the four animals assayed during early heat (Table 1) a fifth may be added. This additional animal was killed less than 45 min. after the start of heat and had a progesterone level of 0·65 μg./100 ml. plasma. The result was not included in the calculations because of a low recovery of radioactivity (9·5%). When the oestrous animals listed in Table 1 are further subdivided the following picture emerges: at 0–1 hr. after onset of heat the mean progesterone concentration was 0·39 μg./100 ml. plasma ($n = 3$), at 3–5 hr. it was 0·26 μg./100 ml. ($n = 1$), at 5–6 hr. it was 0·05 μg./100 ml. ($n = 1$), at 7–12 hr. it was 0·02 μg./100 ml. ($n = 2$), and at the end of oestrus no progesterone was detected ($n = 4$). Histological examination of the ovaries showed that pre-ovulatory follicles were present in all of the receptive animals, except one which had been in heat for 7–12 hr. In this animal, and in the four animals killed just after the receptive period, ovulation (as indicated by rupture points) had occurred (Plate, figs. 1 and 2).

The mean plasma progesterone levels in the groups of animals listed in Table 1 were compared by Duncan's new multiple range test (Edwards, 1963) to ascertain which groups differed significantly. The comparisons made were: (a) early stages of heat v. Day 15, late stages of heat, end of heat (significant at $α = 0·01$); (b) Days 1, 2, 3, 13, 15 v. Day 9 (significant at $α = 0·05$); (c) Day 9 v. early stages of heat (not significantly different).

Immediately after ovulation the plasma values for progesterone were very low. The concentration increased from Day 2 to 9 and started to decline by about Day 11. No plateau was reached during the luteal phase, although there was much over-lapping of values from Day 5 to 11 of the cycle. After Day 11 the progesterone levels declined fairly rapidly. At Day 16, some undetermined number of hours before the onset of oestrus there was an indication of a slight upswing in progesterone concentration (see Table 1).

Two samples at each stage of the cycle (including the onset of heat) were analysed for 20α- and 20β-hydroxypregn-4-en-3-one. The 20α compound appeared in one animal at Day 11 (0·05 μg./100 ml. plasma), but was otherwise undetectable. Endogenous 20β-hydroxypregn-4-en-3-one was never detected.

177

Plasma progesterone fell sharply immediately after ovariectomy at Day 8 of the cycle. The values for individual animals were as follows (μg./100 ml. plasma): 0·26 (intact animal); 0·04 (1 hr. after ovariectomy); 0·03 (4 hr. after ovariectomy); trace (8 hr. after ovariectomy), and undetectable (24 hr. after ovariectomy).

These results suggest that the assay method employed is specific for progesterone and indicate that, under the conditions of the present study, in cyclic guinea-pigs most if not all of the circulating progesterone is of ovarian origin.

Table 1. *Progesterone levels in arterial plasma of cyclic guinea-pigs*

Stage of cycle	No. of animals	μg. progesterone/ 100 ml. plasma (means $\pm$ s.e.)	ml. plasma (range)
Morning of expected oestrus	2	0·02 $\pm$ 0·01	9·8 — 11·6
Early heat (0–5 hr. after onset of oestrus)	4	0·36 $\pm$ 0·07	7·2 — 11·0
Late heat (6–12 hr. after onset of oestrus)	3	0·04 $\pm$ 0·01	6·8 — 11·8
End of heat	4	0·01 $\pm$ 0·01	10·0 — 10·6
Day 1	3	0	7·5 — 10·4
Day 2	3	0·03 $\pm$ 0·01	9·9 — 12·7
Day 3	3	0·13 $\pm$ 0·03	7·4 — 11·4
Day 5	3	0·22 $\pm$ 0·03	8·7 — 11·5
Day 7	4	0·23 $\pm$ 0·03	7·2 — 11·2
Day 9	5	0·30 $\pm$ 0·05	7·8 — 11·5
Day 11	4	0·26 $\pm$ 0·06	6·1 — 11·6
Day 13	3	0·03 $\pm$ 0·03	9·5 — 12·9
Day 15	3	0	10·8 — 13·5

Values of 0 were assigned when levels were undetectable, and calculations were made on this basis. Two samples in which recovery was less than 30 % were excluded from the calculations. The first was an animal in early heat mentioned in Results (p. 508). The second was a Day-7 animal (0·21 μg./100 ml., 20·0 % recovery).

Plotting the time of rupture of the vaginal membrane was found to be useful for predicting (within a few days) when lordosis would be exhibited in the strain of guinea-pigs used. Of the 105 periods of behavioural oestrus periods observed, 59 occurred on the first day of vaginal membrane rupture, 27 within a few days after vaginal opening, and 19 within a few days before vaginal opening. Only one female in the colony displayed constant vaginal opening and atypical lordosis.

At laparotomy the animal with the constantly patent vagina was seen to have very large uterine horns. The ovaries contained large cysts but no vesicular follicles or corpora lutea (Plate, fig. 3). Neither progesterone nor its 20-dihydro metabolites could be detected in this animal.

DISCUSSION

The present study demonstrates that progesterone reaches maximal concentrations in the arterial plasma of cyclic guinea-pigs just before ovulation and again during dioestrus. The ripe follicle may be the source of pre-ovulatory progesterone in this species. Unlike rabbits, in which interstitial tissue releases pre-ovulatory progestins (Hilliard, Archibald & Sawyer, 1963), guinea-pig ovaries contain no large masses of interstitial cells (Loeb, 1923). Degenerating corpora from the previous cycle are not a

33

Endoc. 40, 4

likely source, because the pre-ovulatory levels are at least as high as those measured when even fresh corpora are maximally active. The ovarian structure responsible for releasing progesterone during dioestrus is the corpus luteum. Rowlands & Short (1959) found a fairly high concentration of progesterone in the corpus luteum of the normal cycle and very little in the ovarian residue. The plasma concentrations of progesterone recorded here correlate well with the growth and regression of the corpora (Rowlands, 1956).

A significant adrenal contribution of progesterone in guinea-pigs cannot be ruled out. With a fluorescence technique levels of 3·2 and 0·77 μg./100 ml. peripheral plasma were found in non-pregnant ovariectomized guinea-pigs (Heap & Deanesly, 1966; Heap, Perry & Rowlands, 1967). Failure to detect progesterone 24 hr. after ovariectomy in this study may reflect a higher degree of specificity of the gas chromatographic technique used, or may be attributable to other differences in experimental procedures such as the anaesthetic used during blood collection.

The functional significance of pre-ovulatory progesterone is unknown, but a relation exists between the intensity of the lordosis response as reported elsewhere (Goy & Young, 1957) and the concentrations of progesterone measured here. Plasma progesterone reaches a peak in the first hours of heat ($0\cdot36 \pm 0\cdot07$ μg./100 ml., $n = 4$) when lordosis responses are strongest, declines as heat progresses ($0\cdot04 \pm 0\cdot01$ μg./100 ml., $n = 3$) and lordosis responses become weaker, and becomes almost undetectable ($0\cdot01 \pm 0\cdot01$ μg./100 ml., $n = 4$) when receptivity terminates. Admittedly, the coincidence of a progesterone surge with the onset of female sexual behaviour in intact animals does not necessarily imply a causal relationship. However, there is evidence that: (a) oestrogen followed at a suitable interval by progesterone induces oestrous behaviour in ovariectomized guinea-pigs more reliably than oestrogen treatment alone (Dempsey et al. 1936); (b) this synergistic action of progesterone is dose-dependent (Collins, Boling, Dempsey & Young, 1938); as well as (c) reasonably specific (Hertz, Meyer & Spielman, 1937); and (d) an animal with cystic ovaries and morphological evidence of oestrogenic stimulation did not show normal female behaviour. This suggests that a factor, other than oestrogen, involved in normal oestrus was missing (see this study). Taken together, these data indicate that progesterone may be functionally significant in inducing full oestrus (which includes normal behavioural receptivity) in guinea-pigs. A contrary view derived from a study which did not use behavioural criteria has been expressed by Deanesly (1966).

The possibility that low plasma concentrations of progesterone may facilitate receptivity whereas high levels inhibit it was examined. The luteal phase is inhibitory in the sense that oestrogen–progesterone treatment rarely induces lordosis at Day 4 to 10 (Goy et al. 1966). Yet the present results demonstrate that the highest mean level of progesterone attained during the luteal phase (at Day 9) was $0\cdot30 \pm 0\cdot05$ μg./ 100 ml. ($n = 5$) compared with $0\cdot36 \pm 0\cdot07$ μg./100 ml. ($n = 4$) at the pre-ovulatory period when receptivity begins (the difference is not statistically significant). Further, considerable refractoriness to hormonal stimulation of the lordosis response occurs soon after spontaneous oestrus (i.e. Day 1 to 2, Goy et al. 1966) even though the plasma progesterone levels found in the six animals assayed at that time were low (Text-fig. 1). Although a relation exists between the percentage of females behaviourally unresponsive to oestrogen–progesterone treatment and plasma progesterone concentration from about Day 3 to 15 (Text-fig. 1) it is clear that purely quantitative considerations fail to account for the opposing facilitatory and inhibitory actions of progesterone. The condition of the neural substrate mediating female sexual responses must therefore be considered. Progesterone may be inhibitory when acting on an incompletely oestrogen-primed nervous system, but facilitatory once the priming is adequate (Goy & Phoenix, 1965; Zucker, 1966). A consequence of the facilitatory action of progesterone may be an alteration in the state of the nervous system causing it to become refractory to certain stimuli (Kawakami & Sawyer, 1959; Zucker, 1966).

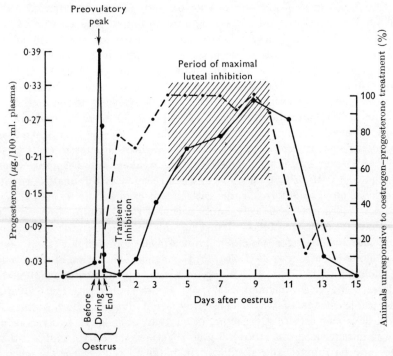

Text-fig. 1. Summary of relationships between progesterone concentration in plasma (———) and the behavioural status of guinea-pigs. For purposes of comparison the data of Goy *et al.* (1966) are expressed in terms of % animals behaviourally unresponsive to oestrogen treatment (- - - -).

A postdoctoral research fellowship from the U.S.P.H.S., (1 F 2 HD 29-006-01) supported one of the authors (H.H.F.). Grants MH 08634 (to R. W. G.) and FR 00163 also supported this work. Mr D. Harvey provided valuable assistance during the final phases of the study. We are particularly grateful to Dr I. Zucker for discussions of the results and to Dr D. Exley and Professor G. W. Harris for helpful comments and encouragement. Publication no. 272 of the Oregon Regional Primate Research Center.

33-2

REFERENCES

Baker, L. N., Ulberg, L. C., Grummer, R. H. & Casida, L. E. (1954). Inhibition of heat by progesterone and its effect on subsequent fertility in gilts. *J. Anim. Sci.* **13**, 648–657.

Boling, J. L. & Blandau, R. J. (1939). The estrogen–progesterone induction of mating responses in the spayed female rat. *Endocrinology* **25**, 359–364.

Bush, I. E. (1961). *The chromatography of steroids*, p. 352 (b). Oxford: Pergamon Press.

Collins, V. J., Boling, J. L., Dempsey, E. W. & Young, W. C. (1938). Quantitative studies of experimentally induced sexual receptivity in the spayed guinea-pig. *Endocrinology* **23**, 188–196.

Deanesly, R. (1966). Pro-oestrus in the guinea-pig: hormonal stimulation of the vaginal epithelium. *J. Reprod. Fert.* **12**, 205–212.

Dempsey, E. W., Hertz, R. & Young, W. C. (1936). The experimental induction of oestrus (sexual receptivity) in the normal and ovariectomized guinea pig. *Am. J. Physiol.* **116**, 201–208.

Edwards, A. L. (1963). *Experimental design in psychological research*. New York: Holt, Rinehart and Winston.

Feder, H. H., Goy, R. W. & Resko, J. A. (1967). Progesterone concentrations in the peripheral plasma of cyclic rats. *J. Physiol., Lond.* **191**, 136–137P.

Frank, A. H. & Fraps, R. M. (1945). Induction of estrus in the ovariectomized golden hamster. *Endocrinology* **37**, 357–361.

Goy, R. W. & Phoenix, C. H. (1965). Inhibitory actions of progesterone. *Am. Zool.* **5**, 725 (Abstr.).

Goy, R. W., Phoenix, C. H. & Young, W. C. (1966). Inhibitory action of the corpus luteum on the hormonal induction of estrous behavior in the guinea pig. *Gen. comp. Endocr.* **6**, 267–275.

Goy, R. W. & Young, W. C. (1957). Strain differences in the behavioural responses of female guinea pigs to α-oestradiol benzoate and progesterone. *Behaviour* **10**, 340–354.

Harris, G. W. & Michael, R. P. (1964). The activation of sexual behaviour by hypothalamic implants of oestrogen. *J. Physiol., Lond.* **171**, 275–301.

Hashimoto, I. & Melampy, R. M. (1967). Ovarian progestin secretion in various reproductive states and experimental conditions in the rat. *Fedn Proc. Fedn Am. Socs exp. Biol.* **26**, 485 (Abstr.).

Heap, R. B. & Deanesly, R. (1966). Progesterone in systemic blood and placentae of intact and ovariectomized pregnant guinea-pigs. *J. Endocr.* **34**, 417–423.

Heap, R. B., Perry, J. S. & Rowlands, I. W. (1967). Corpus luteum function in the guinea-pig; arterial and luteal progesterone levels, and the effects of hysterectomy and hypophysectomy. *J. Reprod. Fert.* **13**, 537–553.

Hertz, R., Meyer, R. K. & Spielman, M. A. (1937). The specificity of progesterone in inducing sexual receptivity in the ovariectomized guinea pig. *Endocrinology* **21**, 533–535.

Hilliard, J., Archibald, D. & Sawyer, C. H. (1963). Gonadotropic activation of preovulatory synthesis and release of progestin in the rabbit. *Endocrinology* **72**, 59–66.

Kawakami, M. & Sawyer, C. H. (1959). Neuroendocrine correlates of changes in brain activity threshold by sex steroids and pituitary hormones. *Endocrinology* **65**, 652–668.

Leathem, J. H. (1938). Experimental induction of estrus in the dog. *Endocrinology* **22**, 559–567.

Lisk, R. D. (1965). Reproductive capacity and behavioural oestrus in the rat bearing hypothalamic implants of sex steroids. *Acta endocr., Copenh.* **48**, 209–219.

Loeb, L. (1923). Types of mammalian ovary. *Proc. Soc. exp. Biol. Med.* **20**, 446–448.

Marshall, F. H. A. & Hammond, J., Jr. (1945). Experimental control by hormone action of the oestrous cycle in the ferret. *J. Endocr.* **4**, 159–168.

Melampy, R. M., Emmerson, M. A., Rakes, J. M., Hanka, L. J. & Eness, P. G. (1957). The effect of progesterone on the estrous response of estrogen-conditioned ovariectomized cows. *J. Anim. Sci.* **16**, 967–975.

Ring, J. R. (1944). The estrogen–progesterone induction of sexual receptivity in the spayed female mouse. *Endocrinology* **34**, 269–275.

Rothchild, I. (1965). Interrelations between progesterone and the ovary, pituitary, and central nervous system in the control of ovulation and the regulation of progesterone secretion. *Vitams Horm.* **23**, 209–327.

Rowlands, I. W. (1956). The corpus luteum of the guinea pig. *Ciba Fdn Colloq. Ageing* **2**, 69–83.

Rowlands, I. W. & Short, R. V. (1959). The progesterone content of the guinea-pig corpus luteum during the reproductive cycle and after hysterectomy. *J. Endocr.* **19**, 81–86.

Runnebaum, B. & Zander, J. (1967). Progesterone in the human peripheral blood in the preovulatory period of the menstrual cycle. *Acta endocr., Copenh.* **55**, 91–96.

Short, R. V. (1967). Reproduction. *A. Rev. Physiol.* **29**, 373–400.

Van der Molen, H. J. & Groen, D. (1965). Determination of progesterone in human peripheral blood using gas–liquid chromatography with electron capture detection. *J. clin. Endocr. Metab.* **25**, 1625–1639.

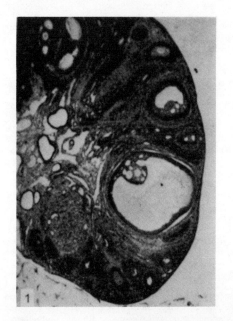

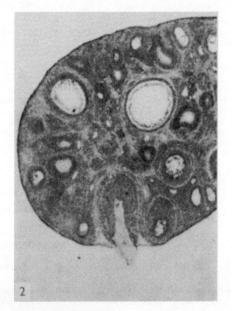

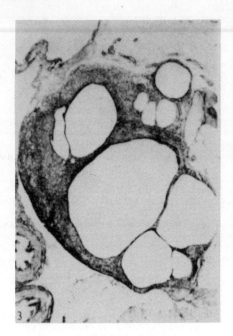

H. H. FEDER, J. A. RESKO AND R. W. GOY   *(Facing p. 513)*

Wiest, W. G. (1959). Conversion of progesterone to 4-pregnen-20α-ol-3-one by rat ovarian tissue *in vitro*. *J. biol. Chem.* **234**, 3115–3121.

Young, W. C., Dempsey, E. W., Hagquist, C. W. & Boling, J. L. (1937). The determination of heat in the guinea pig. *J. Lab. clin. Med.* **23**, 300–302.

Zucker, I. (1966). Facilitatory and inhibitory effects of progesterone on sexual responses of spayed guinea pigs. *J. comp. physiol. Psychol.* **62**, 376–381.

DESCRIPTION OF PLATE

Staining: haematoxylin and eosin. Magnification: × 22.

Fig. 1. Ovary from a guinea-pig killed at the onset of behavioural oestrus. A pre-ovulatory follicle is prominent.

Fig. 2. Ovary from a guinea-pig which had been sexually receptive for 7–12 hr. at the time it was killed. Note follicular rupture point.

Fig. 3. Large cysts and absence of follicular or luteal development in ovary of a guinea-pig with a constantly patent vagina, oestrogen-stimulated uterus, and intermittent display of atypical lordosis. Both ovaries were cystic.

Editor's Comments on Papers 19 Through 24

In primates, a role for estrogens in the stimulation of female behavior was inferred from studies of intact versus ovariectomized females, and in particular from the report by Josephine Ball (Paper 19) of increases in female sexual behavior following estrogen injections. Ball later (Paper 20) described the inhibitory action of progesterone on female primate behavior.

Rather recently, correlational studies with rhesus monkeys (Michael, Saayman, and Zumpe, Paper 21) and humans (Udry and Morris, Paper 22) have indicated correspondence between the phase of the female's cycle and the frequency of copulation. However, in primates, particularly in humans, it has been clearly shown that at least some sexual behavior may continue following ovariectomy. Based upon data such as those of Salmon and Geist (Paper 23) and Waxenberg, Drellich, and Sutherland (Paper 24), an important role has been suggested for adrenal secretions in the maintenance of female sexual behavior. Questions pertaining to the exact chemical compounds capable of stimulating female behavior are of interest, and it is currently not clear whether adrenal androgens, estrogens, or other substances might be capable of maintaining sexual behavior. Similar questions are also being asked regarding the post castrational maintenance of male sexual behavior. In both cases it has also been proposed that these behaviors in primates may be under less strict hormonal control than that which has been demonstrated in other mammals.

Reprinted from *Psychol. Bull.*, **33**, 811 (1936)

19

Sexual Responsiveness in Female Monkeys After Castration and Subsequent Estrin Administration. JOSEPHINE BALL, Johns Hopkins Medical School.

Nine female rhesus monkeys were used for this study. The measurement of sexual receptivity was based on the animals' response to males in several 10-min. mating tests given three times a week over extended periods of time. Most of the animals were observed continuously, except for the hot summer months, for two or three years. Four of them were tested throughout several normal menstrual cycles before castration.

After castration sex interest drops, over varying periods of time, to practically zero.

Injection of estrogenic hormones (Progynon B, Schering and Amniotin, Squibb) raises it to normal.

There seems to be a slight tendency to periodicity after removal of the ovaries. Before this was taken into consideration there was little relation between hormone dosage and amount of reaction. By administering the hormone at monthly intervals this relationship has been improved. The discrepancies encountered on this regime are consistent with the observation of Engle and Hartman that the summer (non-breeding) season reduces response to sex hormones. [15 min., slides.]

Reprinted from *Psychol. Bull.*, **38**, 533 (1941)

20

10:00 A.M.　*Effect of Progesterone Upon Sexual Excitability in the Female Monkey.* JOSEPHINE BALL, Johns Hopkins Medical School.

Because progesterone, an ovarian hormone secreted chiefly after ovulation, has been shown to enhance the effectiveness of estrogens in producing sexual responsiveness in rodents it is not likely to be considered a sex depressant. However, since it prevents ovulation and augments the excretion of estrogens, chronic administration should reduce, rather than increase, their effect on behavior in the castrate, and furthermore, in the intact animal, it should prevent estrous behavior including the homologous preovulatory rise of sex interest which occurs in those infrahuman primates in which this trait has been investigated. These considerations have led to a study of the effect of progesterone upon the sex behavior of the monkey.

Three normal females were injected with 1 to 2 mg. of progesterone daily for the first two weeks following a menstrual cycle in which normal sex behavior had been observed. Ovulation and the accompanying rise of sexual excitability did not occur during the injection period in any case. In two of the animals uterine bleeding occurred three days after the last injection, and this was followed by another normal cycle. In the third female sexual responsiveness rose immediately after the last injection, and ovulation occurred 10 days later.

Sexual excitability produced by estrogens in two castrates was lowered by chronic administration of progesterone.

In seven experiments a single injection of progesterone in various dosages failed to raise the responsiveness of castrates in which sex behavior was not much affected by chronic administration of estrogens.

These experiments show that it is much easier to reduce than to increase sexual excitability in the monkey with progesterone and suggest that this hormone might be tried clinically in cases where medication for temporary reduction of sex tension seems desirable. [15 min., slides.]

Reprinted from *Nature*, **215**, 554–556 (1967)

21

ANIMAL BEHAVIOUR

Sexual Attractiveness and Receptivity in Rhesus Monkeys

THE catarrhine monkeys and the apes alone among animals have a clearly defined menstruation with a cycle lasting about 30 days. They do not show well-circumscribed periods of oestrus (heat), but will copulate throughout the cycle. This has led to the idea that the great development of neocortical mechanisms in the primate has resulted in an emancipation of its brain from the influence of gonadal steroids[1]. Earlier work in both the rhesus monkey[2,3] and the chimpanzee[4-6], however, has indicated that the level of sexual interaction shows some variation with the phases of the menstrual cycle. Furthermore, studies of the rhesus monkey[7,8] have shown that well marked rhythms of mounting behaviour by males occur in relation to the menstrual cycles of their female partners. These rhythms are abolished by bilateral ovariectomy, and so endocrine mechanisms in primate sexual behaviour seem to have been underestimated[9].

Copulation in the rhesus monkey consists of a series of sexual mounts by the male on the female, each with an intromission and thrusting, the series being terminated by a final, ejaculatory mount. The mounting sequence is not haphazard, but results from initiating movements by the male (for example, courtship postures, mounting attempts or clasping), and from initiating movements by the female (sexual presentations and various invitational gestures). Endocrine-dependent changes in the mounting activity of the males could therefore be mediated by changes in this initiating behaviour. This communication attempts to account, in behavioural terms, for the conspicuous decline in male mounting activity that occurs early in the luteal phase of the menstrual cycle.

Mature, intact male and female rhesus monkeys, weighing 8–14 kg, were studied in glass-fronted observation cages for 60 min periods by two observers, in carefully standardized conditions. Between tests animals were housed singly. Four females were used, each was tested with two males for a total of fifteen menstrual cycles involving 176 h of observation, and in eight cycles rhythmic changes in mounting occurred. A system, previously described, was used for scoring components of behaviour at 30 sec intervals giving their temporal sequence[10]. The present report is confined to the way in which the hormonal status of the female influences the outcome of mounting attempts by the male and of sexual invitations by the female.

Fig. 1 shows the changes in each test during four menstrual cycles: (*a*) in the number of mounts made by the male on the female; (*b*) in the number of sexual invitations made by the female to the male, and (*c*) in the proportion of these invitations that stimulate the male to mount. The latter is termed the female success ratio—successful mounting invitations (those followed by a mount) expressed as a percentage of total mounting invitations made by the female. Although rhythmic changes in the number of mounts in a test occurred in relation to the menstrual cycle, with a well marked decline early in the luteal phase, the number of female sexual

invitations remained at about the same level throughout the cycle, and did not decline when male mounting activity did so. The proportion of these invitations that were accepted by the male and resulted in a mount (the female success ratio), however, diminished markedly during the luteal phase when male mounting activity declined. In such cases, type *A*, the decrease in mounting activity was clearly related to the diminished effectiveness of the female sexual invitations, reflecting her diminished value as a sexual stimulus at this time.

This is not, however, the only mechanism responsible for the decline in male sexual activity observed early in the luteal phase. Fig. 2 shows the changes in each test during three further menstrual cycles: (*a*) in the number of mounts made by the male on the female; (*b*) in the number of mounting attempts made by the male, and (*c*) in the proportion of these attempts accepted by the female and followed by a mount. The latter is termed the male success ratio—successful mounting attempts (those followed by a mount) expressed as a percentage of total mounting attempts made by the male. Although the number of mounts declined in the luteal phase of these cycles also, this was not associated with any corresponding decline in the number of male mounting attempts. The

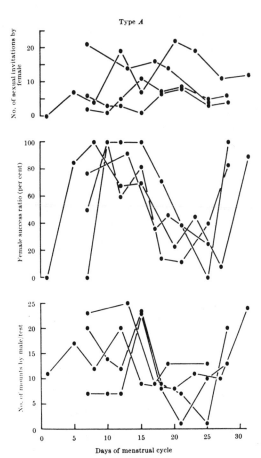

Type *A*

Fig. 1. Although sexual invitations in these females do not decline in the luteal phase of the menstrual cycle, the proportion accepted by the male (female success ratio) does so. This decline in female "attractiveness" correlates with the decline in male mounting (four of five cycles are illustrated).

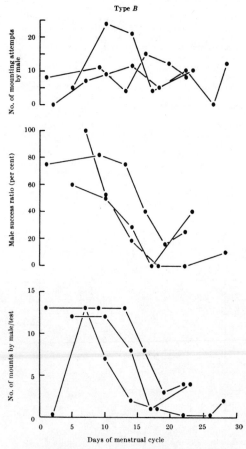

Type *B*

Fig. 2. Although mounting attempts by the male with these females do not decline in the luteal phase, the proportion resulting in a mount (male success ratio) does so, and the number of mounts declines. This is because of increased refusals by the female (three cycles).

one; this may account for its inhibitory effect on ejaculation reported elsewhere[11].

We thank Dr J. Herbert for his assistance and Mr M. Sullivan for his care of the animals. This investigation was supported by grants to R. P. M. from the Medical Research Council, the Bethlem-Maudsley Hospital Research Fund and the National Institute of Mental Health, US Public Health Service.

RICHARD P. MICHAEL
G. S. SAAYMAN
D. ZUMPE

Primate Research Centre,
Bethlem Royal Hospital,
Beckenham, Kent.

Received May 1; revised June 16, 1967.

[1] Ford, C. S., and Beach, F. A., *Patterns of Sexual Behaviour* (Eyre and Spottiswoode, London, 1952).
[2] Ball, J., and Hartman, C. G., *Amer. J. Obstet. Gynec.*, **29**, 117 (1935).
[3] Carpenter, C. R., *J. Comp. Psychol.*, **33**, 113 (1942).
[4] Yerkes, R. M., and Elder, J. H., *Comp. Psychol. Monogr.*, **13**, 1 (1936).
[5] Yerkes, R. M., *Human Biol.*, **11**, 78 (1939).
[6] Young, W. C., and Orbison, W. D., *J. Comp. Psychol.*, **37**, 107 (1943).
[7] Michael, R. P., *Proc. Roy. Soc. Med.*, **58**, 595 (1965).
[8] Michael, R. P., and Herbert, J., *Science*, **140**, 500 (1963).
[9] Michael, R. P., in *Pathology and Treatment of Sexual Deviation* (edit. by Rosen, I.), 24 (Oxford University Press, London, 1964).
[10] Michael, R. P., Herbert, J., and Welegalla, J., *J. Endocrinol.*, **36**, 263 (1966).
[11] Michael, R. P., Herbert, J., and Saayman, G., *Lancet*, i, 1015 (1966).

proportion accepted by the female, the male success ratio, however, conspicuously declined when the number of mounts declined, and this was caused by a corresponding increase in the number of active refusals of male mounting attempts made by these females. In these cases, type *B*, the female had not lost her attractiveness, for the male continued to attempt to mount, and a female refusal mechanism was therefore brought into play during the luteal phase.

By analysing primate sexual behaviour in this way, it was possible to differentiate between the sexual "drive" of the female, on the one hand, expressed by the number of sexual invitations, and her value as a stimulus for the male, on the other, expressed by the female success ratio. Two distinct mechanisms thus seemed to underlie the decline in male mounting activity early in the luteal phase: the first depended on a decrease in the attractiveness of the female, indicated by the failure of her sexual invitations to stimulate mounting; the second depended on a decrease in the receptivity of the female, indicated by an increase in the number of female refusals. Further studies of ovariectomized females given oestrogen and progesterone have shown that these mechanisms can be brought into operation in different females by progester-

188

Reprinted from *Nature*, **220**, 593–596 (1968)

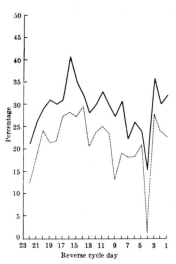

22

Distribution of Coitus in the Menstrual Cycle

J. Richard Udry

Naomi M. Morris

Fig. 1. Percentage of women in the North Carolina sample reporting intercourse and orgasm by reverse day of menstrual cycle. There were forty women, and 73–115 cycles. ———, Intercourse; ...., orgasm.

PERIODICITY is characteristic of the sexual behaviour of lower mammals. In animals with oestrous patterns, the female will accept the male only when she is "in heat", a condition relating temporally to ovulation and the presence of certain gonadal hormones. Primates are different; apes and most monkeys, like humans, have menstrual cycles, and mate at all times during the cycle. Laboratory studies of several species of primates, however, show a periodicity of sexual behaviour within the menstrual rhythm with increased sexual behaviour occurring at mid-cycle when ovulation is believed to occur[1].

Many physiological characteristics are known to vary in women with the phase of the menstrual cycle[2], and it is accepted that gonadal hormones ". . . have a determinative influence on the human psyche . . .", but it is generally believed that this will not show up in sexual behaviour, ". . . because there are other, stronger determinants and because this determinant is weak"[3].

Although it has been demonstrated that the patterns of sexual activity in other primates are under the control of hormones, this is insufficient evidence to impute the same control to humans. Nevertheless, it must give increased credibility to the limited data showing menstrual periodicity in human sexuality. A review of the literature shows that data bearing on the problem come from three sources. The first source is retrospective reports: for example, data from Davis[4] and Terman[5]. Each reported that respondents indicated that they felt increased sexual desire just before and just after menstruation. A comparison made by McCance, Luff and Widdowson[6] between such retrospective reports and subsequent daily recordings of feelings shows no relationship, however, between the results of the two types of reports. This limits the value of the retrospective reports.

The second source of data is the psychoanalytical interpretation of dreams. Benedek's classic monograph[7] presents data obtained from fifteen women, correlated with vaginal smears from which hormone levels were estimated. Her subjects were all under psychoanalytic therapy and most showed histories of complaints related to menstrual functions, which makes this a rather unusual sample. Benedek reports that active heterosexual striving increases with oestrogen production in the first half of the menstrual cycle; that passive-receptive sexuality, a "narcissistic, self-centred state", is associated with the presence of progesterone; and that "the highest integration of the sexual drive" occurs when at the time of ovulation these states overlap. Benedek reports a close relationship between dream content and hormone production. She collected no routine data on overt sexual behaviour. From her analysis one is led to predict a mid-cycle peak in desire, but no inference is made about the actual distribution of coitus.

The third source of data is daily reports of sexual behaviour. McCance, Luff and Widdowson[6] collected menstrual cycle records with daily notations of a number of different feelings and behaviours. Their women were all white, and most had been to college. They had a low reported rate of intercourse—only about half the rates that are usually reported. Their data show a peak frequency at day 8, and another elevated period from day 11 to day 17, calculated on a standardized 28-day menstrual cycle.

We present here data derived from daily reports of the sexual behaviour from two different samples. The first sample consisted of forty premenopausal, married, working, non-white women from North Carolina. Approximately 60 per cent had not graduated from high school, and most were employed in low-status jobs. None was using oral contraceptives.

The sexual data were collected incidentally in connexion with another research purpose. Each woman was paid 50 cents per day for participating in the research, which included the daily deposition in a collection box of a report slip containing yes or no responses to the following items for the previous 24 h: coitus (sexual intercourse or sex relations), orgasm, menstruation. Women were identified and paid by an anonymous code which protected their privacy. Most women delivered slips daily on each of 90 days and there were a few days when information from these women was missing. Data on this sample represent from 93 to 115 cycles.

In this communication the truthfulness of the reports is assumed. A validity check of the responses is reported elsewhere[8]. That the women knew the meaning of the term sexual intercourse or sex relations was assumed. Orgasm was defined for the respondents as follows: a high peak of sexual excitement followed by sudden relaxation. This definition will not satisfy all clinical criteria, but is easily communicable and is broad enough to encompass considerable variation in the experience of women.

The second sample consisted of forty-eight middle class women, most of whom had been to college; they contributed sexual calendars to the files of the Institute for Sex Research, Indiana University. These calendars were obtained through the cooperation of Paul Gebhard. From the institute's total calendar collection we did not include women attending a sterility clinic (whose intercourse schedule may have been prescribed), those past menopause, those not reporting menstruation, those using contraceptive pills, and two who were under continuous psychiatric care.

Unlike the report slips which were collected daily in North Carolina, the institute's calendars were collected yearly or less often. Although women were asked to record daily on the calendars, there is no way of determining how much "filling in" of back data might have been done. The data from these calendars represent from 818 to 882 menstrual cycles.

For the purpose of the analysis, the data from both samples have been organized in the same way. Because the post-ovulatory phase is usually considered less variable than the pre-ovulatory phase in cycles of different lengths, reverse menstrual cycle days were used as reference points. (The lesser variability of the post-ovulatory phase has, however, recently been questioned[9].) Few women reported coitus during the menstrual flow; before 21 reverse days, therefore, the number of observations on non-menstruating women declined rapidly.

Fig. 1 shows the rates of orgasm and intercourse for the North Carolina sample. The sharp rise towards the middle of the cycle, the sharp drop after the middle of the cycle and the second rise towards the end of the cycle in the rate of intercourse are unmistakable.

Fig. 2 shows the rates of intercourse and orgasm for the married women in the institute's sample, and Fig. 3 shows the rates for the single women. There is some overlap in the individuals comprising the married and the single groups, for some of the sexual histories covered periods of more than one marital status. A sharp rise in the rate of intercourse towards the middle of the cycle, followed by a decline in the latter half of the cycle, and a brief recovery before the end of the cycle appear in Fig. 3 much as in Fig. 1. The parallel in the rates of orgasm in the 3 figures is even more striking: in both Fig. 3 and Fig. 1 the peak orgasm rate falls on reverse day 14. Reverse days 4 or 5 show a very low point followed immediately by a return to a higher level. Considering the differences in the samples, the methods of data collection, the levels of the graphs and the sources of orgasm, the similarities between the figures are convincing evidence of some underlying biological mechanism.

For individual women a wide range of variation in the pattern of intercourse and/or orgasm can be seen. Individual curves showing the range of variation are given in Figs. 4 and 5.

The importance of Figs. 1, 2 and 3 lies in their similarities and in their relationship to other known events of the menstrual cycle. The following results of statistical tests of significance are, however, reported. For Fig. 1 the rate of intercourse for day 16 is greater than that for day 4, and the rate of orgasm for day 14 is greater than for day 4 (z test, $P < 0.01$). For Fig. 2 the average rate of intercourse for days 1 to 12 is not different from that for days 13 to 26 (t test, $P > 0.10$); the rate of intercourse for day 6 is lower than the rate for day 20 (z test, $P < 0.02$, greater than 0.01); the rate of orgasm for day 6 is lower than the rate for day 17 (z test, $P < 0.01$). For Fig. 3 the rate of intercourse for day 14 is greater than the rate for day 22 and the rate of orgasm for day 5 is lower than that for day 14 (z test, $P < 0.01$).

These data direct our attention to three questions. (1) What are the controlling hormones? It seems reasonable to hypothesize that oestrogen increases and progesterone decreases the probability of human sexual activity. We consider it premature to answer this question without more direct hormone-behaviour correlations. When data of this sort are obtained, there is every reason to expect that graphs will show even sharper patterns than those that appear when behaviour is correlated only with the reverse menstrual cycle day.

Corroborative evidence concerning the effects of oral contraceptives has been published recently. Grant and Mears[10] state that "loss of libido" is more common with strongly progestogenic compounds containing a low dose of oestrogen than with strongly oestrogenic sequential formulations.

(2) What are the intervening mechanisms by which variations in hormones are translated into variations in the frequency of coitus? Several hypotheses are being explored to answer this question. (a) Variations in hormone concentrations affect the rate of initiation or invitation

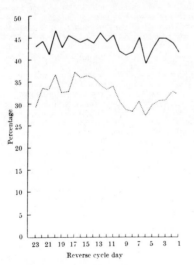

Fig. 2. Percentage of married women in ISR sample reporting intercourse and orgasm by reverse day of menstrual cycle. There were thirty-seven women, and 265–638 cycles. ——, Intercourse; . . . . , orgasm.

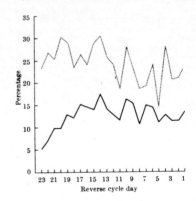

Fig. 3. Percentage of unmarried women in ISR sample reporting intercourse and orgasm by reverse day of menstrual cycle. There were thirteen women, and 77–230 cycles. ——, Intercourse; . . . . , orgasm.

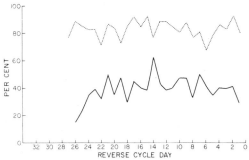

Fig. 4. Rate of intercourse by reverse day of menstrual cycle, for two women in ISR sample. - - -, No. 26 (13-22 cycles). ——, No. 48 (27-55 cycles).

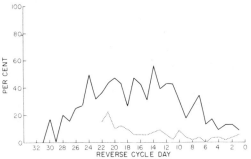

Fig. 5. Rate of orgasm by reverse day of menstrual cycle, for two women in ISR sample. ——, No. 40 (2-23 cycles); - - -, No. 35 (49-54 cycles).

Lachenbruch, William A. Masters, John L. Money and Ethel M. Nash for useful comments.

J. RICHARD UDRY
NAOMI M. MORRIS

University of North Carolina,
Chapel Hill.

Received June 24, 1968.

[1] Michael, Richard P., *Proc. Roy. Soc. Med.*, **48**, 595 (1965); Ford, Clellan S., and Beach, Frank A., *Patterns of Sexual Behaviour*, 203 (Harper and Brothers, New York, 1951).
[2] Southam, Anna L., and Gonzaga, Florante P., *Amer. J. Obst. Gynec.*, **91**, No. 1, 142 (1965).
[3] Maslow, Abraham, H., *Sex Research, New Developments* (edit. by Money, John) Holt, Rinehart and Winston, New York, 1965).
[4] Davis, K. B., *Factors in the Sex Life of 2200 Women* (Harper and Row, New York and London, 1929).
[5] Terman, L. M., *Psychological Factors in Marital Happiness* (McGraw-Hill Book Company, New York, 1938).
[6] McCance, R. A. Luff, M. C., and Widdowson, E. E., *J. Hyg.*, **37**, 571 (1937).
[7] Benedek, Therese, *Psychosexual Functions in Women: Studies in Psychosomatic Medicine* (The Ronald Press Company, New York, 1952).
[8] Udry, J. Richard, and Morris, Naomi M., *J. Marriage and the Family*, **29**, No. 3, 442 (August 1967).
[9] James, William H., *Population Studies*, **57**, No. 1 (July, 1965).
[10] Grant, Ellen C. G., and Mears, Eleanor, *Lancet*, ii, 945 (1967).

from the female; (*b*) variations in hormone concentrations affect the receptivity of the female or her willingness to cooperate when invited; (*c*) variations in female hormone concentrations affect her stimulus value for males, and therefore affect the timing of the sexual approaches she receives from males.

(3) Why does the distribution of the frequency of coitus differ in the two married groups, while the single group matches one of the married groups? The underlying biological similarity of the three groups is suggested by the similarity of the distribution of orgasm. The answer to this question is intricately involved with the previous question. Our present hypothesis is that the differences between groups are sociological phenomena related to differences in the control and initiation of coitus in different groups and by marital status.

In conclusion, variations in the rate of human sexual behaviour are clearly related to the female menstrual cycle. The rates of intercourse and orgasm at one point in the cycle are from two to six times the rates at other points in the cycle. The highest rates of intercourse and orgasm occur about the time when ovulation is thought to occur. Variations from one sample to another suggest that the influence of cyclic female hormones on sexual behaviour is more apparent in some groups than in others. Unexplored social-psychological and sociological processes probably mediate between the hormone variations and overt sexual behaviour.

This work was supported by grants from the US Children's Bureau, the US Public Health Service and the Carolina Population Center. We thank Charles Chase and Sue Gena Ficker for their help, and the late Harry S. McGaughey, jun., and Judson Van Wyk for advice. We also thank Frank A. Beach, William E. Easterling, Paul H. Gebhard, J. F. Hulka, Francis J. Kane, Peter A.

191

Reprinted from *J. Clin. Endocrinol.*, **3**, 235–238 (1943)

Effect of Androgens upon Libido in Women

UDALL J. SALMON, M.D.
AND SAMUEL H. GEIST, M.D.

From the Mount Sinai Hospital, New York, New York

23

URING the course of the therapeutic administration of androgens for various endocrinopathic gynecologic disorders during the past five years, the observation was made that libido was apparently increased in a number of the women being treated (1). Our attention was first drawn to this phenomenon in 1937 by an ovariectomized woman, suffering from the menopause syndrome, who was being treated experimentally with testosterone propionate. During the course of treatment the patient volunteered the information that she had experienced resurgence of sexual desire after a period of quiescence of some ten years. Since then we have been conducting a series of studies to determine the effect of various sterol sex hormones upon the psychosexual reactions of women.

METHODS AND MATERIALS

The present report is based on our observations of a series of 101 women comprising the following groups. *Group A.* Twenty-nine cases of primary frigidity which were being treated for some apparently unrelated gynecologic disorder: dysmenorrhea, 6 cases; menometrorrhagia, 16 cases; polymenorrhea, 3 cases; premenstrual tension, 4 cases. *Group B.* Thirty cases of secondary frigidity occurring *a)* at the menopause, 13 cases; *b)* after bilateral ovariectomy, 7 cases; *c)* after roentgen-ray castration, 7 cases; and *d)* in association with amenorrhea, 3 cases. *Group C.* Forty-two were patients with apparently normal libido who were being treated with androgens for one of the following complaints: dysmenorrhea, 12 cases; menorrhagia, 27 cases; and premenstrual tension, 3 cases.

Received for publication January 8, 1943.
Presented before the twenty-sixth Annual Meeting of the Association for the Study of Internal Secretions in Atlantic City, N. J., on June 9th, 1942.

Types of androgen therapy and dosage used. Fifty-six of the patients were treated with testosterone propionate in sesame oil intramuscularly in doses of 10 or 25 mg. two or three times weekly, the total weekly doses varying from 20 to 75 mg., for periods varying from 6 to 8 weeks; 22 with methyl testosterone in daily doses of 10 to 30 mg. orally for periods varying from 3 to 6 months; 23 received subcutaneous implants of 2 to 8 pellets of testosterone, each weighing 75 mg. Periods of observation following therapy varied from 8 to 38 months.

RESULTS

Group A. Primary Frigidity with Somatic Gynecologic Disorders

Of the 29 cases in this group 21 were treated with testosterone propionate intramuscularly and 8 had pellet implants of testosterone. Of the 21 cases treated with testosterone propionate, 4 reported excessive sexual stimulation. These patients were conscious of a sensation of formication in the vulva which was most marked in the region of the clitoris. The clitoris and prepuce were hypersensitive and in two cases, slightly hypertrophied. Ten of the cases reported complete sexual gratification, 4 reported considerable improvement but were unable to achieve an orgasm and 3 reported no improvement.

Of the 8 cases with testosterone pellet implants 2 reported excessive stimulation (600 and 525 mg., respectively); 4, normal sexual reactions (225 to 450 mg.); and 2, no response (150 and 300 mg.).

Of the 5 failures in the group of 29 cases, 2 could be attributed to marital incompatibilities. In the remaining 3 there was a heightened clitoridal sensitivity without, however, consonant emotional response. These 3 individuals appeared to be normal, well-in-formed, well-adjusted women who professed to be, otherwise, happily married.

Group B. Secondary (Endocrinopathic) Frigidity

In order to evaluate the effect of androgens in the presence of an estrogen deficiency the 30 patients with secondary or endocrinopathic frigidity were divided into three groups and treated as follows: a) one group of 10 received androgens; b) another group of 11, estrogens, and c) the remaining 9 cases, androgens and estrogens simultaneously.

Effect of androgens in estrogen-deficient cases. After periods of treatment varying from 4 to 8 weeks, the total doses varying from 240 to 600 mg. of testosterone propionate, 9 of the 11 patients in this group reported increased libido and increased clitoridal sensitivity, but did not experience coital gratification. One patient reported excessive stimulation and one none at all. The 9 stimulated patients stated that coitus could not be satisfactorily consummated because of dryness and tenderness of the vagina. Following the course of androgens, estrogens were substituted for the testosterone without the patient's knowledge (2.5 mg. of estradiol dipropionate three times weekly). After treatment with estrogens for 10 to 14 days, the patients reported disappearance of the vaginal dryness and tenderness and they were able to perform coitus satisfactorily. This improvement in the subjective vaginal reaction paralleled the appearance of a characteristic estrogenic effect in the vaginal smears.

Effect of estrogens in estrogen-deficient cases. Twelve patients were treated with estrogens (2.5 mg. of estradiol dipropionate three times weekly) until the vaginal smears showed a full estrogenic effect, and the vasomotor symptoms were relieved. Libido was not, however, appreciably affected. Eight patients of this group who, prior to the estrogen therapy, had experienced dyspareunia caused by 'senile' vaginitis and who had in consequence developed a distaste for coitus, were relieved of the coital discomfort after treatment with estrogens, but none of these reported a resurgence of libido. At this point, without their knowledge, androgens were substituted for the estrogens (25 mg. of testosterone propionate

2 or 3 times weekly). After approximately 3 weeks, 4 patients reported restoration of normal sexual reaction; 2, excessive stimulation; 1, slight, and 1, no improvement.

Effect of combined estrogen and androgen therapy in estrogen-deficient cases. The 9 patients in this group were given estrogens and androgens simultaneously, 1 to 5 mg. of estradiol dipropionate in combination with 10 or 25 mg. of testosterone propionate two or three times weekly. In 7 instances, the patients reported a marked improvement after 3 or 4 weeks of treatment. The optimal therapeutic effect was obtained in these cases when the vaginal smear revealed a full estrogenic effect even though the androgens were being given. If sufficient androgen, but insufficient estrogen was given, the patient experienced the stimulating effect of the former but coitus was unsatisfactory and a vaginal orgasm was not achieved, because of the dryness and irritability of the vagina; when, however, an adequate amount of estrogen but an insufficient amount of androgen was given, the vaginal dryness and irritability were relieved but libido was not stimulated.

Group C. Cases with Somatic Gynecological Disorders and Normal Libido

Of the 42 patients in this group 10 were treated with testosterone propionate, 25 mg. twice weekly or 10 mg. three times weekly intramuscularly; 12 with methyl testosterone orally, 20 to 50 mg. daily. Fifteen received implants of testosterone pellets, 75 mg. each, total doses varying from 300 to 600 mg.

Results with testosterone propionate. Of the group treated with testosterone propionate, all but one experienced increase of libido after receiving approximately 200 to 300 mg. With this dose a mild stimulation occurred which persisted for several weeks. Continued administration of the testosterone propionate resulted in a progressive increase of libido, in 4 instances to a distressing degree.

Results with methyl testosterone. Methyl testosterone produced moderate stimulation in doses of 20 and 30 mg. a day after treatment for 2 to 3 weeks. In 6 instances doses of 40 and 50 mg. a day caused excessive stimulation after two weeks of treatment and it persisted for approximately two weeks after discontinuation of the hormone.

Results with pellet implantation. Of the 15 cases in this group receiving pellet implants of testosterone, 3 reported no change (150 mg., 2 cases; 300 mg., 1 case); 5 were moderately stimulated for 4 to 5 weeks (150 mg., 1 case; 225 mg., 1 case; 300 mg., 3 cases). Seven were excessively stimulated for 6 to 8 weeks (450 mg., 3 cases; 600 mg., 4 cases). Subjective evidence of stimulation was reported in 10 of the 12 responsive cases approximately 12 days after the implantation; in 2 cases implanted with 150 mg. increased libido was not noted until approximately 18 days after the implantation.

DISCUSSION

It appears from this study that androgens have a triple action upon the psychosomatic sexual mechanism, causing, *a*) a heightened susceptibility to psychic stimulation; *b*) increased sensitivity of the external genitalia, particularly of the clitoris and *c*) a greater intensity of sexual gratification.

It appears, furthermore, that in the presence of an estrogen deficiency, androgens in the doses used may stimulate the psychic and the sensory components but that culmination in a vaginal orgasm may not be accomplished without the complementary stimulative effects of estrogens upon the vagina. The estrogens cause an increase in the vascularity of the vagina and furnish the physiologic lubricant resulting from the desquamation of the cornified epithelial cells and the cervical secretion which is essential for normal coital sensibility.

It is noteworthy that in 6 of the 13 cases which failed to respond to the androgens, there appeared to be some psychic disturbance to which the frigidity could be attributed. It would seem from these studies that types of frigidity that are predominantly psychogenic in origin are not responsive to androgen therapy. However, the stimulating effects of the androgens are so striking, that even psychogenic types might be benefited, providing one can be reasonably certain that the stimulation

would not *per se* cause undesirable psychic repercussions.

The question arises as to whether this stimulating action of the androgens upon the psychic and somatic components of the sexual mechanism should be considered as a pharmacologic phenomenon or whether one may infer from these observations that in the normal female some endogenous androgens may have a similar effect, acting as a physiologic stimulant or activator of the psychic and somatic components of the sexual mechanism. For several years we have fostered the theory that androgens play an important rôle in the steroid sex hormone physiology of the human female, modifying and balancing the action of the gynecogenic hormones (1, 2, 3). We have based the rationale for the use of androgens in the treatment of certain types of menometrorrhagia (4, 5), dysmenorrhea (6), and premenstrual tension (1, 7) upon this theory. It appears from the studies herein reported that the endogenous androgens in the mature human female may have still another function, namely, to sensitize both the psychic and somatic components of the sexual mechanism so that the individual is emotionally receptive to excitation and somatically sensitive to sexual stimulation.

SUMMARY AND CONCLUSIONS

The effect of androgens upon libido was studied in a group of 101 women who were being treated for some endocrine disorder. The androgens were administered in the form of *a*) testosterone propionate in solution in sesame oil, intramuscularly; *b*) pellets of testosterone, implanted subcutaneously; and *c*) methyl testosterone, orally. All but 13 of the 101 treated women reported some increase in libido; 20 noted excessive stimulation which subsided within 2 to 4 weeks after discontinuation of the androgen therapy. From these studies it appears that androgens have a threefold action, causing *a*) an increased susceptibility to psycho-sexual stimulation; *b*) an increased sensitivity of the external genitalia and *c*) a greater intensity of sexual gratification.

It is suggested on the basis of these observations that endogenous androgens in the normal mature woman may act as the physiologic sensitizer of both the psychic and somatic components of the sexual mechanism.

REFERENCES

1. SALMON, U. J.: *J. Clinical Endocrinology* 1: 162. 1941.
2. GEIST, S. H., U. J. SALMON AND J. A. GAINES: *Endocrinology* 23: 784. 1938.
3. GEIST, S. H., U. J. SALMON, J. A. GAINES AND R. I. WALTER: *J. Am. Med. Assoc.* 114: 1539. 1940.
4. SALMON, U. J., S. H. GEIST, J. A. GAINES ANI R. I. WALTER: *Am. J. Obst. & Gynec.* 41: 991. 1941.
5. GEIST, S. H.: *J. Clinical Endocrinology* 1: 154. 1941.
6. SALMON, U. J., S. H. GEIST AND R. I. WALTER *Am. J. Obst. & Gynec.* 38: 264. 1939.
7. GEIST, S. H., AND U. J. SALMON: *J. Am. Med. Assoc.* 117: 2207. 1942.

Reprinted from *J. Clin. Endocrinol.*, **19**, 193–202 (1959)

THE ROLE OF HORMONES IN HUMAN BEHAVIOR.
I. CHANGES IN FEMALE SEXUALITY AFTER ADRENALECTOMY

24

SHELDON E. WAXENBERG, Ph.D., MARVIN G. DRELLICH, M.D.
AND ARTHUR M. SUTHERLAND, M.D.

Sloan-Kettering Institute for Cancer Research and Memorial Hospital, New York 21, N. Y.

ABSTRACT

Twenty-nine women, mean age 51 years, who had undergone total bilateral oophorectomy and adrenalectomy for metastatic breast cancer were interviewed about changes in their sexual desire, activity and responsiveness at an average period of twelve months after adrenal ablation. Of the 17 patients reporting some sexual desire preoperatively, 14 experienced a decrease after surgery, the majority losing all desire. Of the 17 patients sexually active preoperatively, all reduced their frequency of intercourse postoperatively, almost half stopping entirely. Of the 12 patients responsive in intercourse preoperatively, 11 experienced a decrease after surgery, almost all becoming completely unresponsive. In each of these categories, the number of patients who lost sexual function entirely was statistically significant. Both sexual feelings and activities declined in all but 2 of the patients not already at zero levels among the 20 who obtained remission of metastatic disease after adrenalectomy, as well as in all but 1 of the 9 patients who did not present objective evidence of benefit from the operation.

Neither breast cancer itself nor any previous surgery had as profound an impact on sexual behavior as did adrenalectomy. In agreement with the preponderance of evidence in the literature, changes in a subgroup of 7 patients who had undergone oophorectomy one to five years before their adrenals were removed indicated that ovarian ablation by itself usually does not radically alter sexual behavior. It was concluded that the adrenal glands and, more specifically, the androgens of adrenal origin, play a critical part in maintaining the patterns of sexual behavior in the human female.

INTRODUCTION

THIS investigation explores the hypothesis that sex hormones of adrenal origin have an important influence on the sexual behavior of women. The adrenal glands are a source of both androgens and estrogens and are probably the exclusive source of androgens in women. Five steroids which possess androgenic activity have been isolated by direct chemical extraction from adrenocortical tissue, but none biologically as active as testosterone, the androgen produced by the male testes (1). Although the adrenal androgens are relatively weak substances, they are produced in greater amounts than the adrenal estrogens.

Received July 21, 1958.

193

In contrast to previous studies of effects of androgens in women, which have been concerned with exogenous hormones, this study examines the results of deprivation of endogenous adrenal hormones. Adrenal ablation, or combined oophorectomy-adrenalectomy, is currently resorted to in an effort to remove sources of estrogenic hormones which are believed to promote or sustain the growth of some types of breast cancer. Recent reviews of the results of adrenalectomy in breast cancer patients omit any mention of effects on sexual behavior (2–4).

MATERIAL AND METHODS

Twenty-nine women who had undergone total bilateral oophorectomy and adrenalectomy for metastatic breast cancer were available for study over the course of a year in the Breast Service Clinic of Memorial Center. The patients varied in age from 31 to 72 years. Only 2 were more than 59 years old, and the mean age for the group was 51 years. Sixteen patients were married, 4 were widowed, 6 were divorced or separated, and 3 were single. Breast cancer had been present from two to fifteen years; the median duration was less than five years. Twenty-two patients had undergone simple or radical mastectomy, and 7 had inoperable breast disease. Seven of the patients had undergone bilateral oophorectomy one to five years before adrenalectomy. All the others were oophorectomized at the time of adrenalectomy.

Intensive interviews at two to thirty-nine months after adrenalectomy (mean interval twelve months, median nine months) probed the pre- and postoperative sexual behavior of these women. Since sexual functioning in the human subject is a complex phenomenon, an attempt was made to differentiate 3 important aspects of such behavior: (a) sexual desire, an appetitive state compounded of innumerable somatic and psychologic elements, (b) heterosexual activity, which can be measured objectively in terms of frequency of intercourse, and (c) sexual responsiveness, which entails a subjective evaluation of the degree of participation in and the gratification obtained from sexual activity.

OBSERVATIONS

The changes in sexual functioning which took place between the time of adrenalectomy and the time of the retrospective interviews are summarized in Table 1. Before adrenalectomy, 8 of the patients had no desire, no sexual activity and no responsiveness, and they experienced no changes after the operation. Four women were sexually inactive before adrenalectomy although they felt sexual desire, and 4 others engaged in sexual relations without sexual desire.

Changes in sexual desire

Seventeen of the women reported some level of conscious sexual desire prior to adrenalectomy, whereas 12 were aware of no such feelings. Of the 17 with sexual desire, 3 patients reported no change after the operation, but 14 experienced a noticeable decrease in desire. Of the 14 who reported a decrease, 10 experienced a total loss of desire, and in the remaining 4 it was

TABLE 1. SUMMARY OF CHANGES IN SEXUAL BEHAVIOR AFTER
OOPHORECTOMY AND ADRENALECTOMY

Type of change	Number of patients experiencing change		
	In sexual desire	In sexual activity	In sexual responsiveness
Maintained preoperative level	3	0	1
Decreased from preoperative level	14	17	11
Some decrease	4 ⎱	10 ⎱	2 ⎱
Decrease to zero	10* ⎰	7† ⎰	9* ⎰
Unchanged, zero level	12	12	17
Totals	29	29	29
Percentage of those not at zero level preoperatively who experienced decrease	82%	100%	92%

* Number decreasing to zero significant at .01 level.

† Number decreasing to zero significant at .05 level.

considerably reduced but not absent. The number changing from a state of some desire to no desire is significant at the .01 level.

Changes in sexual activity

Seventeen women were sexually active prior to adrenal surgery. All decreased their sexual activity afterward. Seven discontinued sexual activities altogether, and the other 10 reduced the frequency of intercourse. The number changing from some activity to no activity is significant between the .02 and .05 levels.

Three patients received testosterone after adrenalectomy. Two were in a terminal condition; the third reported some heightening of desire and increase in frequency of intercourse during testosterone therapy.

Changes in sexual responsiveness

Twelve patients reported some pleasurable response in their sexual relations before adrenalectomy. After adrenalectomy, 9 of the 12 reported total loss of responsiveness, 2 experienced a decrease, and in 1 there was no change. The number changing from some responsiveness to none is significant at the .01 level.

A 54-year-old patient, interviewed almost two years after oophorectomy-adrenalectomy which was followed by remission of metastatic breast cancer, illustrates the dramatic changes which can occur:

"My husband is very patient and understanding . . . There was no change in my sex life after the first operation (mastectomy). My feelings were just what they had been before. Immediately after the second operation (combined oophorectomy-adrenalectomy), I noticed a change. There was a loss of interest and desire for sex. There was no pleasure in it whatsoever. Maybe I didn't try, but I just didn't feel any desire for it. Before the operation (combined procedure) I never had any trouble reaching an orgasm. I had one almost every time. There was no change after the breast operation; I would even get desire from reading a book or something like that. After the breast operation, there was no loss of interest on my husband's part either, and there was no special cautiousness on either of our parts. Now my husband is still very considerate, but I have no desire or pleasure in it at all, none at all."

Three patients experienced a subtle shift in emphasis from genital gratifications to the affectionate and tender satisfactions of sexual relationships and to the reassurance to be gained from close body contact.

Relationship between physical condition and sexual changes

The fact that all the patients included in this study had a fatal disease requires that the sexual behavior of those who benefited from adrenalectomy be compared with the behavior of those who did not obtain relief of symptoms or remission of disease. The 4 patients for whom adrenalectomy was without benefit experienced decreases in all sexual manifestations not already at a zero level. The 5 patients who obtained some palliation of symptoms without objective evidence of improvement experienced decreases in all sexual measures which were not already zero, with the exception of 1 woman who maintained her level of desire but became sexually inactive.

There were 20 patients who presented objective evidence of improvement. This proportion of patients benefited by adrenalectomy is greater than usually reported, since this research unavoidably excluded some patients who obtained little or no benefit. In 6 of the 20 improved patients, all sexual variables were at zero levels prior to adrenal ablation. In 12 of the remaining 14 improved patients, there was a postoperative decrease in all variables not already at zero levels; in only 2 of the 14 was there maintenance of some element of the sexual functioning at the preoperative level. Of these 2 women, one maintained sexual desire but had already become sexually inactive, and the other maintained desire and responsiveness but decreased sexual activity after surgery.

Every one of these 20 patients who was sexually active before adrenalectomy reported a decrease in sexual activity notwithstanding observed improvement in her general physical condition. These decreases in sexual functioning reported by this majority of the subjects in the study cannot be attributed to progression of the mammary cancer.[1]

---

[1] Many factors besides general state of health may influence sexual behavior, such as individual sex anatomy, sexual technique, sexual inhibitions, the interpersonal climate

Effects of oophorectomy

In 7 of the 29 cases, the ovaries had been removed sixteen to sixty months before adrenalectomy. Table 2 presents a comparison of the effects of oophorectomy with the effects of subsequent adrenalectomy in this subgroup. All 7 of these patients reported having some sexual desire before oophorectomy. Five were sexually active, and 2 were sexually inactive

TABLE 2. COMPARISON OF EFFECTS ON SEXUAL BEHAVIOR OF
OOPHORECTOMY AND ADRENALECTOMY
(INTERVAL BETWEEN OPERATIONS RANGED FROM 1 TO 5 YEARS)

| Effect | Number of patients experiencing change | | | | | |
| | In sexual desire | | In sexual activity | | In sexual responsiveness | |
	After ovarian ablation	After adrenal ablation	After ovarian ablation	After adrenal ablation	After ovarian ablation	After adrenal ablation
No decrease	2	0	2	0	2	0
Some decrease	4	2	3	2	3	1
Decrease to zero	1	4	0	3	0	4
Unchanged, zero level	0	1	2	2	2	2
Totals	7	7	7	7	7	7

widows. After ovarian surgery, 1 of these patients lost all her sexual desire, 4 experienced a decrease in desire, and 2 reported no change. Two of the women who experienced a decline in desire had undergone oophorectomy

between partners, and the desire for or fear of pregnancy. Most such variables are not likely to have changed fundamentally during the periods in the lives of these women from which comparisons were drawn for this report, namely, the immediate pre-adrenalectomy period versus the post-adrenalectomy period. The patterns of sexual behavior which might have been affected by these variables were largely set down many years previously. Although the impossibility of pregnancy after combined ovarian and adrenal ablative surgery might be expected to increase sexual activity in certain patients; this result was not observed. Nor did any patient indicate that inability to conceive had a directly inhibitory effect.

The patients were suffering from widespread metastatic disease at the time the decision was made to resort to adrenalectomy, and most of them felt better after this operation than they had felt just before the operation. Remissions eventually come to an end and the disease again runs a downhill course, but terminal stage conditions were not used in this comparative study. Trends in sexual functioning throughout the lives of the individual patients were reviewed. The data cannot be presented here in detail, but, briefly, they indicated that neither the discovery of breast malignancy nor any previous surgery had the negative impact on sexual behavior that adrenal ablation had.

combined with hysterectomy. Recent reports on this combined operation indicate that subsequent disturbances in sexuality may be related to irrational fears and distorted beliefs concerning the loss of the uterus and the effects of surgery in the genital area (5, 6). After adrenal ablation, 4 of the patients who had retained some desire after oophorectomy lost all desire and the other 2 patients experienced decreases in desire.

Of the 5 patients who were sexually active at the time of oophorectomy in this subgroup of 7, 2 reported no changes in frequency of intercourse; the other 3 reported decreases, but not discontinuance of intercourse after their ovaries were removed. After adrenalectomy, of the 2 patients unaffected by oophorectomy, one experienced a decrease in frequency of intercourse and the other ceased having sexual relations. The 3 patients whose sexual activity had declined after oophorectomy all experienced further declines, 2 of them discontinuing sexual relations. Changes in sexual responsiveness in this subgroup closely paralleled changes in sexual activity.

There was a decrease to zero of at least one sexual variable in 1 out of 7 patients after ovarian surgery and in 5 out of 6 patients after subsequent adrenal surgery. In none of 5 sexually active patients did all three sexual variables decrease to zero after ovarian ablation; however, this did occur in 2 of the 5 patients after adrenal ablation.

DISCUSSION AND CONCLUSIONS

From the findings of this study, it is apparent that any discussion of the hormonal bases of human female sexuality must concern itself as much with the adrenal glands as with the ovaries. Since the ovaries secrete by far the greater proportion of the endogenous estrogens in the female circulation, their removal or the reduction of their function (as in the menopause) deprives a woman of the major source of estrogens. A survey of recent studies (7) indicates that radical changes in the sexual behavior of women do not follow either surgical castration or the menopause. For example, Filler and Drezner (8) found that among 41 women under the age of 40 whose ovaries were removed, 36 (88 per cent) experienced no change in sexual libido. Kinsey *et al.* (7), reporting on 123 women who had their ovaries removed, confirmed the general opinion that there is no modification of sexual responsiveness or capacity for orgasm which can be clearly identified as due to ablation of the ovaries. Although deprivation of ovarian hormones has little effect on sexual behavior, evidence has been presented in this paper for a close relationship between loss of adrenal hormones and disruption of sexual behavior in women.

The gonads and the adrenals together, activated by the pituitary, may be considered to provide the hormonal basis for human female sexuality; but the major nonreproductive component, the erotic component, is here attributed to the adrenal hormones. The evidence points, furthermore, to the adrenal *androgens* as important hormonal contributors to sexuality in the human female. The adrenal glands are probably the exclusive source of androgens in the intact female. After adrenalectomy, urinary androgen excretion falls to very low levels, whereas after castration or the menopause it is not reduced (1). Dorfman and Shipley state that no vital function has been definitely established for adrenal androgens in the female other than promotion of the growth of pubic and axillary hair (1); nevertheless, the present investigation indicates another important function, namely, that of promoting and sustaining certain aspects of female sexual functioning.

On the behavioral side, the evidence reviewed here indicates that loss of the major source of estrogens does not by itself critically alter the sexual expression of women who have previously achieved good sexual adjustment. The patients in this study, all of whom had undergone both oophorectomy and adrenalectomy, may be considered to have been deprived of their endogenous androgens. It has been shown that their sexual functioning underwent significant alteration in the direction of decrease or total loss of sexual desire and responsiveness, and decrease or cessation of sexual activity. Thus the adrenal glands rather than the ovaries, and the *androgens* supplied by the female adrenals, rather than the estrogens, are implicated in this remarkable decrement in manifest sexuality.

There are certain objections which may be raised to this conclusion. After oophorectomy, estrogens may still be present (9, 10). Since there are apparently enough left to affect the growth of estrogen-dependent breast cancer (though the dependence of cancer growth on these hormones is only presumptive), the influence of estrogens on sexuality cannot be ruled out entirely. Metabolic transformation of massive doses of testosterone propionate to small amounts of estrogen has been shown to occur in women whose gonads and adrenals have been removed (11). Estrogens produced by the adrenal glands or derived metabolically from adrenal androgens may sustain sexual functioning after oophorectomy. Small quantities of estrogens may be sufficient to maintain habitual levels of sexual behavior. If this be true, adrenalectomy may reduce sexual desire by removing the secondary source of estrogens. If not, it must be concluded that loss of adrenal *androgens* accounts for the sexual changes after adrenalectomy.

Evidence from studies on the clinical administration of sex hormones bears most tellingly in favor of the conclusion that it is the androgens which are primarily involved. Estrogens administered to women with gynecologic disorders or cancer do not usually have an effect on sexual desire (12, 13), nor have estrogens proved effective in the treatment of frigidity (1,

14). However, numerous reports dealing with the therapy of gynecologic disorders and cancer have established that exogenous *androgens*, chiefly testosterone propionate, do have a heightening effect on libido in many women, including some who lacked or had only minimal desire before (12, 15–18). One patient in this study who was subsequently given a course of testosterone therapy experienced a noticeable increase in sexual desire and activity.

Various explanations have been advanced regarding the mechanism by which androgenic hormones might influence female sexuality. One attributes the effects to increased vascularization of the vulval region, with sensitization of the clitoris (12, 14). Another regards the metabolic stimulation attendant upon androgen intake to be the essential physiologic change, the increased libido being secondary to the generally increased well-being (7). Still another point of view assigns importance to increased psychic sensitization and susceptibility to psychosexual symbolism under the control of neurohumoral mechanisms in the central nervous system (12, 19). These hypotheses are by no means mutually exclusive.

What of the few exceptional patients in whom ovarian and adrenal ablation did not cause a decrease in sexual desire, and what of those who continued to experience some reduced measure of desire? Defensiveness about the waning of highly prized feminine attributes or the failure of sexual powers may cause some women to report retention of desire when they actually do not have it. Accessory adrenal tissue (20) or possibly food ingested by the patients may supply sex hormones after the combined ovarian-adrenal ablative surgery. Also, after adrenalectomy each patient must have a daily dose of 37.5 to 75.0 mg. of cortisone for preservation of life. The question of whether this cortisone is metabolized to androgenically active substances in significant amounts is controversial on both theoretical and empirical grounds (1, 21–23).

The patients who retained some interest in sexual intimacy with their husbands after adrenal ablation indicated a shift of that interest from the genitally erotic to the more diffusely affectionate gratifications which they derived from the tenderness and reassuring physical closeness in sexual relations. This affectionate component of sexual activity may be sustained not by hormonal but rather by psychologic factors. One patient who before adrenalectomy had dreams of intercourse with accompanying orgasm later had dreams involving physical warmth and closeness to another person without genital excitation.

Patterns of sexual expression are not determined exclusively by hormones, nor is sexual behavior entirely determined by learning and training (7, 24). Sex hormones are important in the proper maturation and maintenance of sexual and reproductive mechanisms in the adult human being, but psychologic factors play a part in setting the levels of intensity and of awareness of the sexual urges and in determining the direction and quality of expression of these urges. The psychologic factors are very intricate. They are developed and modified by the total experience of a person, by relationships with parents and by the social and moral pressures which come to bear on the personality. Disturbances in sexual functioning have been shown in the majority of cases to arise from psychologic causes (14, 25), and they have been favorably influenced by psychotherapeutic intervention. Even when sex hormones are present in intact men and women, there may be no sexual expression because hormonal influences cannot always break through strong psychosocial barriers. On the other hand, even in the state of deprivation resulting from combined oophorectomy-adrenalectomy, sexual feelings and activities may occasionally be sustained by social pressures and by interpersonal needs which are not specifically erotic. Nevertheless, in the majority of cases, deprivation of sex hormones results in loss of interest in the specifically erotic aspects of sexual relationships. The hormones are an essential component among the complex determinants of sexuality.

Under the condition of deprivation of endogenous sex hormones, the adrenal glands have been shown to enter into the determination of human female sexual functioning in a critical manner. There is evidence that adrenal androgens, rather than the estrogens, are the major source or vehicle of this adrenal influence on the sexual behavior of women.

Acknowledgments

This investigation was supported in part by research grants M-1118(C) and M-884 (C2) from the National Institute of Mental Health, U. S. Public Health Service.

Grateful acknowledgment is made to Dr. Norman Treves, Dr. Arthur I. Holleb and to Dr. John Finkbeiner for their assistance in obtaining patients for this study.

REFERENCES

1. DORFMAN, R. I., and SHIPLEY, R. A.: Androgens. New York, John Wiley & Sons, Inc., 1956.
2. GALANTE, M.; FOURNIER, D. J., and WOOD, D. A.: Adrenalectomy for metastatic breast carcinoma, *J.A.M.A.* **163**: 1011, 1957.
3. LIPSETT, M. B.; WHITMORE, W. F., JR.; TREVES, N.; WEST, C. D.; RANDALL, H. T., and PEARSON, O. H.: Bilateral adrenalectomy in the palliation of metastatic breast cancer, *Cancer* **10**: 111, 1957.
4. PERLIA, C. P.; KOFMAN, S.; NAGAMANI, D., and TAYLOR, S. G., III: Critical analysis of palliation produced by adrenalectomy in metastatic cancer of the female breast, *Ann. Int. Med.* **45**: 989, 1956.
5. DRELLICH, M. G.; BIEBER, I., and SUTHERLAND, A. M.: The psychological impact of cancer and cancer surgery. VI. Adaptation to hysterectomy, *Cancer* **9**: 1120, 1956.
6. DRELLICH, M. G., and BIEBER, I.: The psychological importance of the uterus, *J. Nerv. & Ment. Dis.* **126**: 322, 1958.

7. KINSEY, A. C.; POMEROY, W. B.; MARTIN, C. E., and GEBHARD, P. H.: Sexual Behavior in the Human Female. Philadelphia, W. B. Saunders Co., 1953.

8. FILLER, W., and DREZNER, N.: The results of surgical castration in women under forty, *Am. J. Obst. & Gynec.* **47:** 122, 1944.

9. HUGGINS, C.: Endocrine methods of treatment of cancer of the breast, *J. Nat. Cancer Inst.* **15:** 1, 1954.

10. McBRIDE, J. M.: Estrogen excretion levels in the normal postmenopausal woman, *J. Clin. Endocrinol. & Metab.* **17:** 1440, 1957.

11. WEST, C. D.; DAMAST, B. L.; SARRÒ, S. D., and PEARSON, O. H.: Conversion of testosterone to estrogens in castrated, adrenalectomized human females, *J. Biol. Chem.* **218:** 409, 1956.

12. SALMON, U. J., and GEIST, S. H.: Effect of androgens upon libido in women, *J. Clin. Endocrinol.* **3:** 235, 1943.

13. SOPCHAK, A.: Changes in Sexual Behavior, Desire and Fantasy in Breast Cancer Patients under Androgen and Estrogen Therapy. Unpublished doctoral dissertation, Adelphi College, Garden City, N. Y., 1957.

14. PERLOFF, W. H.: Role of the hormones in human sexuality, *Psychosom. Med.* **11:** 133, 1949.

15. LOESER, A. A.: Subcutaneous implantation of female and male hormone in tablet form in women, *Brit. M. J.* **1:** 479, 1940.

16. GREENBLATT, R. B.: Testosterone propionate pellet implantation in gynecic disorders, *J.A.M.A.* **121:** 17, 1943.

17. ABEL, S.: Androgenic therapy in malignant disease of the female genitalia, *Am. J. Obst. & Gynec.* **49:** 327, 1945.

18. FOSS, G. L.: The influence of androgens on sexuality in women, *Lancet* **1:** 667, 1951.

19. BEACH, F. A.: Hormones and Behavior. New York, Paul B. Hoeber, 1948.

20. WHITMORE, W. F., JR.; RANDALL, H. T.; PEARSON, O. H., and WEST, C. D.: Adrenalectomy in the treatment of prostatic cancer, *Geriatrics* **9:** 62, 1954.

21. GALLAGHER, T. F.: Steroid hormone metabolism and the control of adrenal secretion, *in* The Harvey Lectures, Series LII, 1956–1957. New York, Academic Press Inc., 1958, pp. 1–22.

22. MUNSON, P. L.; GOETZ, F. C.; LAIDLAW, J. C.; HARRISON, J. H., and THORN, G. W.: Effect of adrenocortical steroids on androgen excretion by adrenalectomized orchidectomized men, *J. Clin. Endocrinol. & Metab.* **14:** 495, 1954.

23. HOLLIDAY, M. E.; KELLIE, A. E., and WADE, A. P.: The effect of hypophysectomy on the urinary excretion of 17-oxosteroids, *in* Endocrine Aspects of Breast Cancer, ed. by A. R. Currie. Edinburgh, Livingstone, 1958, pp. 224–239.

24. FORD, C. S., and BEACH, F. A.: Patterns of Sexual Behavior. New York, Paul B. Hoeber, 1951.

25. MARMOR, J.: Some considerations concerning orgasm in the female, *Psychosom. Med.* **16:** 240, 1954.

Editor's Comments on Papers 25 Through 28

In the female, as in the male, questions have been tendered regarding the localization of hormone action. Again the techniques utilized have been those of ablation (Brookhart, Dey, and Ranson, Paper 25; Beach, Paper 26) and of direct chemical stimulation (Kent and Liberman, Paper 27). More recent studies have further exploited these techniques and, as in the male, the hypothalamus has been implicated in the control of female behavior. The neocortex, of importance in male behavior, is apparently less essential to the appearance of female sexual responses and may, in fact, inhibit the lordosis response characteristic of the receptive female (see Beach, Paper 30). Information from electrophysiological recordings will undoubtedly affect our future understanding of the hormone and nervous system interaction. The work of Kawakami and Sawyer (Paper 28) is an example of the use of this technique that has influenced current theories of neuroendocrinology as well as behavior.

Reprinted from *Endocrinology*, 28, 561–565 (1941)

25

THE ABOLITION OF MATING BEHAVIOR BY HYPOTHALAMIC LESIONS IN GUINEA PIGS[1]

J. M. BROOKHART, F. L. DEY AND S. W. RANSON

From the Institute of Neurology, Northwestern University Medical School
CHICAGO, ILLINOIS

DISTURBANCES of mating behavior and reproductive functions have been observed in 50 female guinea pigs with hypothalamic lesions (1). Twenty-three out of the 50 animals ran normal cycles so far as the changes in the ovaries, uteri and external genitalia were concerned (2) but they did not come into heat in the sense in which that word will be used in this paper, i.e., they did not show proestrual or estrual behavior and would not accept the male. Animals of this type will be referred to as Group I. In 12 others the genitalia were enlarged, the vaginae remained always open and the ovaries were highly follicular but without coporra lutea (Group II). In 7 the genitalia, including the ovaries, were atrophic and the vaginae remained always closed (Group III). Eight others ran very irregular cycles and showed changes resembling one or the other of the two groups last mentioned.

It is difficult to see how the failure to show estrual behavior could have been due to lack of ovarian hormones in the 23 animals which had normal ovaries. Furthermore, spayed guinea pigs with similar lesions failed to respond to injections of ovarian hormones in amounts greatly in excess of those sufficient to induce estrual behavior in the same spayed animals before the hypothalamic lesions had been placed (3). But before adopting the alternative hypothesis that the lesions in the brain were in themselves responsible for the behavioral abnormalities, it seemed desirable to make further attempts to overcome this abnormality by the injection of hormones. For this purpose a new series of animals was prepared.

METHODS

The animals used in these experiments were young adult female guinea pigs weighing between 400 and 600 gm. Electrolytic lesions were placed in the brain with the aid of the Horsley-Clarke instrument by passing a direct current of 3 ma. for 30 seconds through a unipolar electrode. The coordinates were the same as those used in the earlier experiments already mentioned, the bare points of the otherwise insulated electrodes being placed as nearly as possible 1 mm. above the ventral surface of the brain at the level of the posterior border of the optic chiasma. Three lesions were made in each animal, one in the midline and one a millimeter on either side of the midline. In some animals control lesions were made in the same frontal plane and in the same relation to the midline, but at a level 6 mm. above the ventral surface of the brain and therefore above the level of the hypothalamus.

Some of the animals which received the lesions had been previously spayed, the others were normal. The latter were examined daily as to the condition of the external genitalia and vaginal membrane, and they were separated into 3 groups on the same

Received for publication November 7, 1940.

[1] Aided by grants from the Committee for Research in Problems of Sex of the National Research Council, and from the Rockefeller Foundation.

561

basis as the original 50. The spayed animals had been brought into heat at least once before the placing of the lesions by properly timed injections of estrogen and proges' terone (4). The estrogen[2] was injected in 4 doses of 25 I.U. on hours 0, 24, 48, and 60. The progesterone[3] (0.2 I.U.) was injected on hour 72; heat usually developed in from 4 to 6 hours and lasted 3 to 9 hours. The hormones, dissolved in oil, were injected subcutaneously.

For the determination of heat we have relied upon the methods described (5) modified to suit our particular purposes. For the elicitation of the estrual reflex we have utilized manual stimulation and the stimulation supplied by normal mature males. When attempts were being made to induce heat in the spayed females, the animals were stimulated at hourly intervals for a period of at least 12 hours following the injection of the progesterone. When attempts were being made to bring sterile unspayed animals into heat through the injection of ovarian or pituitary hormones the animals were kept under continuous observation from the time of opening of the vaginal membrane until at least 48 hours after the opening. The pituitary hormones used were follicle stimulating and luteinizing fractions prepared according to the method of Fevold (6) and supplied to us through the courtesy of Dr. W. H. Windle. A centrifuged suspension of fresh ground sheep anterior pituitary was also used in some of the experiments (7). The glands[4] were frozen at the time of removal and kept in a frozen condition until immediately before use. This saline extract was tested on immature female white rats. Four daily injections of 100 mg. eq. of the fresh glands caused canalization of the vaginae and 178 to 229% increase in the weight of the ovaries over the weight of the control ovaries.

RESULTS

None of the animals with hypothalamic lesions came into heat. Some of the un' spayed animals displayed normal cyclic changes in the genitalia (Group I); in others the genitalia hypertrophied and the vaginae remained continuously open (Group II); in still others the genitalia underwent atrophy and the vaginae remained continuously closed (Group III).

Ovarian hormones. These have been used on representatives of each of the 3 types. Five animals of Group I which ran regular cycles so far as the physical changes were concerned, were given subcutaneous injections from 12.5 to 25 I.U. of estrogen daily, beginning 4 days prior to the day upon which the vaginal membrane was ex' pected to rupture. On the first morning after the membrane opened, progesterone (0.2 I.U.) was injected and the animals were tested each hour for at least 48 hours thereafter. None of these animals showed either proestrual or estrual behavior. Since the ovaries in animals of this type have normally developed follicles and corpora lutea (2) it seems probable that they secrete ovarian hormones and that the injected hormones would cause a high concentration of these substances in the blood. The normal ovarian function also speaks in favor of a fairly normal pituitary. It would seem probable therefore, that, following the injections, these animals had in their blood an excess of ovarian hormones and a fairly normal concentration of pituitary hormones. In spite of this they failed to come into heat.

Two animals of Group III whose vaginal membranes had been closed for over a month, and whose external genitalia showed signs of atrophy, were injected with 4 doses totaling 100 I.U. of estrogen followed on the fourth day by 0.2 I.U. of progester' one. The vaginal membranes opened for 2 and 3 days, respectively, but neither of the animals showed proestrual or estrual behavior.

---

[2] Theelin furnished through the courtesy of Dr. Oliver Kamm, Park', Davis & Co.
[3] Prolution furnished through the courtesy of Dr. Erwin Schwenk, Schering Corporation.
[4] Kindly supplied by Dr. Edwin Pike, Armour & Co.

Eight animals of Group II with hypertrophied external genitalia and continuously open vaginae were given injections of progesterone, and in 4 the treatment was repeated a second time. In each case the vaginal orifices closed or became reduced to small slits within 4 days, then reopened after 3 to 8 days, but the animals showed neither proestrual nor estrual behavior. Three animals were killed 2, 3 and 4 days respectively after the vaginae closed following the second injection of progesterone. Luteinization had not occurred in the ovaries of any of these animals. These ovaries were not different from those previously described for this type of animal (2).

Pituitary hormones. It might be assumed that the combined action of some pituitary hormone and the ovarian hormones was a necessary condition for the mating reactions and that by reason of the proximity of the lesion to the hypophysis this

TABLE I

No.	Operation to progest.	Operation to estrus	Interval	Second trial	
				Latent	Duration
	hr.	hr.	days	hr.	hr.
		High Lesions			
1	9	14.0			
2	12	*	10	3.0	6.0
3	12	16.0	10	4.0	6.0
4	12	17.0			
5	12	*	8	5.0	3.5
6	12	*	8	5.0	5.0
		Low Lesions			
7	12	*	10	*	0
8	12	*	9	*	0
9	12	*	10	*	0
10	12	*			
11	12	*	8	*	0
12	12	*	8	*	0

* Failed to come into heat.

pituitary hormone was deficient in amount even in those animals with hypothalamic lesions which showed regular cyclic changes in the genitalia. If the lesions were placed shortly before the animal was due to come into heat the chance for the disappearance of this hypothetical hormone from the blood stream would be decreased. This line of reasoning was tested on a series of 12 spayed animals. Each animal was prepared for the elicitation of the estrual reflexes by the proper injection of estrogen. The animals were then lightly anesthetized with nembutal. In 6, lesions were placed in the usual position 1 mm. dorsal to the ventral surface of the brain, and in 6 others, 6 mm. dorsal to the ventral surface of the brain. The 12 animals were then injected with progesterone, at from 9 to 12 hours after the placing of the brain lesions. As far as could be determined, the general depression resulting from the high lesions was similar in duration to that resulting from the low lesions. The results of these experiments are shown in table 1. None of the animals with the low lesions, i.e., lesions in the hypothalamus 1 mm. above the ventral surface of the brain, came into heat during the period immediately following the operation nor after a second trial with the hormones several days later. On the other hand, 3 of the 6 animals with the high lesions 6 mm. above the ventral surface of the brain and above the level of the hypothalamus showed estrual behavior 14, 16 and 17 hours respectively after the placing of the lesions. The 3 refractory animals were tested again with the hormones some days later, and showed heat of normal duration after the usual latent period.

In so far as one may assume that any hypothetical hypophysial hormone would not have been likely to have disappeared from the blood in 16 or 17 hours following a hypothalamic lesion, the results of this experiment seem to indicate that the failure to show heat is not due to a deficiency of such a hormone. The experiments also show that the hypothalamic lesions have a specific effect not produced by similar lesions 5 mm. higher in the brain.

Luteinizing hormone was given to 5 of the animals of Group II with continuously open vaginae. Each received a single dose of 50 to 100 mg. eq. intraperitoneally. Following the injection the animals were observed every hour until the time of the closure of the vaginal membrane, and no proestrual or estrual behavior was observed at any time. The vaginal membranes closed within 2 days after injection and remained closed for the duration of a normal cycle. Two of these animals were killed for a histological study of their ovaries. It was found that the hormone induced luteinization in the ovaries, although the remnants of the ova and zona pellucida are discernible in some of the luteinized follicles. Two other animals subsequently received injections of chorionic gonadotropic hormone. Although the vaginal membranes closed for the duration of a normal cycle, the animals showed no signs of heat.

Combinations of pituitary and ovarian hormones. Luteinizing hormone, estrogen and progesterone were given to 6 animals, 2 of which belonged to Group II with continuously open vaginae. Three of the 6 were spayed animals with hypothalamic lesions; and one, a spayed animal without a brain lesion, served as a control. They were each given the customary treatment with estrogen. At the time of the last injection of estrogen, at the time progesterone was injected, and again at the end of the normal 4 to 5-hour latent period following the progesterone, luteinizing hormone (0.1 cc. or 82.3 mg. eq.) was supplied by intraperitoneal injection. Hourly observations were made for the following 36 hours, but no indications of estrual or proestrual activity were seen in any of the 5 animals with hypothalamic lesions. The control animal came into heat 5.5 hours after the injection of progesterone, showed estrous reflexes and copulated with a normal male.

Follicle stimulating and luteinizing hormones were given along with estrogen and progesterone to 6 animals, 3 of which belonged to Group I with regular cycles. One belonged to Group II with continuously open vagina, one was a spayed animal with hypothalamic lesions, and another a spayed animal with no brain lesion. Each animal was given a subcutaneous injection of 0.5 gm. eq. of FSH every other day for 6 days. Four subcutaneous injections of estrogen were made at the usual intervals. The pituitary and ovarian hormone injections were so timed in the animals of Group I that the vaginal membranes ruptured after the usual interval following their last openings. At the time of the last injection of estrogen luteinizing hormone was supplied by intraperitoneal injection and another dose was given along with the progesterone. None of the animals with the hypothalamic lesions showed either proestrual or estrual behavior; but the control animal showed estrual reflexes beginning 7.0 hours after the injection of progesterone and lasting for 3.5 hours.

Unfractionated pituitary extract, estrogen, and progesterone were administered to 5 spayed animals, 4 of which had lesions in the hypothalamus and one of which served as a control. Each received intraperitoneally 0.5 gm. eq. of a centrifuged saline extract of fresh anterior pituitary daily for 12 days. During the last 4 days, subcutaneous injections of estrogen and progesterone were injected in the usual manner. None of the experimental animals so treated showed either proestrual or estrual behavior. The control animal came into heat 6.5 hours following the injection of progesterone, and the heat lasted for 3 hours.

DISCUSSION

There is strong evidence that the site of the erogenous action of the gonadal hormones is in the central nervous system, and not in the somatic and visceral structures which are also affected by these substances (8–12). Bard (13) states that it appears reasonable to assume "that an important central mechanism for estrual behavior one which is specifically activated by estrin must be situated somewhere in the territory comprising the preoptic and septal areas, a part of the thalamus, the subthalamus, the hypothalamus and the upper portions of the midbrain." We have shown that properly placed lesions in the hypothalamus abolish estrual behavior in the guinea pig. Because of the close topographic relation of the hypothalamus with the hypophysis and the dependence of the ovaries on the hypophysis it seemed possible that the behavioral abnormality might be dependent on a deficiency of some hypophysial or ovarian hormone. But this idea has received no support from the experiments reported in this paper. Guinea pigs with hypothalamic lesions which have received hypophysial and ovarian hormones in quantities sufficient to overcome any such deficit have not come into heat. It seems therefore highly probable that the lesions seriously damaged the central mechanism for estrual behavior which, according to Bard, is specifically activated by estrin.

Dempsey and Rioch (14) place the integrating mechanism for estrual behavior in the posterior part of the hypothalamus at the level of the mammillary bodies and Bard (13) presented evidence that in the cat it may be situated in the rostral part of the mesencephalon. The lesions in our guinea pigs have been in the anterior part of the hypothalamus. Further research will be required to determine the exact location and boundaries of this central mechanism.

It is important to emphasize, however, that our guinea pigs with hypothalamic lesions were in excellent health and could be kept under observation as long as desired. Some of the animals in the first group of 50 were observed for 6 or 7 months.

CONCLUSIONS

Guinea pigs with properly placed lesions in the anterior part of the hypothalamus failed to show proestrual or estrual behavior and such behavior could not be induced in them by the injection of ovarian hormones alone or in combination with pituitary hormones. The failure of such hormones to bring these animals into heat increases the probability that the behavioral difficulty is due to impairment of a central mechanism for estrual behavior.

REFERENCES

1. Dey, F. L., C. Fisher, C. M. Berry and S. W. Ranson: Am. J. Physiol. 129: 39. 1940.
2. Dey, F. L.: Am. J. Anat. In press.
3. Brookhart, J. M., F. L. Dey and S. W. Ranson: Proc. Soc. Exper. Biol. & Med. 44: 61. 1940.
4. Collins, V. J., J. L. Boling, E. W. Dempsey and W. C. Young: Endocrinology 23: 188. 1938.
5. Young, W. C., E. W. Dempsey, C. W. Haquist and J. L. Boling: J. Lab. & Clin. Med. 23: 300. 1937.
6. Fevold, H. L.: Endocrinology 24: 435. 1939.
7. Schockaert, J. A.: Anat. Rec. 50: 381. 1931.
8. Ball, J.: J. Comp Psychol. 18: 419. 1934.
9. Bard, P.: Am. J. Physiol. 113: 5. 1935.
10. Brooks, C. McC.: Am. J. Physiol. 120: 544. 1937.
11. Root, W. S., and P. Bard: Am. J. Physiol. 119: 392. 1937.
12. Brooks, C. M.: Am. J. Physiol. 121: 157. 1938.
13. Bard, P.: Res. Publ. Ass. Nerv. Ment. Dis. 20: 551. 1940.
14. Dempsey, E. W., and D. M. Rioch: J. Neurophysiol. 2: 9. 1939.

Reprinted from *Psychosomat. Med.*, **6**, 40–55 (1944)

EFFECTS OF INJURY TO THE CEREBRAL CORTEX UPON SEXUALLY-RECEPTIVE BEHAVIOR IN THE FEMALE RAT[1]

FRANK A. BEACH[*]

26

INTRODUCTION

Previous studies have revealed that females of several mammalian species display sexually-receptive behavior after extensive injury to the forebrain. Typical estrous behavior has been observed in female cats following complete removal of the neocortex combined with injury to the rhinencephalon, the basal ganglia and the sensory thalamus (2, 3, 4, 5, 6). Normal mating behavior is shown by female guinea pigs following removal of neocortex, caudate-putamen, hippocampus, septal nuclei and other portions of the forebrain (13). Unilateral hemidecerebration (removal of all central nervous tissue anterior to the mesencephalon), or total decortication fails to eliminate normal mating responses and postcopulatory ovulation in the female rabbit (11). Female rats suffering the loss of more than half of the cerebral cortex are capable of mating with and being impregnated by the male (7, 19). Destruction of nearly all of the neopallium does not disturb the vaginal estrous cycle in the rat, and pseudopregnancy follows stimulation of the cervix in such animals as it does in intact females (12).

Published reports of sexual behavior in brain-operated female rats have been chiefly of the "presence or absence" type (12). There are many kinds of reactions which, although they survive extensive brain injury, are more difficult to elicit after partial decortication (e.g., copulatory responses in the male rat (8)). Still other behavior patterns may function fairly effectively after cortical invasion although close scrutiny reveals that brain injury tends to disrupt the serial nature of the discrete responses which are smoothly integrated in the performance of the unoperated animal. Indeed certain segments of a patterned response may be eliminated by injury to the neopallium without abolishing other elements in

the pattern or seriously reducing its biological effectiveness (e.g., maternal behavior in the male rat (7)).

No investigator has reported intensive, quantitative tests of sexual behavior in female rats before and after removal of the cerebral cortex. The experiment which is to be described was designed to reveal the effects of partial and complete decortication upon each of the several discrete reactions that are coordinated in a pattern commonly termed "sexually receptive behavior."

METHODS

In planning the present work it was anticipated that brain injury might in some way effect the functional activity of the pituitary gland. It is well-known that pituitary secretions are responsible for the elaboration of certain hormones by the ovaries; and that under normal conditions ovarian products in turn are essential to the appearance of sexually receptive behavior. obviate the possibility that cortical lesions might affect mating behavior through interference with the function of the endocrine system, estrus was induced by administration of the appropriate hormone preparation to ovariectomized females. Earlier studies from our laboratory have shown that under the influence of the hormone treatment used in the present experiment castrated females display sexually receptive behavior which is indistinguishable from that shown by intact estrous females (10).

Twenty virgin female rats raised in segregation were ovariectomized at 3 to 4 months of age. To induce estrus in the spayed animals an intramuscular injection of 500 R.U. of estradiol benzoate was followed after a 48-hour interval by an intramuscular injection of 0.1 mg. of progesterone.[2] Sex tests were conducted 16 to 18 hours after progesterone administration. At least two weeks elapsed between tests, and in this length of time the vaginal smear returned to typical diestrous condition, containing only leukocytes.

Beginning 1 to 3 weeks after castration each female was given 3 preoperative tests in which sexual receptivity was measured quantitatively. A vaginal smear was taken before the test. After the female had been

[1] The experiment herein reported was supported by a grant from the Committee for Research in Problems of Sex, National Research Council.

[*] Department of Animal Behavior, American Museum of Natural History, New York, N. Y.

[2] Estradiol benzoate (Progynon-B) was supplied by The Schering Corporation, Bloomfield, N. J. Progesterone (Lutocylin) was made available by The Ciba Pharmaceutical Products Co., Inc., Summit, N.

allowed a 5-minute period of adaptation to the experimental cage (circular, 30″ in diameter) a sexually vigorous male was introduced. If the original male ejaculated before the end of the test a second male was immediately substituted so that throughout the period of observation the female was continuously subjected to the attentions of an active stimulus male.[3] Male and female remained together in the cage for 5 minutes and records made during this period included the number of times the female was mounted by the male, the number of times the female exhibited lordosis (concave arching of the back facilitating intromission), and the number of hopping responses and ear-vibration responses executed by the female. These 3 reactions: hopping, ear-wiggling, and responding to the male's mounting with the assumption of lordosis constitute, by definition, sexual receptivity in our female rats. Except in very rare instances this behavior can be elicited only when the female is in vaginal estrus, and its manifestation may be regarded as a concomitant of the late follicular or early luteal phase of the ovulatory cycle.

In addition to noting receptive reactions the experimenter recorded observations of the character and amount of resistance which the female exhibited. With the exception of purely passive failure to show lordosis, the only form of resistance consisted of kicking backward with the hind feet when the male attempted to mount.

After the third preoperative mating test the females were subjected to unilateral hemidecortication. Cortical tissue was removed by suction while the animal was under deep surgical anaesthesia.

When the animals had approached or exceeded preoperative weight and appeared to be in good health a second series of 3 sex tests was conducted employing the same techniques used in the preoperative series. At the conclusion of these tests 9 females were subjected to a second operation in which an attempt was made to destroy all of the remaining cortex. Following recovery these animals were observed in 3 additional tests.[4]

At the conclusion of the experiment operated brains

[3] Vasectomized males were employed to avoid impregnation of experimental females. Since it was necessary to elicit copulatory attempts from these males even when they were offered nonreceptive females, all males were occasionally injected with testosterone propionate in order to maintain their sexual excitability at a high level. The hormone preparation employed was Oreton, supplied by The Schering Corporation, Bloomfield, N. J.

[4] In some instances females suffering very large bilateral cortical lesions did not completely regain preoperative weight. Such cases are noted in the presentation of experimental results.

were sectioned at 50 micra, stained with thionin, and reconstructed by the method of Lashley (17).

RESULTS

Brain Lesions

Reconstructions of all cortical lesions inflicted in this study are shown in figures I and II. In the first operation the entire neocortex was removed from one hemisphere in the case of 14 females. The fifteenth animal was deprived of 47 per cent of the cortex unilaterally; and the remaining 5 cases sustained unilateral lesions ranging from 29 to 39 per cent of the entire neopallium. Incidental invasion of noncortical structures is described below in connection with other results. Although there was some variation in the extent and locus of the cortical injury there was no relation between these factors and postoperative changes in behavior. Accordingly in the following presentation of results all operated animals have been treated as a homogeneous group.

Effects of Unilateral Cortical Lesions upon Feminine Mating Reactions

Effects upon the lordosis response. In this report the readiness with which a female displayed lordosis is expressed in terms of the number of lordosis responses divided by the number of times the female was mounted. The figure resulting from this procedure is multiplied by 100 and is called the *Copulatory Quotient*, abbreviated to CQ. Thus, $\frac{\text{total lordoses}}{\text{total mounts}} \times 100 = CQ$. The CQ may range from 0 to above 100. Some receptive females show lordosis every time they are mounted, and in addition they may exhibit this reaction in response to other types of physical contact with the male (such as that resulting from the male's exploration of the vaginal region).

Results summarized in Table 1 indicate the effects of removal of 27 to 50 per cent of the neopallium upon the CQ of each of 20 females. Many animals showed some test-to-test variability both before and after operation. Furthermore, if the average CQ for each rat in 3 preoperative tests is compared with the average for the same animal in the 3 postoperative tests, some changes are evident. However, such variation occurred in different directions for different individuals, with the result that after partial decortication 9 females exhibited increase in the CQ, 9 evidenced a decrease in this measure, and in 2 cases there was

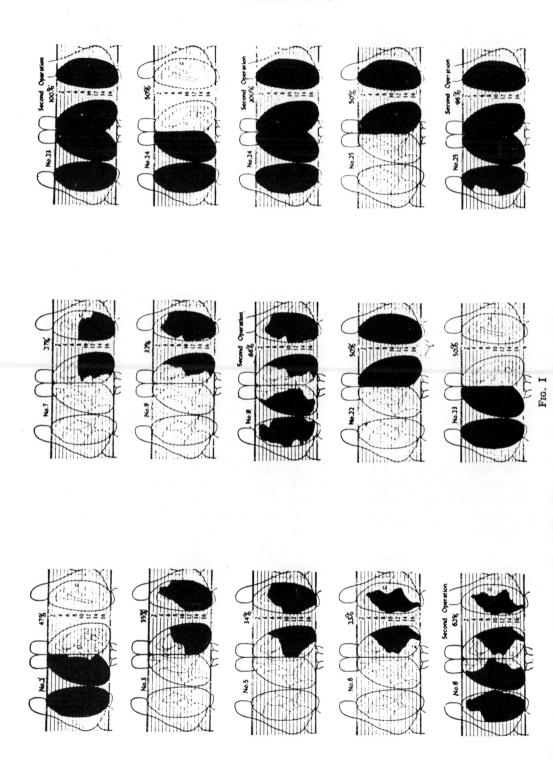

FIG. I

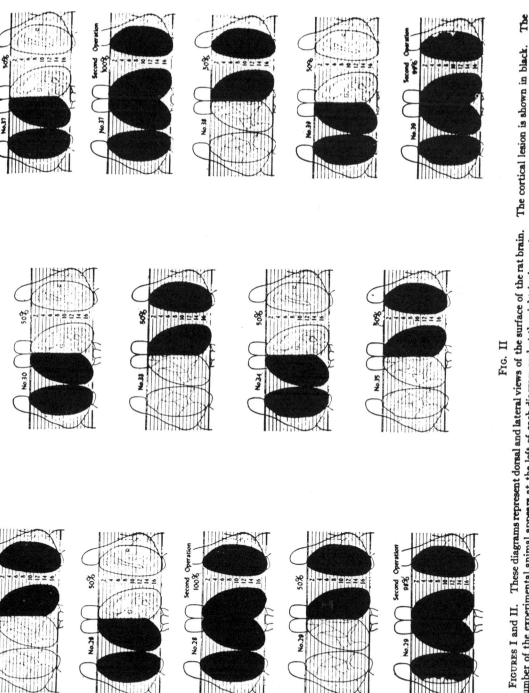

FIG. II

FIGURES I and II. These diagrams represent dorsal and lateral views of the surface of the rat brain. The cortical lesion is shown in black. The number of the experimental animal appears at the left of each diagram and at the right is shown the percentage of neocortex removed.

215

no change. Values shown in Table 1 reveal that before brain operation the average CQ for 20 females was 82; while after operation the group average was 84. The average range for the group (highest CQ to lowest CQ) did not change significantly after operation.

It is to be noted that although 14 animals suffered loss of an entire hemisphere, while 6 cases sustained smaller lesions, there is no indication of any relation between lesion magnitude and postoperative changes in the CQ.

cases in which the response survived. However, before operation 233 ear-wiggle reactions were shown by 20 rats; and after operation the reaction was displayed 415 times by 17 individuals. Seventeen females executed an average of 5.2 ear-wiggles per test before operation, and 8.0 ear-wiggles per test after operation.

Table 2 presents pre- and postoperative records of the ear-wiggle response for all females. Following brain operation the average number of ear-

TABLE 1

Effects of unilateral cortical lesions upon the estrous female rat's tendency to show lordosis

Rat	Per cent of neocortex removed	Average CQ for 3 tests		Highest CQ in any test		Lowest CQ in any test	
		Normal	Operated	Normal	Operated	Normal	Operated
2	47	84	79	125	100	40	56
3	35	67	48	100	100	33	0
5	34	58	97	73	100	43	90
6	39	100	97	100	100	100	92
7	27	105	56	111	80	100	14
9	32	88	93	114	100	62	80
22	50	95	87	100	100	86	92
23	50	82	75	100	100	70	37
24	50	100	100	100	100	100	100
25	50	41	48	86	64	0	37
26	50	84	100	100	100	54	100
28	50	100	100	100	100	100	100
29	50	79	63	100	94	54	6
30	50	71	91	100	100	39	84
33	50	76	82	100	95	61	65
34	50	53	98	100	100	10	94
35	50	94	78	100	96	83	46
37	50	83	81	100	89	67	68
38	50	90	102	100	112	70	95
39	50	90	100	100	100	80	100
Mean	46	82	84	100	96	63	68

TABLE 2

Effects of unilateral cortical lesions upon the ear-wiggle response in estrous female rats

Rat	Average number of ear-wiggles per positive test*		Highest number of ear-wiggles in any test		Lowest number of ear-wiggles in any test	
	Normal	Operated	Normal	Operated	Normal	Operated
2	2.0	0	2	0	0	0
3	1.0	8.0	1	9	0	0
5	5.5	7.3	8	8	3	6
6	5.0	5.0	7	7	0	0
7	3.0	2.0	4	6	2	0
9	2.5	10.0	4	13	1	7
22	7.3	5.0	17	5	1	0
23	5.0	1.0	5	1	0	0
24	10.1	3.7	14	8	4	1
25	2.0	0	2	0	0	0
26	2.5	11.5	4	4	0	0
28	12.7	17.0	18	21	8	12
29	4.5	8.0	7	8	0	0
30	4.0	3.7	4	5	0	3
33	4.7	19.7	7	22	1	18
34	3.5	16.7	6	22	0	10
35	5.0	0	6	0	3	0
37	5.7	18.0	10	22	0	13
38	6.0	19.0	12	28	1	11
39	3.3	1.0	5	1	2	0
Mean	4.8	7.8	7.1	9.5	1.3	4.0

* Including only those tests in which the response occurred at least once.

These data suggest that partial or complete unilateral hemidecortication has no consistent, predictable effect upon the tendency of the female rat in estrous to display lordosis when mounted by the male.

Effects upon the ear-wiggle response. Ear wiggling was shown by each of the 20 females before operation and by 17 cases after partial decortication. This behavior occurred in 72 per cent of the total tests before operation, and was observed in 74 per cent of the postoperative tests for the 17

wiggles per positive test increased in 10 animals, decreased in 9, and was unchanged in 1. However, the 10 females whose scores increased showed an average postoperative gain of 8.7 responses per test; while the 9 rats with lowered scores after operation decreased by an average of only 2.8 ear-wiggles per positive test.

The effects of partial decortication thus appear to have been two-fold. The ear-wiggle response was eliminated in 3 females. The 17 cases in which it survived brain injury displayed the reaction in the same proportion of their tests before and after operation; but during those postoperative tests in which ear-wiggling did occur its average frequency for the group was noticeably increased.

Effects upon the hopping response. Each of the 20 females showed hopping behavior in at least 1 of 3 preoperative tests. This response survived brain operation in 19 animals. A total of 271 hops occurred in 46 preoperative tests; and 464 hops were recorded in 48 postoperative tests. The 19 females which reacted both before and after operation hopped an average of 5.3 times per test preoperatively, and 8.3 times per postoperative test.

Scoring each animal on the basis of average number of hops per positive test before and after cortical injury reveals that 13 cases showed an increase in this behavior, while 7 females exhibited a decrease. As in the case of the ear-wiggle response, the increases in the hopping reaction were of greater magnitude than were the decreases. Thirteen rats exhibited an average postoperative increase of 6.8 hops per test; and the scores of 7 cases decreased by an average of 3.5 hops per test.

Data presented in Table 3 indicate the direction and magnitude of changes in the behavior of individual animals.

Unilateral cortical invasion is thus seen to have eliminated hopping behavior in one individual, decreased its frequency in 6 others, and increased its frequency in the remaining 13 animals.

Effects upon the back-kicking response. Females may resist the male's attempts to explore the vaginal region or to mount by kicking backward with the hind feet. This behavior is shown occasionally by many of the estrous females from our colony. During positive tests the female may back-kick at one moment and permit the male to copulate immediately thereafter.

Before partial decortication each of 20 rats in this experiment showed the back-kicking reaction at least once. After operation the response was exhibited by 16 cases. Those females that back-kicked both before and after operation showed such behavior in 86 per cent of the preoperative and 64 per cent of the postoperative tests. The 16 cases in which the response occurred after operation exhibited an average of 3.7 back-kicks per positive test preoperatively, and 2.5 back-kicks per positive test postoperatively.

There are thus 3 points in evidence to indicate that partial decortication tends to reduce back-

kicking. This behavior was eliminated in 4 of the 20 cases after the operation. For the 16 rats in which the response survived, the proportion of tests in which such behavior occurred was reduced after brain injury. During those postoperative tests in which back-kicking appeared the reaction was executed less frequently than preoperatively.

TABLE 3

Effects of unilateral cortical lesions upon the hopping response in estrous female rats

Rat	Average number of hops per positive test*		Highest number of hops in any test		Lowest number of hops in any test	
	Normal	Operated	Normal	Operated	Normal	Operated
2	2.0	4.0	2	8	0	1
3	2.5	1.5	3	2	2	0
5	3.0	6.3	5	10	1	3
6	4.0	2.7	6	1	2	5
7	3.7	3.0	4	3	3	0
9	2.5	3.3	3	4	2	3
22	14.0	5.5	29	10	2	0
23	8.0	4.7	15	7	0	2
24	10.7	6.7	15	11	3	3
25	4.0	3.0	9	3	0	0
26	4.7	17.0	7	4	1	0
28	11.7	18.7	21	23	6	15
29	5.0	3.0	5	3	0	0
30	6.0	6.7	6	5	0	3
33	4.0	14.3	8	20	1	11
34	4.0	22.3	7	36	0	9
35	6.7	0	13	0	3	0
37	1.0	2.0	1	2	0	0
38	4.0	20.7	6	28	1	8
39	6.0	18.3	11	1	2	0
Mean	5.1	8.3	8.8	9.0	1.4	3.1

* Including only those tests in which the response occurred at least once.

Effects of Complete Decortication upon Feminine Mating Reactions

Bilateral cortical lesions were inflicted in the case of 9 females. Seven of these animals were deprived of 97 to 100 per cent of the neopallium. In considering the effects of bilateral injury we have included only those data gathered from observations of these females. They will be referred to as "complete decorticates" although, as can be seen from reconstructions presented in Figures I and II small neocortical remnants remained in 3 cases.

Effects upon the lordosis response. Table 4 reveals some of the effects of partial and of complete decortication upon the estrous female's tendency to exhibit lordosis. The average CQ (Copulatory Quotient) for the group of 7 females was remarkably constant throughout the experiment, varying less than 1 point after each operation.

The rank-order of the 7 individuals was likewise relatively unaffected by cortical loss. Female 25 ranked seventh in the preoperative tests (lowest average CQ). After removal of the neocortex from one hemisphere No. 25 still ranked seventh; and complete decortication did not alter her rank. Animal 29 ranked sixth as a normal, and retained this rank after each operation. Number 23 ranked fifth, and 37 fourth throughout all 3 test series.

TABLE 4

Effects of hemi- and complete decortication upon the estrous female rat's tendency to show lordosis

Rat	Average CQ in 3 tests as:			Range of CQ in 3 tests as:		
	Normal	Hemidecorticate	Decorticate	Normal	Hemidecorticate	Decorticate
28	100	100	117	100–100	100–100	100–150
24	100	100	89	100–100	100–100	53–113
39	90	100	103	80–100	100–100	100–108
37	83	81	85	67–100	68– 89	75–100
23	82	75	75	70–100	37–100	38–100
29	79	63	55	54–100	6– 94	31–100
25	41	48	48	0– 86	37– 64	0– 83
Mean	82.1	82.4	81.7	67– 98	64– 93	57–108
				D = 31	D = 29	D = 51

The remaining 3 animals maintained approximately the same rank orders after each operation. Comparing preoperative CQ with CQ after total decortication we find the rank-order coefficient of correlation to be +.94.

Group means presented in Table 4 indicate that complete decortication had little effect upon the lordosis response as measured by the average CQ; but close inspection of daily test results reveals that in several animals loss of the neopallium was followed by definite peculiarities in the female's tendency to exhibit lordosis. These postoperative changes are difficult to describe quantitatively but seem to have sufficient importance to be reported in qualitative terms.

Rat No. 23 did not exhibit an increase in CQ after decortication. However, as a complete decorticate this animal showed lordosis when a pipette was inserted in the vagina as the vaginal smear was being taken. Although she was smeared before all pre- and postoperative tests, female No. 23 never responded to the pipette in this fashion before complete decortication.

In all tests as a normal and as a hemidecorticate rat No. 24 showed a CQ of 100. After complete decortication her CQ varied from 53 to 113. In one test this female exhibited lordosis when the male nosed at her vaginal region without mounting. At the time of another test female No. 24 displayed lordosis when the pipette was introduced into the vagina to secure the vaginal smear; but immediately thereafter when placed with a male she exhibited a low CQ of 53. In this particular test rat No. 24 was very slow to show lordosis and kept moving about the cage, carrying the male upon her back when he mounted. When at length the male did succeed in copulating, the female held the lordosis position for many seconds after the male had dismounted. Before complete decortication rat No. 24 never showed lordosis when the smear was taken, never failed to give lordosis when mounted by the male, and never maintained lordosis after the male had dismounted.

After complete decortication female No. 28 exhibited lordosis when the male's paws touched her back,—a reaction which had not occurred in previous tests. In addition this animal stiffened into the lordosis position when picked up in the experimenter's hand. If she was placed gently upon the table the female remained in the lordosis pose for several seconds after the hand was withdrawn. When the male dismounted after copulation female No. 28 held lordosis for several seconds. In one test lordosis was maintained without interruption while the male copulated 4 times. None of the reactions described above were shown by this individual prior to removal of the neocortex.

Female No. 29 showed a reduction in the CQ following complete decortication. However, as a complete decorticate this animal exhibited lordosis when the pipette was inserted in the vagina and, upon occasion, when the male investigated without mounting.

As complete decorticates female No. 37 displayed lordosis when the male explored; and rat No. 39 responded to the insertion of the pipette with the assumption of lordosis.

The protocol excerpts cited above suggest that in some postoperative tests 6 of the 7 decorticated females gave evidence of increase in the readiness to display lordosis.[5] Four animals showed lordosis when the pipette was inserted in the vagina. Four exhibited this response when the male explored without mounting. One rat stiffened into lordosis in the experimenter's hand. Such changes in behavior might reasonably be expected to result in an increase in the CQ, but mean values presented in Table 4 indicate that no such increase occurred. Some insight into the reasons for this apparent contradiction may be derived from a comparison of the average range of CQ scores before and after decortication.

Table 4 includes the lowest and highest CQ recorded for each female in 3 tests before and 3 tests after operation. Preoperatively the average of the lowest CQs for the group was 67; while the mean of the highest CQs was 98. The difference between these two values, that is the *average range*, was thus 31 points. Hemidecortication had no pronounced effect upon this measure. However, after complete decortication the average of the lowest CQs was 57 (10 points below preoperative levels); and the average of the highest CQs was 108 (10 points above preoperative levels). The average range after complete decortication was therefore 51 points,—20 points greater than the preoperative average range.

Loss of the cerebral cortex is thus seen to have increased the variability of the female's tendency to show lordosis. Although in one test after operation a female might exhibit a CQ higher than any shown preoperatively, the same animal in a second postoperative test was not unlikely to display a lower CQ than had been recorded for her before operation. This lack of consistency extended to behavior displayed at different times in the same test. For example, female No. 24 exhibited lordosis when a pipette was inserted in the vagina. Placed with a male immediately afterward she was mounted and palpated 17 times, but lordosis occurred only 8 times ($CQ = 53$).

Effects upon general activity. The effects of decortication upon general activity will be considered at this point because postoperative changes in activity frequently exerted a marked effect upon the CQ. Several experimenters have described increases in "spontaneous" activity in the rat

[5] It is of interest to note that the only animal which failed to show any increase in the tendency to exhibit lordosis was female No. 25 who displayed the lowest CQ of the group in all pre- and postoperative tests.

following injury to various portions of the cerebrum (9, 16, 18).

Four of the 7 decorticated females whose sexual behavior is described here exhibited pronounced increases in activity in postoperative tests. This hyperactivity took the form of running rapidly around and around the perimeter of the circular testing cage. Often a decorticate female made as many as 25 complete circuits of the cage without stopping (approximately 65 yards). After complete decortication this behavior was exhibited by females No. 23, 25 and 37 in all 3 tests, and by No. 39 in 2 of the 3 tests.

Normal female rats in estrous are usually quite active, frequently hopping away at the male's approach. The male appears to be stimulated by this purely temporary retreat, and follows the female closely. After a few hops the normal, estrous female stands still, allows the pursuing male to mount and responds with the assumption of lordosis. The decorticated females showed no such cooperative behavior. As hyperactive females displayed the continuous running the males usually pursued vigorously, and frequently clasped the female momentarily. Under such conditions the hyperactive rats showed no response to the male's clasp, but persisted in their circumlocutions of the cage, often dragging the clasping male for some distance before he was shaken loose. Some males became quite adept at maintaining the clasp and at the same time keeping up with the female by prodigous hops of the hind legs. Even this exhausting persistence failed to elicit lordosis on the part of the female.

The impression gained from repeated observations of this behavior was one of a new and impelling drive superimposed over an original one. The tendency to show lordosis did not seem to have been eliminated nor even necessarily weakened; but appeared to be temporarily, completely overridden by the extremely powerful drive toward hyperactivity. In the course of a single test highly receptive behavior often alternated with periods of continuous running during which the lordosis response was totally suppressed. For example, one female (No. 37) stood stationary in the cage and showed lordosis to every mount while the male copulated 10 times. After the tenth copulation the female suddenly began circling the cage and continued to do so without interruption for the remainder of the observation period. While she was running the male attempted to mount several times and was unable to induce the female to halt or even to depart from her stereotyped path around the circumference of the cage.

Both phases of this female's behavior represent marked deviations from the normal pattern. Intact females, no matter how receptive, never remain in the same spot permitting repeated copulations. Almost invariably after a completed copulation the female moves away. The male then pursues and another copulation ensues. On the other hand, the intact estrous rat does not run around the cage continually, paying no attention to the male's attempts to mount. If a normal female is nonreceptive she may run away from the male, but his mating attempts are actively resisted by back-kicking, rolling over on the back, or even by assuming the offensive and launching a vigorous attack. Furthermore the unoperated, nonreceptive female runs away only when the male approaches, while the decorticates reported herein persisted in their running regardless of the male's behavior.

The hyperactivity shown by 4 decorticate females tended to reduce their CQ scores because they failed to exhibit lordosis when mounted during periods of heightened activity. This reduction in the CQ was however counterbalanced by the increased ease with which lordosis could be elicited when the females were inactive. The net result was therefore an increased variability of the measure in question.

Extreme inactivity was observed to alternate with hyperactivity in the case of female No. 37 described above. Another animal (No. 28) was completely inactive at all times. In every test after decortication this female's only movement about the cage occurred when she was shoved forward by the force of the male's copulatory thrusts.

Effects upon the ear-wiggle response. In Table 5 are presented data indicating the effect of complete decortication upon the estrous female's tendency to display the ear-wiggle response. This reaction was exhibited by all females before and after removal of the cerebral cortex. Ear-wiggling appeared in 71 per cent of the tests before operation, and in 95 per cent of those given after decortication. During those tests in which the response occurred it appeared an average of 6.6 times per test before, and 11.2 times after operation. These scores suggest that the ear-wiggle reaction was evoked more easily in decorticated than in intact females.

Additional evidence of the increased ease with which this item of estrous behavior was elicited after removal of the neocortex is found in the daily records of 2 females. Rat No. 28 showed ear-wiggling postoperatively when picked up by the experimenter. Female No. 29 exhibited this response after operation when a pipette was inserted in the vagina to collect the smear. Before complete decortication neither female showed ear-wiggling save when with the male.

Although females tended to display the ear-wiggle more readily after removal of the cerebral cortex this item of estrous behavior lacked, in some decorticates, normal integration with the other behavioral elements which make up the mating pattern of the unoperated female. Normal, receptive females in our colony display ear-wiggling in close conjunction with the hopping response,— usually showing the ear vibration immediately after a series of several hops. Frequently the female hops, poses and ear-wiggles, and, as the male mounts, the ear movement is continued during the actual copulation. Appearance of the ear-wiggle is an almost infallible sign of high sexual receptivity.[6] Following decortication some females often exhibited the ear-wiggle response at times when other aspects of the receptive pattern were lacking.

In several instances rat No. 25 was mounted by the male, failed to display lordosis, and then executed the ear-wiggle when the male dismounted. This same animal displayed repeated periods of hyperactivity during which the male pursued her around the periphery of the cage. As she continued to run the female ear-wiggled almost continuously. Female No. 29 showed a marked postoperative increase in ear-wiggling (Table 5), but always exhibited this response immediately after the male dismounted rather than before and during the mount as is the case with the normal female. Rat No. 39 displayed no postoperative changes in the ear-wiggle response during tests with the male; but following decortication this rat showed ear-wiggling when a pipette was inserted in the vagina.

These data suggest that the actual frequency of ear-wiggling rose postoperatively but that the concatenation of this response with other elements in the copulatory pattern was in some instances

[6] Although this relationship is very close in animals of our stock it is apparently less so among female rats from some other colonies (1).

disrupted. This indication is strengthened by the fact that whereas in normal females there is a close relationship between ear-wiggling scores and willingness to show lordosis, the degree of this relationship is considerably reduced by decortication. Comparison of the rank order of the 7 females rated

sented in Table 5, the remaining 5 animals tended to exhibit this response in a higher proportion of tests after operation, and to hop somewhat more frequently in positive tests following decortication. However, in 3 cases the average frequency increased and in the remaining 2 it decreased. The

TABLE 5

Effects of complete decortication upon the estrous female rat's tendency to exhibit ear-wiggling, hopping and back-kicking

Response	Per cent of the group showing the response		Rat	Per cent of tests in which the response appeared		Average frequency of response for tests in which it appeared	
	Normal	Decorticate		3 tests as normal	3 tests as decorticate	3 tests as normal	3 tests as decorticate
Ear-wiggling	100	100	23	33	67	5.0	6.5
			24	100	100	10.3	7.0
			25	33	100	2.0	15.3
			28	100	100	12.7	9.0
			29	67	100	4.5	27.3
			37	67	100	8.5	11.0
			39	100	100	3.3	3.3
	Average.............			71	95	6.6	11.2
Hopping	100	71	23	67	100	8.0	4.7
			24	100	100	10.7	7.0
			25	67	100	6.0	13.3
			28	100	0	11.7	0
			29	33	100	5.0	6.3
			37	67	0	1.0	0
			39	100	100	6.0	10.0
	Average for cases that showed the response.................			76	100	6.9	8.3
Back-kicking	100	14	23	67	33	2.5	1.0
			24	67	0	7.5	0
			25	100	0	2.3	0
			28	67	0	3.0	0
			29	100	0	5.3	0
			37	33	0	1.0	0
			39	33	0	4.0	0
	Average for cases that showed the response.................			67	33	3.6	1.0

as to CQ and average number of ear-wiggles per positive test reveals a rank order correlation of $+.79$ between these two variables before operation, and a coefficient of $+.41$ after removal of the cerebral cortex.

Effects upon the hopping response. Removal of the neocortex eliminated hopping in 2 of the 7 females. As shown by the group averages presented in Table 5, the remaining 5 animals tended to exhibit this response in a higher proportion of tests after operation, and to hop somewhat more frequently in positive tests following decortication. However, in 3 cases the average frequency increased and in the remaining 2 it decreased. The average for all 5 females rose only slightly after operation. These data can only be taken to suggest that decortication may totally eliminate hopping or may be followed by moderate increases or decreases in frequency of the response.

Two of the 5 females in which hopping reactions survived loss of the neocortex displayed definite postoperative abnormalities in the execution of the response. As described above the normal female in estrous hops away from the male and then halts and permits copulation. Hopping thus appears as a prelude to coition and in all probability serves as one source of precopulatory stimulation contributing to the male's state of sexual excitement. Two decorticate females tended to display hopping at what might be termed "inappropriate" times; and in their behavior this reaction was not normally integrated with the other elements in the mating pattern. Scores presented in Table 5 show that after decortication rat No. 25 hopped more than twice as frequently as she had before operation. However in postoperative tests this female rarely executed the hopping response until immediately after the male had dismounted. Many copulations were followed by hopping on the part of the female but only infrequently did the behavior precede the male's mount as it commonly does in the case of the intact female. In one test this decorticate rat was mounted by the male 3 times and failed to show lordosis in each instance; but after every unsuccessful copulatory attempt the female hopped as soon as the male released her.

Female No. 29 exhibited postoperative hopping behavior closely comparable to that reported for the animal described above. After decortication No. 29 showed an increase in the frequency of the hopping reaction (Table 5). However, the vast majority of her hopping responses occurred immediately after the male had dismounted.

Continued observation of both of these decorticate females indicated that the frequency of "precopulatory" hopping was markedly decreased, and that the appearance of this reaction usually depended upon strong and direct stimulation which occurred when they were clasped and palpated by the male.

Effects upon the back-kicking response. Values shown in Table 5 reveal that removal of the cerebral cortex had a pronounced tendency to abolish back-kicking. Although this response was exhibited by each of the 7 females prior to operation it was shown by only one animal after decortication, and in that case occurred a single time in but one test. Six of 7 decorticates thus showed no active resistance to the male's sexual advances.

The scores of the seventh animal, female No. 23, reveal a marked postoperative decrease in the frequency of back-kicking. In this rat decortication was followed by the appearance of a peculiar type of aggressive behavior. As soon as the male

was placed in the testing cage the female customarily approached and attempted to bite him. Severe wounds were often inflicted. These attacks were entirely unprovoked and apparently had no relationship to the male's copulatory behavior nor to the female's receptive responses. The female did not attack the male when he was attempting to mate nor immediately after a completed copulation. To all intents and purposes the female's aggression was entirely dissociated from her sexual behavior and the attacks had no apparent effect upon her copulatory scores.

In Table 6 we have summarized the qualitative descriptions of deviations from the normal, sexually receptive pattern which appeared in the behavior of the decorticate females.

Some evidence was obtained indicating that decorticated females were less capable of stimulating the sexual advances of the male 'than were normal females. Males were placed in the cage with a decorticate female and an intact female, both fully in estrous and completely receptive. Under these conditions males showed a marked tendency to discriminate against the brain-operated female and to mate almost exclusively with the intact animal. These incidental observations were not made under controlled conditions and therefore deserve only the briefest mention.

Effects of Cortical Injury upon Body Weight

First operation. In Table 7 are presented data showing the effects of unilateral brain injury upon the body weight of 14 of the 20 females whose behavior has been reported above. Weight records of the remaining 6 cases are not available but there is no reason to assume that they would have differed significantly from those presented here. It will be noted that by the time the first postoperative test was administered 11 of the 14 individuals exceeded their preoperative weight. Three animals still weighed less than they had at the time of operation. The weight of unoperated animals frequently varies as much as 10 per cent and it is therefore safe to assume that when postoperative testing was initiated all 20 animals were in good health insofar as body weight reflects this condition. Between the first and last postoperative test 8 animals increased in weight, 5 decreased, and one showed no change. All of the weight losses were small. The average weight for the group increased after operation and increased again during the testing series. These data seem to indicate that the animals described were in good health following recovery after the first brain operation. Inspection of test results fails to show any difference between the behavioral performance of those animals which lost weight as compared with those which gained.

means in Table 8 indicates that at the time of the second brain operation the average weight was slightly above that recorded prior to the first operation. Following the second operation the average weight decreased somewhat, and at the beginning of the second period of postoperative tests it was

TABLE 6

Abnormalities appearing in the execution of separate elements in the mating behavior of the estrous female rat, or in the integration of these elements into a functional pattern, following removal of the neocortex

Rat	Lordosis response	General activity	Ear-wiggle response	Hopping response
23	Gave lordosis in response to pipette. When male mounted female rolled over and stiffened into lordosis while lying on her side.	Circled cage rapidly in all tests, male following. In one test made 25 circuits without stopping.	No apparent abnormalities.	No apparent abnormalities.
24	Gave lordosis in response to pipette. Held lordosis for several seconds after male dismounted.	Ran about cage dragging male on her back.	No apparent abnormalities.	No apparent abnormalities.
25	(See notes on activity.)	Often failed to stand still when mounted by male. Circled cage repeatedly with male following. 9 circuits without stopping. Repeated 4 times in one test.	Ear-wiggled continuously while circling cage during hyperactive stage. Ear-wiggled immediately *after* male dismounted, even though female may not have shown lordosis.	Hopped immediately *after* male dismounted, even though female may have shown no lordosis.
28	Assumed lordosis in experimenter's hand when picked up and held it when put down. Lordosis whenever male mounted or put paws on her back. Held lordosis for long periods after male dismounted. Often permitted 3 or 4 copulations without relaxing from lordosis assumed upon the first mount.	Remained in spot where she was placed by experimenter. Only movement about cage occurred when female was shoved by male. Frequently received 6 to 10 copulations without moving a step.	Ear-wiggled when picked up by the experimenter.	No apparent abnormalities.
29	Gave lordosis in response to pipette. Showed lordosis when male investigated without mounting.	No apparent abnormalities.	Ear-wiggled *after* male dismounted. This occurred 27 times in one test.	Hopped *after* male dismounted.
37	Gave lordosis when male investigated without mounting.	In one test stood motionless for 10 copulations, then circled cage repeatedly. In all tests circled cage many times without stopping.	No apparent abnormalities.	No apparent abnormalities.
39	Gave lordosis in response to the pipette.	Circled cage repeatedly at a rapid run in 2 tests.	Ear-wiggled in response to the pipette.	No apparent abnormalities.

Second operation. Records of the body weights of 7 of the 9 animals sustaining extensive bilateral brain injury are summarized in Table 8. The weight records of the remaining 2 cases are not available, but there is no reason to assume that they would have differed significantly from those of the other animals. Inspection of the group

still below preoperative levels. During the course of the tests following the second operation the average body weight decreased slightly. At the time of the beginning of the postoperative testing schedule 5 animals were below and 2 were above the weight recorded at the time of the second operation. In 2 individuals the loss exceeded the limits of normal variation, being greater than 10 per cent of the body weight (Nos. 23 and 37). In the remaining 3 cases the loss was less marked. During the course of the postoperative tests 3 females increased, 3 decreased in weight and one did not change. Only one of these changes exceeded 10 per cent of the total body weight (No. 23). A

TABLE 7

Effects of unilateral cortical injury upon body weight

Rat	Weight before operation	Weight at first postoperative test	Weight at last postoperative test
22	249	232	228
23	234	254	262
24	233	239	250
25	257	258	271
26	246	259	249
28	210	212	219
29	340	306	345
30	255	246	255
33	195	256	239
34	225	240	242
35	235	273	269
37	245	255	249
38	210	249	252
39	260	261	261
Mean	242	253	256

TABLE 8

Effects of bilateral cortical injury upon body weight

Rat	Weight before first operation	Weight before second operation	Weight at start of tests after second operation	Weight at end of tests after second operation
23	234	255	216	193
24	233	245	250	238
25	257	260	257	258
28	210	215	221	216
29	340	330	310	315
37	245	255	224	224
39	260	253	248	261
Mean	254	259	246	243

careful inspection of test results reveals that the mating behavior of females which were below preoperative weight was closely comparable to that of those animals which regained their original weights.

However, the data at hand do suggest a possible reason for weight loss in the 2 cases where it was greatest. Females No. 23 and 37 showed the most

marked loss in weight following the second operation; and neither of these animals gained weight in the course of the testing series. Both individuals were noticeably hyperactive during postoperative tests. It seems likely that the frequent bursts of constant activity occurring during the periods of observation were symptomatic of a general level of increased activity which persisted while the animals were in their living cages. It is quite possible that this hyperactivity was responsible for the failure of these animals to regain weight lost immediately after operation.[7]

Effects of Incidental Invasion of Noncortical Structures

First operation. It will be recalled that the first operation inflicted upon each female was unilateral, being restricted to either the right or left cerebral hemisphere. In 18 of the 20 operated cases the fibers of the corpus callosum directly underlying the cortical lesion were destroyed. In the remaining 2 females only medial remnants of the corpus callosum were left intact. There is no reason to assume that removal of those parts of the corpus callosum which are subadjacent to destroyed cortical areas should exert any effect beyond that produced by the cortical damage alone.

The hippocampus of one hemisphere was completely destroyed in 7 cases. In 6 rats only the dorsal or the lateral portions of the hippocampus were removed, and in the remaining 5 females the hippocampus was uninjured. Comparison of the behavioral records of 7 individuals lacking the hippocampus and 5 individuals with this structure intact indicates that hippocampal injury under these conditions has no detrimental effect upon feminine mating responses. There is on the contrary some suggestion that unilateral destruction of the hippocampus combined with neocortical injury may be followed by greater increase in the frequency of certain estrous reactions than brain loss which is confined to the neopallium. This effect however is merely suggested by the few data available.

The corpus striatum was unilaterally invaded in all operations. In 5 rats the striatal lesion was slight and was restricted to the lateral or dorsal

[7] Several earlier studies have shown that certain types of cortical injury may result in marked increase in general activity coupled with a rise in the amount of food ingested daily, and at the same time with a definite loss in body weight (9, 14, 15).

portions of the head of the caudate nucleus of one hemisphere. Twelve females sustained injury involving more than half of one corpus striatum. In the majority of these cases the loss was confined to the caudate. Comparing the test scores of females with minor striatal injury to those with more extensive invasion of this structure reveals that major unilateral lesions do not have a depressing effect upon the reactions considered. On the contrary animals suffering major invasions of the caudate tended to show greater postoperative increases in the frequency of ear-wiggling and hopping than did those cases suffering slight injury to the basal ganglia. Here again as in the case of the hippocampus the small group differences cannot be regarded as reliable. Increases in the frequency of estrous reactions were not the invariable sequelae to extensive striatal damage. Some individuals suffering the greatest injury to the basal ganglia failed to show postoperative increase in any of the measures employed, whereas others with very slight caudate injury exhibited marked increases.

In 18 of the 20 operated females the thalamus was intact. One rat suffered moderate invasion of the medial geniculate nucleus and the final case sustained a lesion involving nearly all of the dorsal thalamus and epithalamus anterior to the posterior commissure. The postoperative behavior of the 2 females with thalamic injury did not differ in any way from that of those individuals in which the thalamus was spared.

Second operation. Nine rats were subjected to a second operation in which part or all of the remaining neocortex was removed. The total lesion after the second operation ranged from 62 to 100 per cent of the neopallium.

The corpus callosum was completely destroyed in all of these animals.

The hippocampus of both hemispheres was intact in 4 individuals. This structure was intact on one hemisphere and partially or completely destroyed on the opposite hemisphere in 2 cases. It was completely destroyed bilaterally in 3 cases. Careful inspection of test results fail to show any differences between the postoperative behavior of females in these 3 groups. The data point strongly to the conclusion that hippocampal injury combined with removal of the overlying neopallium has no effect beyond that produced by cortical injury alone.

The 9 bilaterally operated females included 3 cases in which invasion of the corpus striatum was confined to minor injury of the lateral aspects of the caudate. In 2 rats striatal invasion was slight on one hemisphere and much more extensive on the opposite hemisphere. The remaining 4 cases suffered loss of 25 to 50 per cent of the striatum on both hemispheres. The postoperative behavior of females with extensive bilateral striatal injury differed in no observable manner from that of other individuals suffering very slight invasion of this subcortical region. There is no suggestion in our findings that injury to the basal ganglia and the cortex has any effect in addition to that to be expected from purely cortical removal.

The thalamus was not invaded in 6 of the 7 bilaterally operated females. In the seventh case the medial geniculate nucleus sustained lateral injury which was apparently without effect upon postoperative behavior. In most cases the pyriform cortex was intact. Six animals suffered varying amounts of destruction of this region, but in no case did such invasion produce any detectable change in behavior.

DISCUSSION

Results of the present investigation are in agreement with the findings of earlier workers in showing that neocortical tissue is not essential to mating behavior in the female rat in estrous. The normal estrous pattern includes 3 discrete responses, namely lordosis, ear-wiggling and hopping. The frequency with which these acts occur does not tend to be reduced by injury to the cortex. In fact in some animals there is a distinct postoperative increase in response frequency. Sexually receptive behavior in the form of back-kicking to prevent the male's attempts to mount is greatly reduced or eliminated following decortication.

The survival of the female rat's mating behavior despite removal of the cerebral cortex stands in marked contrast to the complete loss of copulatory reactions which follow decortication in the male of this species. Male rats from our colony do not mate, and show little or no interest in the receptive female following destruction of the neopallium (8). In the male this type of brain injury always results in marked reduction in the capacity for sexual arousal. This clear cut sexual difference in the importance of the cortex to the occurrence of mating behavior is of definite theoretical interest.

The evidence presented in the present report should not be taken to indicate that the neocortex does not participate in the mediation of the normal female's copulatory behavior. Although the separate receptive reactions survive extensive cortical destruction, the integration of these items of behavior into a smooth-flowing pattern of concatenated responses is often partially disrupted after brain operation. In the intact female the cortex is at least partially responsible for the sequential timing of the separate estrus responses. In the absence of the cortex, although none of the discrete reactions are lost, the biological effectiveness of the pattern as a whole is reduced. Poorly timed execution of the disintegrated elements in some cases actually interferes with the male's attempt to achieve intromission, and in other cases appears to reduce the "stimulation value" of the female as a sexual object.

SUMMARY

Methods

Twenty virgin female rats raised in segregation were ovariectomized at 3 to 4 months of age. Estrous was induced in these animals by the injection of estradiol benzoate and progesterone. Each animal was observed in mating tests with a sexually vigorous male during 3 successive periods of induced estrous. Records taken during every test, included the number of times the female was mounted by the male, the number of times the female exhibited lordosis (concave arching of the back facilitating intromission), the number of hopping responses and ear-vibration responses executed by the female, and the number of times the female kicked backward in an attempt to resist the male's sexual mounts.

After the third preoperative mating test part or all of the neocortex was removed from one hemisphere. Following postoperative recovery estrous was induced by hormone administration and a second series of 3 sex tests was administered to each female. At the conclusion of these tests 9 rats were subjected to a second operation in which part or all of the remaining cortex was destroyed. Only the records of 7 females deprived of 97 to 100 per cent of the cortex were subjected to extensive analysis. Following recovery these animals were observed in 3 additional sex tests during induced estrous.

Results

Unilateral removal of 29 to 50 per cent of the neocortex in 20 females. The tendency to exhibit lordosis (as measured by the Copulatory Quotient) showed no consistent change. Ear-wiggling was eliminated in 3 individuals, decreased in 6, increased in 10, and remained the same in one animal. Increases in the frequency of ear-wiggling were of greater magnitude than were decreases. Hopping was eliminated in one female, decreased in 8 cases, and increased in 11. Increases were more extensive than decreases. Back-kicking was eliminated in 4 rats and reduced markedly in the remaining 16 individuals.

Destruction of 97 to 100 per cent of the neocortex in 7 females. Average CQ (Copulatory Quotient) scores were relatively unaffected, but the female's tendency to show lordosis became more variable. In the same test a female would display lordosis when the male investigated without mounting; and 2 minutes later the same animal might fail to show lordosis in response to several successive copulatory attempts by the male. Six of the 7 decorticates displayed lordosis in response to stimuli which had not elicited this reaction prior to operation (pipette in the vagina, grasp of experimenter's hand, male's preliminary investigations, etc.). Two females frequently maintained the lordosis position long after the male had dismounted, and one animal often permitted several successive copulations during the same lordosis. Five of 7 rats experienced periods of hyperactivity, repeatedly circling the perimeter of the cage at a fast run. During such periods males were rarely able to force a lordosis. This fact contributed to the increased variability of the CQ scores.

Ear-wiggling increased in 4 of the 7 females, decreased in 2 and remained the same in one rat. Increases were of greater magnitude than were decreases, and the average frequency of this response for the group rose from 6.6 per positive test before operation to 11.2 after decortication. Two rats displayed ear-wiggling in response to stimuli which had not elicited the behavior preoperatively (pipette in the vagina, and grasp of the experimenter's hand). In 2 individuals although the ear-wiggle reaction increased in frequency after decortication, it lacked normal integration with remaining elements of the receptive pattern, appearing only after the male dismounted instead of before and during the copulation.

Hopping was eliminated in 2 cases, decreased in 2 and increased in 3. In 2 females postoperative hopping usually occurred immediately following the male's dismount and rarely appeared before the mount as it does in the pattern of the normal female.

Back-kicking was eliminated in 6 of 7 females and the frequency of its occurrence was markedly reduced in the seventh individual.

Incidental observation suggested that completely receptive decorticate females were sexually less stimulating to the male than were intact females.

Records of changes in the bodily weight of the operated females indicate that after operation the animals were in good health.

Careful study of incidental invasions to noncortical structures appeared to support the conclusion that postoperative changes in behavior were primarily if not exclusively the result of neocortical damage.

REFERENCES

1. BALL, J.: A test for measuring sexual excitability in the female rat. Comp. Psychol. Monogr., 14: 1, 1937.
2. BARD, P.: On emotional expression after decortication with some remarks on certain theoretical views. Part II. Psychol. Rev., 41: 424, 1934.
3. BARD, P.: Oestrual behavior in surviving decorticate cats. Proc. Amer. Physiol. Soc., Amer. J. Physiol., 116: 4, 1936.
4. BARD, P.: Central nervous mechanisms for emotional behavior patterns in animals. Res. Publ. Ass. nerv. ment. Dis., 19: 190, 1939.
5. BARD, P.: The hypothalamus and sexual behavior. Res. Publ. Ass. nerv. ment. Dis., 20: 551, 1940.
6. BARD, P., AND RIOCH, D. McK.: A study of four cats deprived of neocortex and additional portions of the forebrain. Bull. Johns Hopkins Hosp., 60: 73, 1937.
7. BEACH, F. A.: The neural basis of innate behavior. I. Effects of cortical lesions upon the maternal behavior patterns in the rat. J. comp. Psychol., 24: 393, 1937.
8. BEACH, F. A.: Effects of cortical lesions upon the copulatory behavior of male rats. J. comp. Psychol., 29: 193, 1940.
9. BEACH, F. A.: Effects of brain lesions upon running activity in the male rat. J. comp. Psychol., 31: 145, 1941.
10. BEACH, F. A.: Importance of progesterone to the induction of sexual receptivity in spayed female rats. Proc. Soc. exp. Biol. & Med., 51: 369, 1942.
11. BROOKS, C. McC.: The role of the cerebral cortex and of various sense organs in the excitation and execution of mating activity in the rabbit. Amer. J. Physiol., 120: 544, 1937.
12. DAVIS, C. D.: The effect of ablations of neocortex on mating, maternal behavior and the production of pseudopregnancy in the female rat and on copulatory activity in the male. Amer J. Physiol., 127: 374, 1939.
13. DEMPSEY, E. W., AND RIOCH, D. McK.: The localization in the brain stem of the oestrous responses of the female guinea pig. J. Neurophysiol., 2: 9, 1939.
14. FRANZ, S. I.: On the functions of the cerebrum. The frontal lobes. Arch. Psychol., 1: 1, 1907.
15. FULTON, J. F., JACOBSEN, C. F., AND KENNARD, M. A.: A note concerning the relation of the frontal lobes to posture and grasping in monkeys. Brain, 55: 524, 1932.
16. LASHLEY, K. S.: Studies of cerebral function in learning. Psychobiol., 2: 55, 1920.
17. LASHLEY, K. S.: Brain Mechanisms and Intelligence. Chicago, Univ. Chicago Press, 1929.
18. RICHTER, C. P., AND HAWKES, C. D.: Increased spontaneous activity and food intake produced in rats by removal of the frontal poles of the brain. J. Neurol. and Psychiat., 2: 145, 1939.
19. STONE, C. P.: Effects of cortical destruction on reproductive behavior and maze learning in albino rats. J. comp. Psychol., 26: 217, 1938.

Reprinted from *Endocrinology*, **45**, 29–32 (1949)

INDUCTION OF PSYCHIC ESTRUS IN THE HAMSTER WITH PROGESTERONE ADMINISTERED VIA THE LATERAL BRAIN VENTRICLE[1]

GEORGE C. KENT, JR. AND M. JACK LIBERMAN

From the Department of Zoology, Physiology and Entomology,[2]
Louisiana State University

BATON ROUGE

27

FOLLOWING estrone priming, the ovariectomized golden hamster responds to injections of progesterone in adequate doses by exhibiting psychic estrus as do also certain other common laboratory rodents, a fact first established by Frank and Fraps (1945) and subsequently verified by Kent and Liberman (1947) while studying vaginal smears associated with induced mating. While the threshold dose of progesterone necessary to effect psychic estrus when administered via a subcutaneous route has not been established, Frank and Fraps observed no mating among animals subjected to subcutaneous injection of 0.02 mg. of progesterone 24 to 48 hours after estrogen priming, but found that 0.05 mg. of progesterone was as effective in inducing mating as was 10 times this quantity. The purpose of the present paper is to report that doses of progesterone too small to effect psychic estrus when administered subcutaneously to ovariectomized, estrogen-primed females may be entirely adequate in inducing typical mating responses when introduced directly into the ventricles of the brain. In the present experiments no attempt was made to determine the minimum dose of progesterone necessary to effect psychic estrus by either the subcutaneous or the ventricular route. The possible significance of the findings will be discussed briefly.

MATERIALS AND METHODS

Animals utilized in the present experiment had been ovariectomized two weeks (in a few cases, slightly longer) before receiving the priming regimen of estrone. At the time of the initial estrone injection the vaginal smear was typical of castrate animals and none had exhibited estrous cycles or mating responses in the presence of males subsequent to ovariectomy. The regimen of estrone consisted of a series of injections of 50 μg estrone daily for 6 days, a regimen sufficient to produce, on or before the sixth day, typical vaginal estrous smears devoid of leucocytes and composed of epithelial cells with polyhedral squamous cells predominating.

Received for publication March 7, 1949.

[1] Aided in part by a grant from the American Association for the Advancement of Science through the Louisiana Academy of Sciences.

[2] Contribution no. 91.

Twenty-four hours after the last dose of estrone the animals were tested for mating responses. Of 63 animals employed to this point, one exhibited a mating response as a result of the effects of the estrone, and this animal was discarded. Both Frank and Fraps, and Kent and Liberman have shown that the effect of estrone in the induction of psychic estrus is erratic, and the exhibition of psychic estrus by this animal is not unexpected at the dosage level herein utilized. Two animals were subsequently lost in trephining.

Sixty animals were utilized in three groups as follows: Fifteen animals received subcutaneously 0.025 mg. progesterone (crystalline synthetic) in 0.05 cc. of olive oil previously neutralized with sodium bicarbonate and extracted with ether; 30 animals received 0.025 mg. progesterone in 0.05 cc. oil directly into the right lateral ventricle; 15 animals received 0.05 cc. oil alone, directly into the right lateral ventricle. Several additional animals were injected via the ventricle with oil containing India ink to determine the distribution of the granules and therefore the distribution of progesterone within the ventricular system. These latter animals were autopsied immediately after injection. The ink granules were found in the contralateral ventricle, in the hypothalamic region of the 3rd ventricle, and in the iter. It is presumed that, in the test animals, the progesterone underwent a similar dissemination.

Injection into the ventricle was via a trephine hole placed 6 hours earlier in the right parietal bone half way between the coronal and lambdoidal sutures and 1.5 mm. lateral to the sagittal suture. The animals were ambulatory and alert less than 10 minutes after trephining. A 26 gauge intradermal needle attached to a 1 cc. tuberculin syringe was directed ventrad and slightly laterad through the hole, the meninges, and the cerebral cortex while the animal was under light ether anesthesia.

The brains of the test animals receiving progesterone via the ventricles were fixed in ammoniacal alcohol after observations had been completed at the end of twenty-four hours and were sectioned transversely by hand. The pathway of the needle through the cerebral cortex and into the ventricle was observed in the sections with the aid of low power lenses. In most cases the needle penetrated the right lateral ventricle and the postero-dorsal-most extent of the right inferior horn. Trauma was probably not avoided entirely in the adjacent ventricular walls. In no instance had the needle been directed mediad, and as far as could be ascertained by gross examination the thalamus suffered no direct damage.

OBSERVATIONS

Of the 15 animals which received subcutaneously 0.025 mg. progesterone in oil all gave negative responses when subsequently tested for mating reactions. Testing commenced 20 minutes after the injection and a response was considered negative when the female failed to exhibit lordosis after 4 hourly exposures to a male of approximately 10 minutes each, followed by a single exposure 10 hours later. Antagonism on the part of the female necessitated the removal of the male before 10 minutes had elapsed in all except one instance. In previous experiments utilizing larger doses (0.1 mg.) of progesterone the present authors (1947) were able (after 3 hours) to elicit 33 positive responses in 33 trials. Frank and Fraps described the latent period of response to progesterone as not more than 2 or 3 hours.

Of the 15 animals receiving 0.05 cc. oil alone into the ventricle, all exhibited negative mating responses.

Of the 30 animals receiving 0.025 mg. progesterone in oil into the ventricle, all exhibited positive mating responses (lordosis and mating) at the end of 1 hour. Sixteen animals in this group were tested for mating responses 1 hour after injection and were positive at that time. One animal was tested at 50 minutes, 4 at 30 minutes, 2 at 25 minutes, 1 at 20 minutes, and 6 at 10 minutes, and all except one of the six in the last group were positive when tested. The animal which refused to exhibit lordosis 10 minutes after injection was positive when re-tested 5 minutes later. Lordosis was exhibited from 15 seconds (1 animal) to 2 minutes (7 animals) after introduction of a male into the cage with the test female for the first time after injection, most of them responding in 30 to 90 seconds. When tested at subsequent hourly intervals the reaction time to the male invariably decreased until, on the third hour, the female typically exhibited lordosis 10 seconds after exposure. The reflex following intraventricular injection was, in our experience, particularly strong, although no objective measurements were made. In one instance in which the male was withdrawn from the cage before having mounted a lordotic female the latter remained in rigid lordosis for 40 minutes.

Observation of vaginal smears 24 hours after the animals received 0.025 mg. progesterone via the ventricles revealed that these animals did not exhibit a vaginal response equivalent to that elicited in previous experiments (Kent and Liberman, 1947) in which the animals received, subcutaneously, a larger dose (0.1 mg.) of progesterone. The latter animals within 24 hours exhibited a condition reminiscent of typical diestrous smears of unoperated females, characterized by the presence of large numbers of leucocytes. In the present experiments utilizing smaller doses of progesterone leucocytes appeared in small numbers only and the smears apparently were tending toward an ultimate castrate condition. Jones and Astwood (1942) have shown in the rat that, if the dosage of progesterone is adequate, diestrus will follow an estrous smear even in the presence of excess estrogen.

Numerous investigations chiefly on the guinea pig have indicated a likelihood that one or more hypothalamic nuclei may be indispensable elements in the complex neural and hormonal integrating mechanism which ultimately brings about mating behavior at a specific stage of the estrous cycle in rodents (Bard, 1940; Brookhart et al., 1940, 1941; Dey et al., 1940, 1942; Dey, 1943). In the present experiment the following facts seem significant: 1. A dose of progesterone too small to effect psychic estrus when administered subcutaneously into the body wall will, if injected directly into the neurocoele of the forebrain of animals pretreated with estrone, facilitate the exhibition of a strong mating reflex. 2. The latent period of response to the progesterone is relatively short (10 minutes or longer). On the basis of these observations it is suggested that, in the intact animal, progesterone may act (in a manner not understood) directly upon one or more nervous centers causing females to enter a physiological state known as psychic estrus. This state is manifest by the exhibition of postural reflexes initiated by adequate environmental stimuli chief of which, in a natural environment, are the presence and behavior of the male. Such an interpretation precludes a role of any other endocrine organ or secretion (with the exception of estrone the role of which has not been elucidated) in the exhibition of mating activity on the part of the female hamster once progesterone, elaborated by a preovulatory follicle (perhaps under the stimulus of prolactin in the rat as suggested by Everett in 1944) affects the proper nervous centers. Further integrating mechanisms resulting in lordosis are, therefore, envisaged as entirely nervous in nature.

SUMMARY

Fifteen ovariectomized golden hamsters pretreated with estrone received subcutaneously 0.025 mg. progesterone in oil and, as expected on the basis of previous observations of the size of dose necessary to effect psychic estrus by subcutaneous administration, exhibited no subsequent mating reaction when presented with males. Thirty additional animals received 0.025 mg. progesterone in oil directly into the right lateral ventricle, and all exhibited positive mating responses within one hour or less. Fifteen animals received an equivalent amount of oil alone directly into the ventricle and these exhibited no mating responses. The progesterone is believed to have been distributed in the contralateral ventricle, in the hypothalamic region of the 3rd ventricle, and in the iter. On the basis of these experiments it is suggested that progesterone may act directly upon one or more cranial nuclei facilitating, in an unknown manner, the manifestation of reflex mating activity in the female hamster, thus effecting the physiological state known as psychic estrus. In the present experiment vaginal smears typical of diestrus did not occur subsequent to progesterone administration although in earlier experiments utilizing larger doses of progesterone diestrual smears were observed.

REFERENCES

BARD, P.: *Research Publ., A. Nerv. & Ment. Dis.* 20: 551. 1940.
BROOKHART, J. M., F. L. DEY AND S. W. RANSON: *Proc. Soc. Exper. Biol. & Med.* 44: 61. 1940.
BROOKHART, J. M., F. L. DEY AND S. W. RANSON: *Endocrinology* 28: 561. 1941.
DEY, F. L.: *Endocrinology* 33: 75. 1943.
DEY, F. L., C. FISHER, C. M. BERRY AND S. W. RANSON: *Am. J. Physiol.* 129: 39. 1940.
DEY, F. L., C. R. LEININGER AND S. W. RANSON: *Endocrinology* 30: 323. 1942.
EVERETT, J. W.: *Endocrinology* 35: 507. 1944.
FRANK, A. H., AND R. M. FRAPS: *Endocrinology* 37: 357. 1945.
JONES, G. E. S., AND E. B. ASTWOOD: *Endocrinology* 30: 295. 1942.
KENT, G. C., JR., AND M. J. LIBERMAN, *J. Exper. Zool.* 106: 267. 1947.

Reprinted from *Endocrinology,.* **65**, 652–668 (1959)

NEUROENDOCRINE CORRELATES OF CHANGES IN BRAIN ACTIVITY THRESHOLDS BY SEX STEROIDS AND PITUITARY HORMONES[1]

28

M. KAWAKAMI AND CHARLES H. SAWYER[2]

*Department of Anatomy, University of California at Los Angeles, and
Veterans Administration Hospital, Long Beach, California*

ABSTRACT

A neurophysiological study has been made in rabbits of the effects of sex steroids and pituitary and placental hormones on sex behavior and pituitary function as related to changes in thresholds of two opposing cerebral systems: the *EEG arousal threshold* involving the brain stem reticular formation (I), and the *EEG after-reaction threshold* involving the rhinencephalon and hypothalamus (II). In the estrous or estrogen-primed female rabbit, progesterone at first lowers both thresholds for a few hours and subsequently elevates them to supranormal levels until withdrawal or rebound brings them down again. The early phase is related to estrus and a lowered threshold of pituitary activation, whereas the elevated thresholds correlate with anestrus and pituitary inhibition. Testosterone, certain pituitary and placental hormones, seasonal influences and even prolonged treatment with female sex steroids affect the two thresholds differentially. Relating the effects on thresholds to behavior and to known thresholds of pituitary activation, one may conclude that changes in I are more closely related to sexual behavior, whereas alterations in II are correlated with pituitary thresholds for the release of ovulating hormone. It is concluded that both the facilitatory and inhibitory effects of sex steroids on pituitary activation, as well as the influences on behavior, are mediated by means of altered thresholds of cerebral activity.

T HE mechanisms by which steroid hormones exert their effects on animal behavior and on the release of pituitary hormones are poorly understood. Beach (1) proposed several years ago, on indirect evidence, that the sex steroids must alter thresholds within the nervous system. However, as recently as 1957, in a symposium on the effects of steroid hormones on the nervous system, Goldstein (2), after reviewing recent progress in Beach's laboratory, admitted that "we have little information on the relationship between hormone action and neural functioning," and that "not much more is known about the neural-hormonal interaction than was known in 1940."

A pioneer attempt to localize the behavioral influence of a steroid within the brain was made by Kent and Liberman (3) in 1949. They showed that sub-threshold doses of progesterone induced estrous behavior in the ovari-

---

Received April 27, 1959.

[1] Supported in part by a grant (B1162) from the National Institutes of Health.

[2] Fellow of the Commonwealth Fund 1958–59.

ectomized estrogen-primed hamster if the steroid was injected into the lateral cerebral ventricle. The work has been criticized as not adequately controlled (4), but has not been refuted. More recently Fisher (5) and Harris, Michael and Scott (5) have effected sexual behavior in rats and cats, respectively, by injecting tiny amounts of steroids directly into the brain substance, in doses which were ineffective when administered systemically.

Another approach has been made by Faure (7), who has described the influence of hormones on an olfacto-bucco-ano-genital syndrome in rabbits. The behavioral manifestations have been related to electroencephalographic (EEG) and sensitivity changes in certain rhinencephalic nuclei. The after-reaction to coitus in the rabbit (8) resembles Faure's syndrome behaviorally and electroencephalographically, and it can be induced in a similar manner, i.e. by pituitary hormones and by direct electrical stimulation of the brain (9). Ishizuka *et al.* (10) have reported changes in the EEG activity of sympathetic and parasympathetic regions of the rabbit hypothalamus resulting from treatment with large doses of estrogen and progesterone, respectively.

The biphasic effects of progesterone on sexual behavior and on the threshold of pituitary activation in the estrous female rabbit (11) appeared to offer a unique system for study of threshold changes in cerebral activity by neurophysiological methods, inasmuch as dramatic changes were observed within a few hours. The present work takes advantage of these relationships in unanesthetized rabbits with electrodes chronically implanted in various parts of the brain, by studying alterations induced by sex steroids on 2 types of neural thresholds: (a) the EEG arousal threshold to direct stimulation of the midbrain reticular formation (12), and (b) the EEG after-reaction threshold to low frequency stimulation of hypothalamic or rhinencephalic nuclei (9). In general the threshold changes in the 2 systems parallel each other. Where they diverge it would appear that the arousal threshold is the more closely related to sexual behavior while the after-reaction threshold is the more intimately linked to pituitary activation. Part of the results have previously been referred to in abstracts and reviews (13–17).

METHODS

Large New Zealand white rabbits, ranging in weight around 4 kg., were employed. Their care, the methods of implanting permanent electrodes in the brain and the types of hormone preparations used have been described (8, 9, 18). Methods of assessing thresholds are presented below (in Results).

RESULTS

Characteristics of EEG arousal and EEG after-reaction. Figure 1A illustrates the types of EEG arousal changes in the frontal and limbic cortex of the rabbit brain after direct electrical stimulation of the reticular forma-

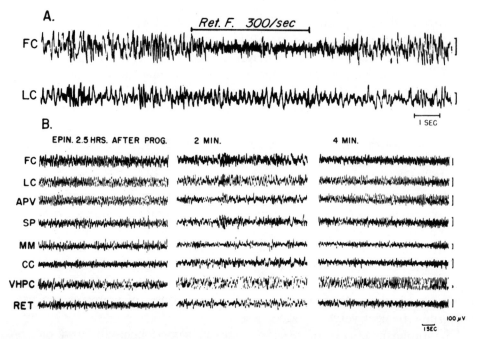

Fig. 1. EEG records of arousal (A) in the frontal cortex (FC) and limbic cortex (LC) stimulated by an electrode in the midbrain reticular formation. Stimulation parameters, 2 V, 300/sec.; 0.5 millisecond square waves, for 5 sec. Note the induction of desynchrony in FC and the synchrony (6/sec.) in LC. (B) EEG after-reaction sequence induced by 25 μg. epinephrine intravenously. Description in text. This type of after-reaction is also induced by stimulating hypothalamic or rhinencephalic nuclei at a frequency of 5/sec. APV, nucleus paraventricularis anterior; SP, septum; MM, medial mammillary nuclei; CC, corpus callosum; VHPC, ventral hippocampus; RET, midbrain reticular formation.

tion at a frequency of 300/sec. The lowest voltage to induce this change promptly when the stimulating current is applied for 5 sec. is known as the EEG arousal threshold. It is ordinarily slightly lower than the threshold of behavioral arousal, during which the rabbit raises its head or looks around.

Figure 1B illustrates a typical EEG after-reaction, induced in this case by 25 μg. epinephrine after progesterone. Within 2 minutes sleep waves appeared, and, by 4 minutes, the "hyperaroused" EEG which we are calling the phase of hippocampal hyperactivity. The first and last records in B show the differences between normal arousal and the "hyperactivity" phase, in which the hippocampus and projections from it reveal a high amplitude 8/sec. rhythm. The EEG after-reaction threshold is the lowest voltage, which, on application for 30 sec. at a frequency of 5/sec. and a wave duration of 0.5 msec., will induce the characteristic EEG after-reaction. It was found in many rabbits that, at threshold, the sleep spindles of the after-reaction were induced almost immediately by electrical stimulation. In other instances a waiting period of half an hour was instituted between stimulations. As mentioned later, the phenomenon can be induced by epinephrine only when the electrical threshold is very low (e.g. <0.05V).

EEG arousal and EEG after-reaction are mutually antagonistic phenomena. The after-reaction manifestations, both behavioral and electroencephalographic, are impossible to elicit in a noisy, brightly lighted room. Conversely, when the rabbit, in after-reaction, has reached the stage of hippocampal hyperactivity, it can be aroused to a behaviorally alert condition only with great difficulty. In one representative animal in which repeated measurements were made of arousal thresholds during wakefulness (A), sleep (B), and the after-reaction stage of hippocampal hyperactivity (C), the thresholds of behavioral arousal (which were easier to differentiate than the shift from hippocampal hyperactivity to EEG arousal) were: A, 0.3–0.4V; B, 0.4–0.6V; and C, 0.8–1.25V. Thus it took twice as great a stimulus to "wake" the rabbit from the depths of the after-reaction as it did from sleep or the sleep-spindle stage of the after-reaction.

Intraventricular experiments. Influenced by the findings of Kent and Liberman (3), our first experiments tested in acute preparations the effects of progesterone injected into the lateral cerebral ventricle. The rabbits were immobilized with curare and maintained by artificial respiration in the stereotaxic apparatus for periods up to 30 hours. The experiments were conducted before the techniques for stimulating the EEG after-reaction had been developed, and only the EEG arousal thresholds were assessed.

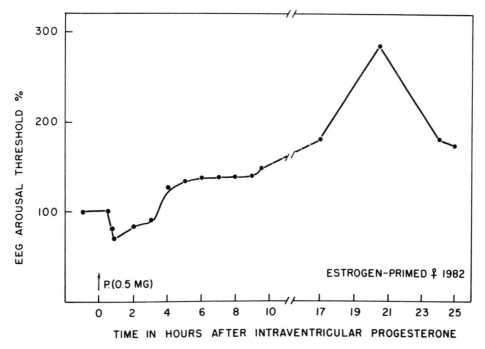

Fig. 2. EEG arousal curve in acute rabbit preparation in which progesterone in oil was injected directly into the lateral cerebral ventricle.

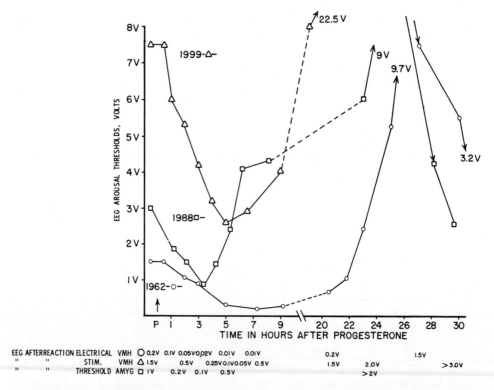

FIG. 3. Typical biphasic threshold changes in 3 estrogen-primed ovariectomized rabbits following subcutaneous injection of 2 mg. progesterone in oil. The data at the bottom are the EEG after-reaction thresholds to electrical stimulation of the ventromedial nucleus of the hypothalamus (VMH) or the amygdala (AMYG).

Representative of 5 progesterone experiments in which the rabbit survived 24 hours or more is the case illustrated in Figure 2. Within less than an hour after the intraventricular injection of 0.5 mg. progesterone in 0.05 ml. sesame oil, the arousal threshold had dropped some 30%, but by 4 hours it had risen above the initial level. By the next day it was 2 or 3 times as high. Although these results were consistently biphasic and were not induced in oil-injected controls, the secondary rise was suspect because of the probable deterioration of the acute preparation. Furthermore, it was impossible to obtain behavioral correlates from such animals. In view of these limitations the chronic preparation with implanted electrodes was adopted, and subcutaneous administration of the steroid was instituted to take advantage of its wider applicability in repeated experiments.

Subcutaneous progesterone in estrogen-primed rabbits. The effects of progesterone (2 mg., s.c.) on EEG arousal thresholds and related EEG after-reaction data in 3 representative "chronic" female rabbits are summarized in Figure 3. To avoid confusion the after-reaction threshold curves are not included, but they would parallel the respective arousal threshold curves very closely. Related to these threshold changes each rabbit was most highly estrous and gave the EEG after-reaction response to coitus, to vaginal stimulation and to epinephrine most readily within its period of most reduced thresholds. The duration of this period was only 3 hours in #1988 but 20 hours in #1962. Conversely, during the periods of most highly elevated thresholds, around 24 hours, the rabbits were anestrous, and the EEG after-reaction could not be elicited by vaginal stimulation or epinephrine. There was a suggestion of a "rebound" from the progesterone inhibition in rabbit #1988: at 30 hours the rabbit responded to vaginal stimulation with an EEG after-reaction; its EEG arousal curve had also recovered at this stage.

In Figure 3 the EEG arousal stimuli were all applied directly to the midbrain reticular formation. Figure 4 shows that the thresholds of arousal from stimulating the caudate nucleus or the dorsomedial thalamic nucleus undergo similar changes after treatment with progesterone. The EEG after-reaction threshold with the septum as the stimulation site reveals parallel alterations. Behaviorally, in addition to her receptivity of the

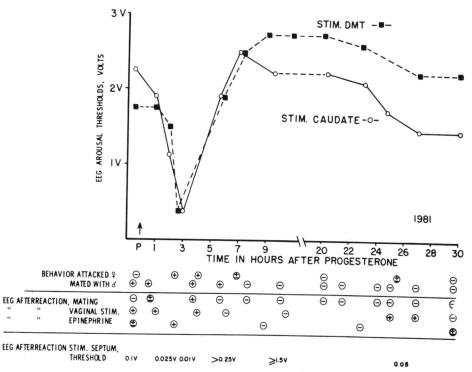

Fig. 4. EEG arousal threshold changes in an estrogen-primed female rabbit following progesterone (2 mg., s.c.) treatment, and related behavioral and EEG after-reaction data. The arousal (300/sec.) stimuli were applied to the caudate nucleus and the dorsomedial nucleus of the thalamus (DMT), while the after-reaction (5/sec.) stimuli were applied to the septum.

male, this animal's reaction toward an anestrous female rabbit is also tabulated. The positive epinephrine EEG after-reaction response at 3 hours is the very one shown in Figure 1B, and the positive responses to coitus and vaginal stimulation were those illustrated in Figures 2 and 3 of an earlier paper (8). The response to septal stimulation in this rabbit has also been illustrated (Fig. 6 in Ref. 9). The "rebound" from progesterone after 24 hours permitted positive EEG after-reaction responses to vaginal stimulation and a weakly positive response to epinephrine. During their periods of minimal thresholds after progesterone a few rabbits gave EEG

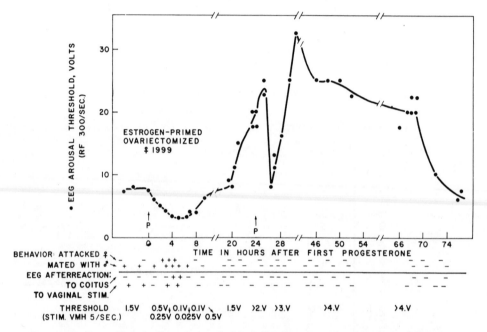

FIG. 5. Prolonged elevation of thresholds following second treatment with 2 mg. progesterone 24 hours after the first injection. Further explanation in text.

after-reaction response to painful stimuli or strong cabbage odor blown into a nostril.

Estrogen-progesterone in male rabbits. In 2 intact (non-castrate) vigorous male rabbits the estrogen-progesterone sequence had no effect on thresholds of arousal or of the EEG after-reaction. In both rabbits the EEG arousal thresholds remained in the 2 volt range throughout while the EEG after-reaction thresholds, to stimulating the ventromedial hypothalamic nucleus in one and the olfactory bulb in the other, remained constant at 4 and 2 volts, respectively. The only observable effect of the female steroids was inhibition of sexual aggressiveness.

The second of the 2 males was castrated, and 4 months later the estrogen-progesterone treatment was repeated. At this time is showed biphasic EEG arousal and after-reaction curves typical of the intact or ovariectomized female. Within 5 hours after progesterone the arousal threshold had dropped 50% and its after-reaction threshold 80%; they rose secondarily to supranormal levels. This castrate male would still mount estrous females prior to his own estrogen treatment, after which he completely lost interest.

Extended treatment with progesterone. To study the effects of 2 successive daily injections of progesterone, the experiment summarized in Figure 5 was conducted. The results during the first day were typical to those already presented. After the second injection of progesterone, at 24 hours, there was a brief reduction in EEG arousal threshold, but the curve remained above the starting level on day 1, and neither coitus nor an after-reaction to vaginal stimulation was observed. Following this brief depression the thresholds stayed highly elevated for at least 40 additional hours while the rabbit remained anestrous and unresponsive to vaginal stimula-

TABLE 1. CHANGES IN SEXUAL BEHAVIOR, EEG AROUSAL AND EEG AFTER-REACTION THRESHOLDS DURING EARLY PREGNANCY

Hours after coitus	Sex behavior	EEG arousal threshold (RF 300/ sec.)	EEG after-reaction		
			To coitus or vag. stim.	Threshold (VMH 5/sec.)	Response to hormones
0	Estrous	0.5 V	+	>0.25 V	
5	Estrous	0.4 V	+	0.1 V	
40	Anestrous	1.0 V	−	>3.0 V	
150	Anestrous	0.8 V	−	>4.0 V	LH 15 U. i.p.　　　(±) Oxytocin 150 mU. (−) 　　　　500 mU. (−) 　　　2000 mU. (±)

tion. The EEG after-reaction threshold changes were even greater, relatively, than the changes in arousal threshold; the after-reaction was never actually attained after 20 hours.

A natural experiment in extended progesterone treatment was provided by a spontaneously estrous female rabbit that copulated with an intact male while her thresholds were being studied. The behavioral and threshold data during early pregnancy are summarized in Table 1. Again it is apparent that the changes in after-reaction threshold are of considerably greater magnitude than the range of arousal threshold levels. The threshold of after-reaction to hormonal stimulation was high at 150 hours post-coitum: it took 2 units of Pitocin to evoke even a brief after-reaction response (see below).

Time relationships of the EEG after-reaction. Figures 6 and 7 illustrate not only the relative changes in EEG arousal and after-reaction thresholds but also the latency and the duration of the phases of the EEG after-reaction responses to coitus, vaginal stimulation and electrical stimulation. It is apparent that the latency of the spindle bursts is shortest and the dura-

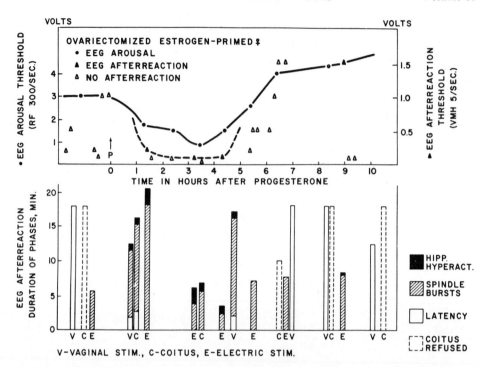

F<small>IG</small>. 6. Thresholds and EEG after-reaction data during the first 10 hours after progesterone treatment of an ovariectomized estrogen-primed rabbit. Open triangles represent unsuccessful attempts to elicit the EEG after-reaction.

tion of hippocampal hyperactivity longest during the period of minimal thresholds after progesterone. The threshold of the after-reaction is relatively much more depressed than the arousal threshold, and in the intact (non-castrate) female it may be lowered by exogenous estrogen alone (Fig. 7). In the non-ovariectomized rabbit the threshold of EEG arousal is also lowered somewhat by estrogen alone, and both thresholds are depressed still further by progesterone. In the 2 rabbits included in Figures 6 and 7, the latency of spindle bursts after electrical stimulation was practically zero throughout, but the hippocampal hyperactivity generally started earlier and continued somewhat longer when elicited during the period of minimal thresholds than at other times. In the ovariectomized rabbits coitus did not usually occur after treatment with estrogen alone, and the EEG arousal and after-reaction thresholds usually both remained relatively high until progesterone treatment.

Progesterone in the absence of estrogen. The effects of treating an unprimed ovariectomized rabbit with progesterone alone are summarized in Figure 8. This rabbit, though anestrous, had an initially low threshold of EEG after-reaction, and a positive response was obtained with vaginal stimulation prior to treatment with progesterone. Progesterone did not affect the EEG arousal threshold, but within a few hours it had elevated

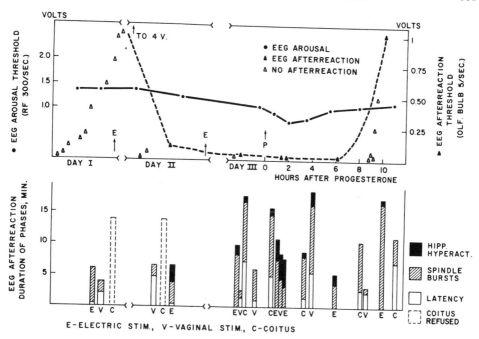

FIG. 7. Thresholds and EEG after-reaction data following estrogen and progesterone treatment of an intact (non-castrate) female rabbit.

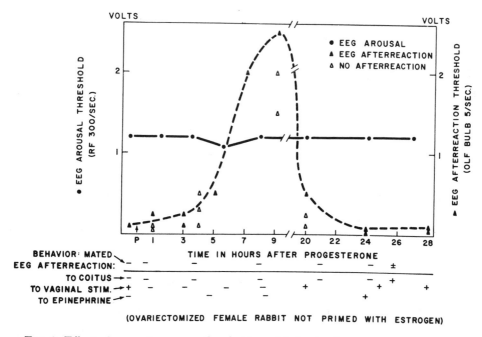

FIG. 8. Effect of progesterone on thresholds and behavior in an anestrous ovariectomized rabbit. Note the failure of progesterone alone to affect the arousal threshold as compared with its marked effect on the after-reaction threshold.

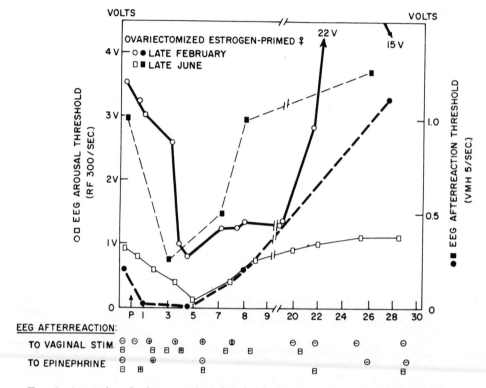

Fig. 9. Arousal and after-reaction thresholds in 1 ovariectomized estrogen-treated rabbit during February and June. Note the low EEG after-reaction threshold in February as compared with June. Coital stimulation in June did not evoke an after-reaction.

the after-reaction threshold several hundred per cent. By 24 hours the EEG after-reaction threshold was again low and at this time positive after-reaction responses were obtained not only to vaginal stimulation but also to epinephrine and even to a questionable coital performance. This latter facilitation would appear to be a definite instance of rebound from progesterone inhibition (19).

Seasonal variations in thresholds. Thresholds in several individual rabbits were measured during 2 or more seasons and certain consistent changes were observed. Figure 9 illustrates some of these variations with data on one ovariectomized female rabbit during late February and late June. During February the arousal threshold was initially rather high, and the drop and subsequent rise to very high levels after progesterone were both precipitous. The EEG after-reaction threshold was very low and for some hours after progesterone a positive response could be obtained with vaginal stimulation. In June the latter threshold was relatively high, never dropping below 0.25 V, and for only a brief period could an after-reaction to vaginal stimulation be achieved. The EEG arousal threshold was comparatively low and its secondary rise took it only as high as the initial level.

During the first 10 hours after progesterone in June the rabbit would mate, but coitus did not évoke an after-reaction. Coitus was not tested during February in this animal. In the same rabbit in late November the thresholds were intermediate between the levels in February and June.

In general between January and May the secondary rise after progesterone took the rabbit's arousal thresholds to about 300% of the initial high levels, whereas between June and December the secondary rises merely restored the initial low thresholds. On the other hand, EEG afterreaction thresholds from February through May dropped (after progesterone) to levels in the 0.01–0.1 V range, while those from June to December seldom dropped below 0.5 V.

Effects of gonadotropins and neurohypophysial hormones on thresholds. All of the preparations which induced the EEG after-reaction (9), including HCG, PMS, LH, LTH, oxytocin and vasopressin, lowered the threshold for electrical stimulation of the after-reaction for a few hours. Neither estrogen nor progesterone was necessary for the response, and there was little if any effect on the EEG arousal threshold. The EEG after-reaction could be elicited by these preparations equally well in castrate males and females. The latter remained anestrous in the absence of the steroids, and vaginal stimulation was almost completely ineffective in evoking the afterreaction even during the periods of minimal threshold. During these periods, both in castrate males and females, epinephrine evoked the afterreaction response.

Figure 10 is representative of this group of experiments except that the female rabbit was estrogen-primed and responsive to coitus and vaginal stimulation. Shortly after treatment with HCG (100 I.U., i.v.) there occurred a "spontaneous" after-reaction, following which the after-reaction threshold to electrical stimulation remained depressed for about 4 hours. The EEG arousal threshold was slightly lowered, an effect not found in the absence of estrogen.

Dosages of the other preparations which evoked "spontaneous" EEG after-reactions and lowered the threshold were as follows: PMS, 100 I.U., i.v., or 300 I.U., i.p.; LH, 5–10 Armour units, i.p.; prolactin, 75–200 I.U., i.p.; Pitocin, 100 mU., i.p.; and Pitressin, 200 mU., i.p. On the other hand, FSH in doses up to 25 Armour units, i.p., failed to lower the EEG afterreaction threshold; indeed it appeared to elevate it in more than one instance. Similarly ACTH, in dosages up to 20 U., i.p., failed to lower the threshold.

Effects of testosterone on thresholds. The failure of estrogen and progesterone to alter the EEG thresholds in the intact male rabbit implies that male sex hormone may counteract the effects of the female steroids on the nervous system. Yet testosterone has been reported to maintain estrous behavior in ovariectomized female rabbits (19). It was therefore of interest to study the effects of the androgen on the female's behavior and on associ-

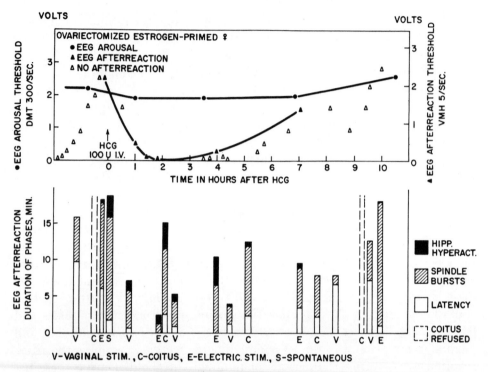

FIG. 10. Effect of human chorionic gonadotrophin (HCG) on EEG after-reaction threshold. In the anestrous rabbit HCG exerts a similar effect on the after-reaction threshold, but the slight depression in arousal threshold and the after-reaction in response to coitus or vaginal stimulation would not have occurred in the absence of the estrogen.

ated thresholds. Two different effects were observed in estrogen-primed ovariectomized females, depending on the dosage of testosterone (Fig. 11). With high dosage (5 mg. testosterone propionate, s.c.), estrus was maintained in 2 cases for 50 hours or more, though coitus never resulted in an after-reaction, and the EEG arousal threshold was maintained at a low level. When the arousal threshold eventually rose (72–76 hours in Fig. 11) the rabbit became anestrous (exhaustion of testosterone?). In 2 similar animals receiving lower dosages of testosterone (0.2 mg. and 2.5 mg.) estrous behavior was maintained only a very few hours. The onset of anestrus could be correlated with a sharp rise in the EEG arousal threshold to some 400% of the initial value by 24 hours the impression was gained that the weak doses of testosterone were adequate to counteract the effect of estrogen on behavior but not enough to maintain estrous behavior on their own.

In the male, sex drive could not be correlated with arousal thresholds. Four intact male rabbits were each treated with 5 mg. testosterone propionate and their thresholds followed for 48–72 hours. In none was the EEG arousal threshold altered, and with a single exception, in several dozen

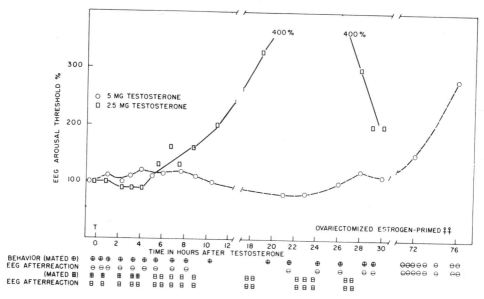

FIG. 11. Effects of high and low dosages of testosterone propionate on arousal thresholds and estrous behavior in ovariectomized estrogen-treated rabbits. Note that coitus following testosterone did not evoke an after-reaction.

instances, no EEG after-reaction followed ejaculation. The effect of the exogenous testosterone on EEG after-reaction thresholds to electrical stimulation was variable and unpredictable. In one vigorous buck there was apparently no effect; the after-reaction threshold remained high, i.e. 2–3 V. In two less vigorous males the EEG after-reaction threshold dropped to the 0.1 V level within 7 hours and remained low for the next 2 days, during which they would mount estrous females only haltingly or not at all. In the least vigorous of the four the threshold remained high for 30 hours after testosterone, during which not a single mounting occurred. Suddenly this buck's behavior changed; it became vigorous and potent and it mounted and ejaculated repeatedly during the next 40 hours, during which its EEG after-reaction threshold dropped to, and remained at, the low 0.1–1.0 V level. Equally suddenly it became impotent at 70 hours, and its threshold rose again to 3 V.

DISCUSSION

The present results confirm by a quantitative electrophysiological method the biphasic progesterone effect proposed by Sawyer and Everett (11)—facilitation followed by inhibition. During the early post-progesterone phase, in which pituitary activation is facilitated, the rabbit is in estrus, and both arousal and after-reaction thresholds are depressed. During the later phase, in which the release of pituitary hormones is inhibited, the rabbit is anestrous, and both thresholds are elevated. Both curves may show a facilitatory withdrawal effect, a rebound to lower thresholds, after 24 hours.

Pregnancy or prolonged treatment with progesterone will maintain both thresholds at elevated levels, the after-reaction threshold being relatively much higher than the arousal threshold. In keeping with the hypothesis that lowered arousal thresholds correlate positively with estrus and lowered after-reaction thresholds correlate with pituitary activation, it is of interest that pregnant animals will occasionally mate, but do not usually release pituitary gonadotropin in response to coitus (20). Thus ovulation is prevented during pregnancy unless the after-reaction threshold is lowered by estrogen, an interpretation consistent with the findings of Mayer and Klein (21). It would appear that both the facilitatory and the inhibitory effects of sex steroids on the adenohypophysis may be mediated by influences on the brain.

A close relationship between threshold of pituitary activation and threshold of the EEG after-reaction was observed in the following instances in which the arousal threshold was in less close agreement: (1) Large doses of testosterone, which block release of ovulating hormone (22), elevate the after-reaction threshold while the arousal threshold is maintained at a low level and the rabbit remains in heat; (2) "Spontaneous" ovulation in estrogen-treated rabbits (23) occurs more commonly during winter and spring, when the after-reaction threshold is especially low, than during summer and fall, when this threshold is elevated disproportionately relative to the arousal threshold.

On the other hand, sex behavior is unrelated to the after-reaction threshold but closely related to the arousal threshold. When placental and pituitary hormones were administered to ovariectomized rabbits untreated with sex steroids, the EEG after-reaction threshold dropped but estrus was not achieved. The behavioral effects of the sex steroids, on the other hand, clearly point to a close relationship of sex behavior to arousal thresholds. Even in the testosterone experiment, estrous behavior correlated with a lowered arousal threshold, and anestrus supervened as the arousal threshold climbed. The higher percentage of estrous rabbits during the spring breeding season would be difficult to relate to absolute levels of EEG arousal thresholds at that season, but perhaps the degree of *change* in threshold in response to the steroids is more important than the absolute level. Both the primary depression and the secondary elevation in arousal threshold after progesterone injection are greater in the spring than in the summer and fall.

The present results are consistent with the proposal of Kent and Liberman (3) that progesterone can induce estrous behavior by a direct action on the brain. The influence of the steroid on an EEG arousal threshold, which is both stimulated and recorded via electrodes implanted in the brain, is most readily explained by an action of the hormone on the central nervous system. The centrally stimulated EEG arousal response is immediate, and independent of nerves, peripheral organs, or fields of nervous influence.

When the EEG after-reaction is induced by electrical stimulation it may occur too soon to have been mediated by pituitary hormones released in response to the electrical stimulation. Nevertheless, the electrical threshold is subject to modification by pituitary hormones as well as the sex steroids. At lowered thresholds, by whatever process attained, either electrical stimulation or epinephrine can trigger the mechanism, although they do not themselves appear to affect the thresholds per se. The steroids and pituitary hormones seem to exert their effects only on thresholds, but if the latter are lowered sufficiently, "spontaneous" triggering occurs. Estrogen and progesterone, in addition to affecting the central mechanism (i.e. the threshold to direct electrical stimulation) lower the threshold to peripheral (e.g. vaginal) stimulation, a feature not achieved by pituitary or placental hormones in the absence of estrogen. Placental gonadotropins and oxytocin have been reported by Faure (7) to lower the threshold for electrically evoking the olfacto-bucco-ano-genital syndrome, which has certain features of the EEG after-reaction.

In its broader implications the present work provides examples of definite influences of hormones on neurophysiological thresholds, a problem which has long interested psychologists (1, 2, 4). The areas in the brain which are affected by steroids are so extensive as to suggest that the whole nervous system is influenced primarily and localized systems of integrated behavior secondarily. Sex steroids affect simultaneously in parallel or opposite directions the midbrain reticular system, which probably includes a mammillary body mating-behavior "center" (24) and the rhinencephalic-hypothalamic system, which includes the basal tuberal gonadotropic "center."

In a recent essay on "translation of bodily needs into behaviour" Dell (25) proposes that a "vigilant" condition of the brain stem reticular system provides a non-specific excitatory state for appetitive behavior, and that elements in the internal environment induce this condition. Consummatory behavior then produces changes which depress reticular activity and vigilance. The results of the present study are consistent with this scheme and they suggest that, in female sexual behavior in the rabbit, internal secretions are responsible for conditioning not only the vigilant appetitive state but also the post-consummatory behavior, which involves a rhinencephalic-hypothalamic circuit.

REFERENCES

1. BEACH, F. A.: *Hormones and Behavior*, Hoeber, New York, 1948.
2. GOLDSTEIN, A. C.: in *Hormones, Brain Function and Behavior*, p. 99, H. Hoagland, ed., p. 99. Academic Press, New York, 1957.
3. KENT, G. C. AND M. J. LIBERMAN: *Endocrinology* **45**: 29. 1949.

4. LEHRMAN, D. S.: in *L'Instinct*, p. 475, Masson and Company, Paris, 1956.
5. FISHER, A. E.: *Science* 124: 228. 1956.
6. HARRIS, G. W., R. P. MICHAEL AND P. P. SCOTT: in Ciba Foundation Symposium on *Neurological Basis of Behavior*, p. 236, J. and A. Churchill, Ltd., London, 1958.
7. FAURE, J.: *Rev. Path. Gen. et Physiol. Clin.* No. 690, p. 1029; No. 691, p. 1263; No. 692, p. 1445, 1957.
8. SAWYER, C. H. AND M. KAWAKAMI: *Endocrinology* 65: 622. 1959.
9. KAWAKAMI, M. AND C. H. SAWYER: *Endocrinology* 65: 631. 1959.
10. ISHIZUKA, N., K. KURACHI, N. SUGITA AND N. YOSHII: *Med. J. Osaka Univ.* 5: 729. 1954.
11. SAWYER, C. H. AND J. W. EVERETT: *Endocrinology* (in press 1959)?
12. MORUZZI, G. AND H. W. MAGOUN: *EEG Clin. Neurophysiol.* 1: 455. 1949.
13. KAWAKAMI, M. AND C. H. SAWYER: *The Physiologist* 1: 48. 1957.
14. KAWAKAMI, M. AND C. H. SAWYER: *Fed. Proc.* 17: 83. 1958.
15. KAWAKAMI, M. AND C. H. SAWYER: *Anat. Rec.* 133: 398. 1959.
16. SAWYER, C. H.: in *Reticular Formation of the Brain*, p. 223, H. H. Jasper, ed., Little, Brown and Company, Boston, 1958.
17. SAWYER, C. H.: in Conf. on *Endocrinology of Reproduction*, p. 1, C. W. Lloyd, ed., Academic Press, New York, 1959.
18. SAWYER, C. H. AND J. E. MARKEE: *Endocrinology* 65: 614. 1959.
19. KLEIN, M.: in *L'Instinct*, p. 287, Masson and Company, Paris, 1956.
20. MAKEPEACE, A. W., G. L. WEINSTEIN AND M. H. FRIEDMAN: *Endocrinology* 22: 667. 1938.
21. MAYER, G. AND M. KLEIN: *C. R. Soc. Biol. Paris* 140: 1011. 1946.
22. ZONDEK, B. AND J. SKLOW: *Endocrinology* 28: 923. 1941.
23. SAWYER, C. H.: *Endocrinology* (in press, 1959).
24. SAWYER, C. H.: *Anat. Rec.* 124: 358. 1956.
25. DELL, P. C.: in Ciba Foundation Symposium on *Neurological Basis of Behavior*, p. 187, J. and A. Churchill, Ltd., London, 1958.

Hormones and the Behavior of Both Sexes

III

Editor's Comments on Papers 29 Through 32

The following papers deal with problems of both male and female behavior and are grouped primarily for that reason. The first three papers (Bard, Paper 29; Beach, Paper 30; Pfaff, Paper 31), however, share, in addition, an attempt to identify the loci of the neural control of sexual behavior. These problems have been dealt with earlier in regard to both males and females, and the review by Beach summarizes many important studies concerned with these questions. The Pfaff study is offered as a representative of the use of an even more recent technique to identify the site of hormone localization in the nervous system. Questions regarding the site and nature of hormone receptors are of contemporary interest within all areas of endocrinology.

The editor is again forced to apologize for the limited treatment of problems of sex differences and sexual behavior in humans. The importance of hormones in determining observed sex differences in human behavior remains equivocal. The paper by Money (Paper 32) is, however, offered here as one of the earliest, albeit comparatively recent, attempts to present some of the available information related to the physiological origins of human sexuality.

Reprinted from *Res. Publ. Assn. Res. Nervous Mental Diseases*, **20**, 551–579 (1940)

29

CHAPTER XIX
THE HYPOTHALAMUS AND SEXUAL BEHAVIOR[1]

PHILIP BARD, Ph.D.

INTRODUCTION

MANY of the forms of overt sexual activity characteristic of different species of animals constitute stereotyped modes of response. In this respect they are allied to other forms of emotional display and to such co-ordinated responses as those which serve to maintain the body temperature relatively constant in the face of environmental heat and cold. Each of these is a reaction pattern in which certain items of behavior occur together or in definite sequence to form a complex act directed toward a specific end. Any such patterned response may be regarded as the expression of the activity of some particular mechanism within the central nervous system, and its very occurrence invites the neurologist to determine the site, arrangement and mode of activation of the essential central machinery.

The fact that the hypothalamus contains a neural mechanism necessary for the ready and full expression of anger (Bard, 1934, 1939; Ranson, 1934; Ranson and Magoun, 1939) and is probably requisite for the type of behavior which denotes fear (Bard, 1934, 1939) vaguely suggests that this ancient part of the brain may also be concerned with sexual behavior. There is also the fact, discussed by Brooks (1940), that in certain species at least, the hypothalamus exerts an influence on the gonadotropic functions of the hypophysis. But one must distinguish between a central nervous mechanism which indirectly controls the output of gonadal hormones and one which under the influence of these humoral agents is capable of fashioning overt sexual behavior. Granting the existence of the first there is no compelling reason to suppose that the second is its cerebral bedfellow. As a matter of fact the evidence which bears

---

[1] Studies of the neural aspects of sexual behavior which have been carried out in the Johns Hopkins Physiological Laboratory and which are described, in some instances for the first time, in this Chapter have been aided by grants from the Committee for Research in Problems of Sex, National Research Council.

551

directly on the question whether the hypothalamus plays a significant part in the direct nervous control of sexual behavior is not extensive. The relation of this part of the diencephalon to estrual behavior has been explored, but the results are not entirely decisive. In the present state of our knowledge sound strategy calls for a flank attack. I shall therefore devote considerable attention to an examination of the management of sexual activities by other levels of central integration.

Studies of patterned responses of the type under consideration have shown that certain ones are dependent on the functional integrity of one or another circumscribed portion of the brain. The essential neural mechanism thus delimited may be spoken of as the center for that particular behavior pattern. Excellent examples of centers of this kind are the hypothalamic mechanisms concerned in body temperature regulation and in the expression of anger. In each of these cases it is found that although destruction of the essential locus permanently abolishes the total response, some or all of the items of response which compose the pattern may still be elicited in isolation or in groups of a few. This fact indicates that an outstanding feature of the center is its capacity to weld together individual responses to form a complex reaction pattern. The center may also serve as a focus where afferent impulses or changes in the blood act to evoke or repress the total response. Furthermore, with the center intact it is found that other parts of the central nervous system normally exert modifying influences either directly upon it or at some point along the pathways over which it controls the peripheral motor neurones. Such in outline is the nature of the central neural mechanisms responsible for complex motor performances which have the same general biological significance as has sexual behavior. We may assume as a working hypothesis that the more complex expressions of sexual excitement have a similar central representation.

The study of the neural bases of sexual behavior gains much interest from the fact that among vertebrates nearly all specific sexual responses depend on the presence of gonadal hormones in the blood. In all mammals which have been studied, prepuberal castration precludes the development of any typical sexual activity, while removal of gonads after puberty either abruptly puts an end to the behavior (*e.g.*, the estrual activities of the females of most species) or causes it to diminish gradually in incidence and vigor. These facts show that the central integrating mechanisms are under the direct or indirect influence of specific chemical substances. Indeed, in animals which show a periodicity in mating, the existence of the central machinery which produces the more or less complicated patterns of sexual response is revealed only during the breeding season; at all other times this neural mechanism lies dormant, apparently refractory to any kind of external stimulus. Therefore one of the chief problems confronting the neurophysiologist investigating sexual behavior concerns the way in which gonadal hormones activate the central mechanisms.

The question whether ovarian and testicular hormones act directly on the central nervous tissue or indirectly through afferent paths from peripheral structures which are affected by the hormones is one which is most difficult experimentally to answer. The decisive experiment would involve the severance of all afferent connections of the central nervous system, a procedure which is of course not feasible. The most marked peripheral effects of the ovarian hormones are found in the genital tract. Josephine Ball (1934) demonstrated that normal sex behavior occurs in rats after removal of the uterus and vagina, thus showing that afferent impulses from these organs are not essential for mating activity. It is not difficult to denervate this part of the body and at the same time to rid the erogenous zone which includes and surrounds the external genitalia of all afferent innervation. In a series of experiments on female cats it was found (Bard, 1935) that after abdominal sympathectomy and extirpation of the sacral cord the animals, on entering into heat spontaneously or as a result of administration of an estrogen, showed feline estrual behavior (crouching and treading) in typical fashion; they lacked only the specific estrual response to vaginal stimulation (the afterreaction). Similarly, Root and Bard (1937) showed that the sexual aggressiveness of male cats is maintained after removal of the sacral and lower lumbar spinal segments, a procedure which renders the penis, the testes and scrotum, the perineum and the surrounding skin completely anesthetic and analgesic. When placed with a female in heat such a male will grasp her by the neck, mount his foreparts, execute copulatory movements with lumbar muscles and develop a full erection of the penis. The erection is mediated by efferent fibers which pass through the hypogastric nerves. If ablation of the lower spinal segments is supplemented by removal of the abdominal sympathetic chains, the capacity for erection is abolished, but sexual aggressiveness appears unimpaired. These experiments demonstrate conclusively that in cats the activation of the central mechanisms responsible for the patterned sexual behavior of both sexes is not dependent on afferent impulses from the genital organs or from the erogenous zones surrounding the external genitalia. It does not follow that this activation is wholly independent of all afferent impulses.

VARIETIES OF MAMMALIAN SEXUAL (MATING) BEHAVIOR[2]

Very little information is at hand concerning the central management of sexual behavior in man and other primates. Brooks (1940) has mentioned the few clinical findings which suggest that portions of the brain, especially the hypothalamic region, may exert some specific control. Experimental investigations of this subject have been confined almost wholly to work on cats, rabbits, guinea pigs and rats. Among these different animals the mating activities of the males differ only in certain details; they consist chiefly in mounting, clasping, various forms of palpation, copulatory movements and intromission. On the other hand, the patterns of response shown by the females of these species differ. For this reason and also because much more work has been done on the central control of estrual behavior than on the nervous management of any other form of reproductive activity descriptions of expressions of sexual excitement in the females of certain species will be given in some detail to serve as a background against which one must view the effects of experimental modifications of the central nervous system.

Estrual behavior—cat. For experimental investigations of estrual activities the female cat is unquestionably the subject of choice; no other common laboratory animal exhibits estrus so conspicuously. In normal and in ovariectomized cats this form of behavior can be evoked by administration of estrin (estradiol benzoate in oil is particularly efficacious) and when thus produced it is identical with that which occurs normally as a result of hypophysial activation of the ovaries (Bard, 1939).

Feline estrual behavior can conveniently be divided into (i) *courtship activities*, which apparently serve to attract the male, and (ii) the *after-reaction* which shortly follows intromission or any other mechanical stimulation of the distal vagina. The courtship activities include playful rolling, excessive rubbing, a curious low vocalization (the estrual call) and crouching and treading. The estrual crouch is a most specific posture. Resting on chest and forearms with pelvis raised somewhat and with tail elevated and turned to one side, the animal tends to execute treading movements of the hindlegs. It is in this posture that the male is accepted. Many estrual cats crouch and tread quite spontaneously in the absence of any specific external stimulus. Others do not show this behavior unless they are in the presence of a male or receive some stimulation of the external genital region. Some females

---

[2] Here as elsewhere in this paper, the term "sexual behavior" is used in a somewhat restricted sense to include only copulatory activities and the associated mating behavior. Other forms of reproductive activity such as parturition and maternal behavior are not considered. For accounts of what little is known of the neural bases of the complex reaction patterns of maternal behavior the reader is referred to the recent communications of Beach (1937, 1938) and Stone (1939).

crouch and tread on being petted or when the nape of the neck is grasped in imitation of the action of the male who on mounting holds the female in this way. Intromission or other vaginal stimulation is usually accompanied by an augmentation of treading and by a loud cry or growl; some animals utter a piercing shriek. The after-reaction varies somewhat. In most animals it consists in more or less frantic rubbing, squirming, licking and rolling. Some cats only lick themselves and squirm on their sides. A few never do more than lick their paws and rub against nearby objects. In its most excessive form the after-reaction, while executed with adroitness, appears convulsive in nature. It is typical of cats in heat. Artificial vaginal stimulation of anestrous cats causes angry vocalization and various forms of struggling and is followed merely by retreat and licking of the vulva. Cats which normally engage in much rubbing may rub after such stimulation, but obviously this is not an after-reaction. In a series of more than one hundred ovariectomized anestrual cats only three instances of behavior resembling the after-reaction have been seen. Each of these females rolled or squirmed moderately after vaginal stimulation, but since each of them indulged spontaneously in a certain amount of such activity they need not be regarded as exceptions to the rule that the after-reaction is strictly a response of estrus.

It is important to distinguish between these estrual responses and certain reactions which can be elicited by genital stimulation in female cats not in heat. During anestrus, when they will not accept the male and are in fact sexually most antagonistic, female cats, if not frightened, apprehensive or restive, will elevate the pelvis and tail and sometimes execute light stepping movements on extended hindlegs when the vulval, perineal or anal regions are gently rubbed or tapped. They do not, however, lower the foreparts and partially flex the hindlegs and so fail to assume the crouch typical of estrus. Further, unilateral stimulation of this kind applied to these zones causes deflection of the tail to the opposite side, lateral turning of the pelvis to the stimulated side and flexion of the ipsilateral hindleg with extension of the contralateral. The tail response is obtained by the slightest tactile stimulation when the cat is standing or lying on its back; the reactions of the pelvis and hindlegs are best obtained when the animal is resting quietly in the dorsal decubitus. These responses are most assuredly not estrual. They are not even peculiar to the female, for all of them can be obtained in male cats by stimulation of the corresponding skin areas.

Rabbit. In contrast to the cat, the female rabbit shows little or no postural or phasic activity which serves to entice the male. If receptive she permits the male to mount and, when mounted, she responds chiefly by raising her pelvis and tail. If she fails to give this simple reaction, intromission does not take place. There is no overt after-reaction, but in this species ovulation normally occurs only as a result of the strong sexual excitement attending coitus. It has been shown by Brooks (1937) that while this ovulatory response is not dependent on genital stimulation it fails to follow mounting and intromission if the female is rendered incapable of cooperating in the sexual act. Thus in the rabbit the occurrence of ovulation furnishes an additional indication of the normality of the mating behavior of the female.

Guinea pig. Young, Myers and Dempsey (1935) have described the well-defined pattern of reflex behavior which the female guinea pig exhibits only during estrus. It consists in opisthotonus, elevation of the pelvis and vulvar dilatation; frequently it is accompanied by a guttural purring vocalization.

This relatively simple response occurs when the male mounts, but it can be evoked by rubbing or touching the skin or hair of the perineal or dorsal lumbar regions.

It appears from the work of Young and his collaborators (see Dempsey, Hertz and Young, 1936) that this estrual response and sexual receptivity can be induced by estrin in only a certain percentage of ovariectomized guinea pigs. If a small preliminary injection of estrin is followed 36 to 48 hours later by progesterone receptivity invariably follows after a brief interval. In this respect the guinea pig apparently differs from other animals; in the rabbit progesterone actually inhibits mating (Makepeace, Weinstein and Friedman, 1937).

Rat. In the presence of a male the female in estrus tends to crouch in a tense manner for brief periods and to dart about. When mounted she responds very much as does the guinea pig. If not in heat she avoids the male, but if he becomes aggressive she actively refuses him.

LEVELS OF FUNCTION GOVERNING SEXUAL BEHAVIOR

In considering the part played by a subdivision of the brain in the control of any form of patterned behavior it is necessary to examine the extent to which different levels of central function mediate or integrate individual items of the total response. Unless this be done one is apt to gain a distorted view of the nature of the central control. In the present case this procedure will give some indication of the manner in which the intact central nervous system fashions some of the complex modes of behavior which are specifically directed toward the continuation of the species.

The spinal level

It has long been known that reflex erection of the penis and even seminal ejaculation can be evoked by genital stimulation after supralumbar transection of the spinal cord. These responses appear to be most easily obtained in the spinal dog and spinal man. Sherrington (1900, pp. 852–853) pointed out that in the dog, after spinal transection above the lumbar segments, responses due to contractions of skeletal muscles can be obtained by genital stimulation. He described these as follows:

"Touching the preputial skin evokes bilateral extension at the knees and ankles, and to a less extent at the hip-joints; accompanying this there is depression of the tail. If through the skin more posteriorly the glans penis be pressed, the posterior end of the body is curved downwards, pushing the penal bone forward. These reflex movements suggest themselves as belonging to the act of copulation, and this suggestion is strengthened by failure to obtain similar movement on touching the analogous parts of the spinal bitch."

Dusser de Barenne and Koskoff (1932, 1934) have described a posture of flexor rigidity of the hindlegs which can be obtained in male cats immediately after spinal transection. The adequate stimulus is set up only by placing the animal in the prone position with thighs abducted and hindquarters in contact with a supporting surface. During the mating periods, the rigidity was often accompanied by prolonged priapism, and chiefly for this reason the response was interpreted as a spinal copulation reflex. Such activity, however, falls far short of effective mating behavior. It is apparent that supraspinal levels are prepotent in the elaboration of the sexual behavior of male carnivores.

It is stated (Fulton, 1939) that uterine contractions and vaginal secretion may be evoked in the spinal bitch by manipulation of the "sexual skin," but so far as I am aware no one has observed the occurrence in spinal dogs of specific estrual responses involving skeletal muscle.

Dempsey and Rioch (1939) tested the spinal reflexes of female guinea pigs both before and after injection of estrin and progesterone. Eight animals with cords cut at levels ranging from T5 to L2 were studied in the chronic condition, and it was concluded that "No change could be found in the spinal reflexes which was correlated with the endocrine condition of the animal, and in no case was any reflex obtained which was at all suggestive of the responses which are normally obtained at oestrus." Equally negative results were obtained by Bromiley and Bard (1940) in a study of chronic spinal cats which were tested before and after injections of estrin.

Recently Maes (1939) has claimed that typical components of estrual behavior are elicitable in the decapitate cat, but only if the animal is in heat. The reactions which he adopted as tests for "heat" were: raising the pelvis and "treading" in response to tapping the perineum, and contralateral deflection of the tail when the perineal region is tapped on one side of the midline. His selection of the tail response as "another reflex apparently characteristic of heat" is an unhappy one, for, as already pointed out, it can readily be obtained not only in normal anestrual female cats, but also in males. Maes studied thirteen decapitate cats. Their "sexual reflexes" were tested with the animal held in a crouching position. Nine of the animals were in either spontaneous or induced estrus and showed "a complete positive response to tapping the perineum." The four animals not in heat when decapitated showed negative responses as regards "treading" and tail movements, but raising the pelvis was obtained in three of them. Maes concluded that "some components at least of the sexual behavior are short arc reflexes, comparable to the scratch reflex, which can be elicited independently of the higher centers, but the occurrence of which depends strictly on hormonal conditions."

In a rather extensive study of the responses to genital stimulation which are obtainable in spinal and decerebrate cats Bromiley and Bard (1940) examined a number of decapitate preparations. Their results are not in accord with the interpretations of Maes. We found that the decapitate female cat, whether fully in heat or wholly anestrous, shows in the dorsal decubitus alternate movements of the hindlegs and dorsal flexion of the pelvis when the vulva or perineum is gently rubbed. Also, if in either preparation the tactile stimulus is applied to one side of the midline, the tail moves to the opposite side, the pelvis turns toward the stimulated side, the ipsilateral hindleg flexes while the contralateral extends and in this asymmetrical attitude both legs may execute alternate movements of flexion and extension at thigh, knee and ankle. All of these reactions have been obtained in decapitate males, but the tail deflection is less marked than in females, and sometimes it is absent. If the decapitate cat is placed on its chest and belly with the limbs in the normal resting or squatting position, rubbing the vulval, perineal or anal regions induces some elevation of the pelvis and vigorous alternating rhythmic movements of the hindlegs which closely resemble the treading movements seen in normal female cats which are in heat. It is this response which Maes observed, but we cannot agree that it depends on a state of estrus, for we have repeatedly obtained it in a series of anestrual decapitate females and even in decapitate males. In fact one of the best examples of this spinal "treading" was found in an animal which had been spayed thirty-nine months before and had never been treated with estrin; its genital tract was markedly atrophic. The decapitate cat does not give any specific response to mechanical stimulation of the vagina; the most common effect is flexion of the hindlimbs.

On the basis of the experiments just described I believe it can be stated with some confidence that neither estrin nor any bodily alteration produced by estrin can cause the spinal cord, after separation from the brain, to mediate any reflex suggestive of normal estrual behavior which cannot also be evoked in the anestrual spinal animal. This is not the equivalent of saying that estrin has no effect on spinal reflexes. Herren and Haterius (1931) have reported that in rats the Achilles reflex time is longer during normal estrus than during the dioestrum, shorter in ovariectomized animals before injection of estrin than after treatment. It may be that some of the reflexes of decapitate cats mentioned above are altered quantitatively by estrin; it is my impression that deflection of the tail on unilateral genital stimulation is somewhat brisker in females which have been decapitated during estrus.

It is not easy to assay the significance of those reflexes of the spinal animal which resemble certain components of the estrual behavior of normal individuals. They cannot properly be called estrual, for their occurrence is not dependent on a state of estrus. But as non-specific items of response they may be used by prespinal levels of integration in fashioning the estrual pattern of the normal animal. Some of the spinal responses to genital and anal stimulation, such as lateral deflection of the tail, may be parts of a cleaning reaction.

Bulbar, pontile, and lower mesencephalic levels

The decerebrate animal as usually prepared is one in which there remains of the neuraxis rostral to the spinal cord only the medulla, pons and lower mesencephalon. It therefore may serve as a basis for judging responses which are integrated at these supraspinal levels. The preparation, however, is dominated by the tonic activity of a bulbo-spinal postural mechanism which tends to fix the animal in the attitude of decerebrate rigidity. Phasic responses are evoked with far less readiness than in a spinal or decorticate animal; righting and walking cannot be elicited. These facts must be kept in mind in examining the capacity of a decerebrate animal to yield sexual behavior involving alternating or rhythmic movements.

Dempsey and Rioch (1939) decerebrated seven guinea pigs, four at the intercollicular level, three by the anemic method; although all were in estrus as a result of injections of estrin followed by progesterone, none gave any response "which even remotely resembled oestrous behavior." They obtained the same negative results in three estrous cats which had been decerebrated at the intercollicular level. All their animals showed marked decerebrate rigidity.

Bromiley and Bard (1940) have studied, with reference to genital and possible sexual responses, a series of twenty-eight female cats which were decerebrated at various levels between the point of exit of the third nerves and the trapezoid body. Nine of the animals were in full behavioral estrus at the time of decerebration, nineteen (of which ten had been spayed) were completely anestrual. To vaginal, vulval, perineal and anal stimulation the responses of all twenty-eight were essentially the same. We have convinced ourselves that the reactions of estrual and anestrual preparations of this kind are qualitatively identical. The state of estrus confers on them no additional reflexes or patterns of behavior which can be detected by ordinary means.

In the course of this work we found that all decerebrate female cats, whether wholly anestrual or fully in heat, regularly show a most striking collapse of the extensor rigidity of the forelegs when the distal vagina or its orifice is mechanically stimulated (Fig. 244).[3] The gentle insertion of a pipette or rod is the adequate stimulus. This response is specifically vaginal in origin; it cannot be obtained by stimulation of anus, rectum, external genitalia, urogenital sinus (vestibule) or urethra (neck of bladder), and no trace of it was elicitable in a series of twelve decerebrate male cats by any kind of genital stimulation. The inhibition of the extensors of the forelegs is tonic in the sense that it is maintained as long as the vaginal stimulation is kept up. During this time the shortening reaction cannot be detected; one can passively flex and extend a foreleg without encountering any resistance, but the moment the stimulus is withdrawn the member shows a marked shortening reaction, regains an attitude of rigid extension and again strongly resists passive flexion. Moderate flexion of the hindlegs almost invariably accompanies the foreleg collapse, but these limbs retain fair extensor tone. The response imposes on the standing decerebrate animal an attitude which is very suggestive of the estrual crouch of the normal female cat. As a part of the courtship activities the crouch does not depend on vaginal stimulation; it is even independent of the afferent con-

---

[3] This response was briefly reported a year ago (Bard, 1939). Since that time many more observations have been made.

nections of the vagina (Bard, 1935). Therefore Bromiley and I have been a little hesistant in concluding that the crouching attitude evocable in the female decerebrate cat denotes the existence of a bulbospinal mechanism capable of executing the major postural adjustment of feline estrus. To obtain further insight we investigated the effects of vaginal and other genital

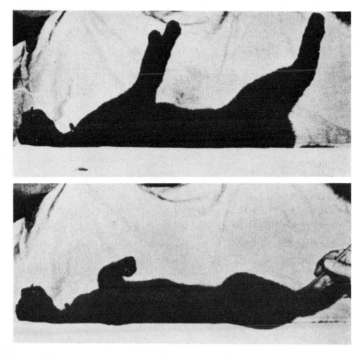

Fig. 244. Two photographs which illustrate the effect of vaginal stimulation on the attitude of the limbs of a decerebrate cat. The upper picture shows the animal before stimulation; the lower was taken a minute later, just after the tip of a rod had been inserted in the distal vagina. Note the complete collapse of the extensor rigidity of the forelimbs and the partial relaxation of that of the hindlimbs. When the cat is held in the standing position this response results in a crouching posture. (From unpublished experiments of Bromiley and Bard).

stimulation in ten decerebrate bitches. Although all exhibited good tonic neck and labyrinthine reflexes, we could obtain not the slightest trace of such a reaction to vaginal stimulation in any of them. This is perhaps significant in view of the fact that a crouch is not a part of the estrual behavior of the female dog.

Vaginal stimulation in the decerebrate cat, even when the

extensor tone is minimal, does not lead to any special phasic activity of the legs. On the other hand, vigorous rubbing of the vulval and perineal areas occasionally induces, if the animal is in the standing position, a sort of wobbling movement of the body and pelvis which is a little suggestive of estrual treading. Actual stepping does not occur, but there is some alternate flexion and extension of the hindlegs and alternate contraction of the lumbar muscles on the two sides. This response, which was obtained in only six of our twenty-eight decerebrate females—and has not been seen in males—is not at all depend-

Fig. 245. Photograph of a decerebrate cat showing the responses of hindlegs and tail to unilateral stimulation of the external genital region. The left side of the vulval and perineal areas is being rubbed with a rod (blurred in the photograph). (From unpublished experiments of Bromiley and Bard).

ent on estrus. It seems to be the result of the alternate bilateral occurrence of the same set of responses which are elicitable in decapitate or spinal cats on unilateral stimulation of the external genital region. These reactions are easily obtained in most decerebrate cats, males and females alike, especially when they are placed on their backs and a light mechanical stimulus is applied just to one side of the midline within the area immediately surrounding the external genitalia. They are: partial flexion of the ipsilateral hindleg, an increment in the extension of the contralateral hindleg, turning of the pelvis to the stimulated side and deflection of the tail to the opposite side. With the exception of the pelvic turning—which was obscured by the position of the animal—all these reactions can be seen in Fig. 245. The threshold and the vigor of the responses vary somewhat from preparation to preparation, the limb flexion being more marked the less the extensor rigidity, but they are not influenced by estrin. Their elicitation in males excludes all possibility that they are directly connected with estrus.

In discussing the neural mechanisms of estrual behavior Maes (1939) has suggested that "Dempsey and Rioch's failure to obtain sexual reflexes from decerebrate cats may be explained by the extensor rigidity which follows the mesencephalic transection, rhythmic reactions (as treading) and reflex raising of the pelvis requiring for their performance a normal balance of muscle tone." This was a reasonable argument, but it has lost much of its cogency in view of our finding that genital stimulation in decerebrate cats may evoke a reaction which, though suggestive of treading, is entirely independent of the sexual state of the animal.

There does not seem to be any positive evidence for the occurrence of sexual behavior of any kind in decerebrate male animals. It is a curious fact that it is extremely difficult to obtain any marked degree of reflex penile vasodilatation in decerebrate cats and dogs. Bromiley and I were never able to induce by manipulation of the organ anything more than the merest suggestion of erection in any of our decerebrate cats and Crystal has had the same experience. Further, this refractory condition is evidently not due to any overactivity of the vasoconstrictor mechanism, for removal of the sympathetic chains in abdomen and pelvis does not render the decerebrate cat any more prone to show an erection (Crystal).

Sexual behavior in decorticate animal

The evidence already brought forward indicates that after truncation of the central nervous system at any level below the middle third of the mesencephalon only the merest fragments of normal sexual behavior are elicitable. Further, it seems clear enough that, in the cat at least, estrin exerts its specific effect on the nervous system at some point above the lower portion of the midbrain. Attention must therefore be directed to higher levels of central integration.

There is evidence that in several mammalian forms the full pattern of mating behavior can be elaborated after removal of cerebral cortex. In a study of four chronically decorticate cats Bard and Rioch (1937; see also Bard, 1934) secured the first proof that a decorticate mammal is capable of showing typical estrual behavior. Their cat 103, in which all neocortex, part of the rhinencephalon, a little of the striatum and the rostro-lateral half of each thalamus had been surgically extirpated, spontaneously went into heat and displayed all the essential features of estrual behavior.

⁻In an attempt to secure further information concerning the estrual capacities of animals without cortex Bard (1936, 1939) prepared four additional cats. During preoperative observation periods estrus was induced by administration of estradiol benzoate and a basis was thus established for judging the postoperative responses to this substance. In each cat all neocortex was ablated. Considerable portions of rhinencephalic cortex were removed in three and in one animal this part of the brain was totally ablated. The striatal complex was partially removed in all; in the animal lacking olfactory cortex it was almost wholly ablated. The significant details of the observations made on these animals have been presented elsewhere (Bard, 1939). Here it will suffice to state that after removal of all cortex, much of the striatum and with functional loss—through secondary degeneration—of the greater part of the thalamus, the female cat when given estrin is capable of displaying typical estrual behavior. Courtship activities even occur spontaneously, the male is readily accepted and after-reactions follow intromission or any mechanical stimulation of the vagina. When not under the influence of estrin such animals do not display any trace of estrus, show anger when mounted and do not permit intromission. The after-reactions of all these cats, though distinct and sometimes briefly frantic, were usually delayed and lacked the vigor and nicety of execution which characterized them before the cerebral ablation.

Brooks (1937) has shown that female rabbits continue to mate and ovulate after removal of the neocortex and destruction of the olfactory bulbs. Three of his animals were completely decorticate (rhinencephalic cortex as well as neocortex removed); each exhibited typical mating behavior, but only one ovulated as a result of coitus. Rabbits lacking all cortex do not eat spontaneously and it is practically impossible to feed these herbivores adequately by artificial means; the ovulatory failures were doubtless due to poor nutritive condition.

Dempsey and Rioch (1939) obtained, by administration of estrin and progesterone, typical estrous responses in ovariectomized guinea pigs after removal of neocortex, hippocampus and most of the caudate and putamen.

It is also established that the mating activities of the female rat are independent of the neocortex, but totally decorticate individuals of this species have not been examined. Recently C. D. Davis (1939) has shown that after bilateral removal of all neocortex female rats still have normal estrous cycles, mate, become pregnant and give birth to normal young. Such animals also show the phenomenon of pseudopregnancy which is undoubtedly dependent on the central nervous system. In confirmation of the work of Beach (1937, 1938) and Stone (1939), Davis found that ablation of all neocortex precludes the display of maternal behavior.

Less information is available concerning the relationship of the cerebral cortex to the mating activities of male animals. Brooks (1937), extending the earlier experiments of Stone (see Stone, 1939), determined that the mating behavior of male rabbits is not interfered with by ablation of all neocortex, but is permanently abolished if the olfactory bulbs are also removed. Yet both Stone and Brooks found that mere removal of the olfactory bulbs in no way diminishes the sexual aggressiveness of this animal. Brooks' experiments indicate that the male rabbit, in contrast to the female, requires for the excitation of mating activity either the specific olfactory influence of the rhinencephalon or the neocortical mediation of visual, auditory or somesthetic sensibility. But the results of his investigation strongly suggest that the actual execution of the male's mating activity is effected by subcortical mechanisms.

Diencephalic and upper mesencephalic levels

The experimental results brought together in the foregoing section permit the conclusion that the cerebral cortex is not essential for the excitation and execution of the estrual behavior patterns of cats, rabbits and guinea pigs. Also, it has been shown that if in addition to the cortex much of the striatum is removed the capacity to show estrual behavior is not impaired. Further, the experiments on long surviving decorticate animals demonstrate that the responses of these preparations do not depend on those thalamic nuclei which project to the cortex, for even when the dorsal diencephalon is surgically spared these parts of the brain stem undergo a profound secondary degeneration. When this group of facts is combined with the evidence that true estrual behavior cannot be evoked in spinal, bulbospinal or low mid-brain preparations it appears reasonable to proceed on the assumption that an important central mechanism for estrual behavior, one which is specifically activated by estrin, must be situated somewhere in cerebral territory comprising the preoptic and septal areas, a part of the thalamus, the subthalamus, the hypothalamus and the upper portions of the midbrain.

Dempsey and Rioch (1939) have found that after recovery from operation female guinea pigs with unilateral and bilateral lesions involving the septal nuclei, anterior commissure, fornix and the medial half of the tuberculum olfactorium show normal sexual responses when injected with estrin and progesterone. After more extensive destructions involving the preoptic area and the rostral part of the hypothalamus their guinea pigs failed to show estrual responses, but were in such poor condition that the authors could not attribute these negative results to the destruction of any neural mechanism essential to sexual behavior.

Results obtained in a few acute experiments led Dempsey and Rioch to suggest that in the cat and guinea pig "the anterior limit of the mechanism which controls sexual behavior lies between the intercollicular level and the anterior limit of the mammillary bodies." The brain stem of an ovariectomized guinea pig which had been treated with estrin and progesterone was cut through along a transverse plane passing from the rostral edge of the anterior colliculus to a level just in front of the mammillary bodies. Seven hours later, "unmistakable oestrous reflexes were observed when the lumbar and perineal regions of the animal were stimulated." Then a cut was made from the same point dorsally which struck the base just behind the mammillary bodies. Thereafter no estrual responses could be obtained. A similar acute experiment was carried out on a cat which, as a result of injection of estrin, exhibited good estrual behavior. After removal of all brain tissue lying above a transection which sloped from a point 2 to 3 mm. rostral to the anterior colliculi to a point just ahead of the optic chiasm the animal responded to manual stimulation of the external genitalia by raising and rotating its pelvis, flexing its forelegs and extending its hindlegs. It thus assumed what appeared to be an estrual crouch. On vaginal stimulation it crouched on its forequarters and executed "treading" movements with extended hindlegs. The same responses were obtained after a second transection had been made from the same dorsal level to a point just rostral to the mammillary bodies. But after a third transection at the intercollicular level no comparable reactions could be evoked. The failure of these workers to obtain any estrual responses in guinea pigs and cats after primary decerebration at the intercollicular level has already been mentioned.

The effects on estrual behavior of restricted lesions involving the hypothalamus. Impressed by the probability that in the cat the essential central mechanism for estrual behavior must lie somewhere in the brain stem between the region of the anterior commissure and the middle of the mesencephalon, Magoun and Bard recently undertook a study of the effects of lesions restricted to the diencephalon and upper mesencephalon. The animals used were ovariectomized female cats which had received one or more treatments with estrin (estradiol benzoate in oil) and whose estrual capacities were therefore well established. Although this work has just begun we have already secured a few results which seem to bear rather directly on the question at issue and therefore will be presented in preliminary fashion.

The electrolytic lesions were made under nembutal anesthesia with the aid of the Horsley-Clarke instrument in the manner described by Ranson (1934). After varying periods of survival, the brains were fixed in formalin, washed, soaked in 20 per cent alcohol, and cut serially at 50 or 75 micra by a frozen section method (Marshall, 1940). Alternate sections at varying intervals through the series were stained by the thionin and Weil methods. The extent of the lesions was determined microscopically, drawn with the aid of a projector, and in some instances photographed (Fig. 246, 247 and 248).

Ranson (1934) has reported that female cats with "large bilaterally symmetrical lesions in the tuber, lateral to or behind the infundibulum" mate, give birth to full sized litters and nurse and care for the kittens quite normally. His animals went spontaneously into heat as a result of hypophysial activation of the ovaries. We have had several cats with somewhat comparable lesions whose responses to administration of estradiol benzoate were precisely the same as before operation. An example is cat E86 which was sacrificed 10 weeks after the lesions were made. Throughout this survival period she tended to be slightly lethargic and she showed a considerable,

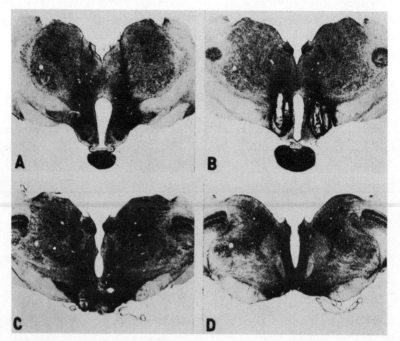

Fig. 246. Photomicrographs of four transverse sections through the brain stem of cat E86. Thionin. A, B, C, D are at intervals of 1 mm. In this figure and in Figs. 247 and 248 a small circular perforation marks the right side of each section (Magoun and Bard).

but by no means complete loss of ability to regulate body temperature in the face of environmental heat and cold (she was tested over the range, 8°–40°C.). Four sections through her brain stem are shown in Fig. 246 and the extent of the lesion may be further judged by a brief description of the histological findings.

Cat E86. The bilateral lesion commences in the anterior tuberal region (Fig. 246A) and reaches its full size in the posterior part of the tuber (Fig. 246B). Here it does not completely destroy the hypothalamus and is somewhat asymmetrical, so that a small amount of the right lateral hypothalamic area and the extent of the left periventricular area remain intact. The ventromedial part of the tuber is spared bilaterally.

The mammillary region (Fig. 246C) is also incompletely destroyed. The lateral part of the right lateral hypothalamic area, the dorsal part of the left posterior hypothalamic nucleus, and the ventromedial part of both medial mammillary nuclei, chiefly the left, remain intact.

The lesion rapidly decreases in size in the posterior mammillary region, and except for gliosis in the most anterior part of the interpeduncular area (Fig. 246D), the midbrain is intact.

Cat E86 was given estradiol (2,000 R.U.) on the seventh postoperative day. Three days later she exhibited mild courtship activities and showed a good after-reaction. This behavior, however, was not quite as pronounced as that which the same dose of estrin had produced before operation. Five weeks later, after a period of anestrus, another injection of the same amount of estrin led to a display of sexual behavior which was the equal of any seen in this animal before the hypothalamic injury.

In several cats extensive hypothalamic damage, resulting in loss of all temperature regulation and rather marked lethargy, has definitely curtailed the capacity to show estrual behavior. E77 was given estrin every three days during her survival of two weeks, but she never showed crouching and treading in response to vulval or vaginal stimulation; on the ninth postoperative day she exhibited a brief and mild but perfectly distinct after-reaction. Cats E79 and E91 also lived only two weeks after lesions which destroyed large portions of the hypothalamus, but neither showed the slightest trace of any form of estrual behavior. In the case of E79 the negative result may be attributable to the fact that she was not given estrin until the seventh postoperative day (the dose was repeated on the tenth day). E91 was fully in heat when the lesions were made and she received estradiol on the day of operation and on the fourth and tenth days, but she appeared to be in rather poor condition throughout her survival.

In this group of animals with very large hypothalamic destruction is cat E85 which was studied intensively during its survival of just seven weeks. The extent of the damage to her brain stem can be appreciated by inspection of Fig. 247 and perusal of the following account of the lesion.

Cat E85. The bilateral lesion in this animal commences in the supraoptic part of the hypothalamus and destroys all the caudal portion of this region

(Fig. 247A) except a narrow ventral strip and the most lateral part of the right lateral hypothalamic area.

At the tuberal level (Fig. 247B) the hypothalamus is completely destroyed except for a fragment of the right lateral hypothalamic area and for a narrow area on either side of the anterior part of the median eminence, through which the ventral supraoptic region retains connection with the infundibular stalk.

At the mammillary level (Fig. 247C) the lesion destroys the hypothalamus and involves the medial part of the basis pedunculi on the left. On the right side the lesion extends to the basis pedunculi but does not appreciably invade

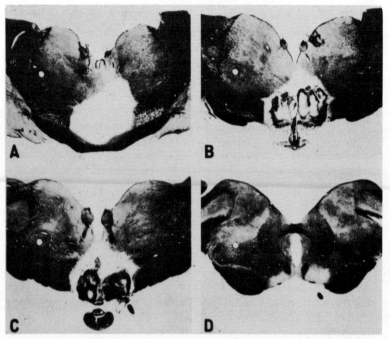

Fig. 247. Photomicrographs of four transverse sections through the brain stem of cat E85. Weil stain. A, B, C, D are at intervals of 1.5 mm. (Magoun and Bard).

it; in the lateral margins of the glial scar, close to the ventral surface of the brain, there are present a few large myelinated fibers, some of which have irregularly swollen and globulated sheaths. Fibers may be recognized in this position in the normal brain and appear to us to be related to the basis pedunculi rather than to the lateral hypothalamic area. No neurons remain in the lateral hypothalamic area at the mammillary level, but on the right side the small number of neurons normally found scattered within the medial part of the basis pedunculi remain intact; on the left side they are destroyed. No part of the mammillary bodies or premammillary areas remain.

In addition to the massive hypothalamic destruction, the lesion involves the ventromedial part of the thalamus, more on the left than the right; the brachium conjunctivum and striatal connections contained in the medial part of the subthalamus are invaded, especially on the left side.

The lesion ends at the transition to the midbrain, but softening and gliosis involve the anterior part of the interpeduncular area (Fig. 247D) and an area of softening extends backward for a short distance beneath the aqueduct. The remainder of the midbrain is intact.

The functional loss produced by this huge lesion is well indicated by the fact that E85 was wholly unable to maintain a normal body temperature when that of the environment fell below 28°–30°C. Just before the observations on this animal were terminated she was tested in a cold room at 8°–10°C. Her rectal temperature fell to 26°C. in 35 min. and to 23.5°C. after an hour. She failed to shiver or show piloerection and at no time did she engage in restless movements. She was not studied in a hot box, but on one occasion when the room became overheated her temperature rose to 42.8°C.; she did not pant or sweat. Although this animal was always lethargic, she never appeared acutely sick and she could not be described as cachectic. At autopsy infection was found in all cranial sinuses, but the lungs and gastrointestinal tract were normal. She had been kept in good nutritive condition by forced feeding. During the week preceding the operation E85 responded to a single injection of 2,000 R.U. of estradiol benzoate by exhibiting good estrual behavior and she accepted a male. Beginning on the day the lesions were made she was given 2,000 R.U. of the estrogen every third day for a period of 34 days. Throughout this time her responses to vulval and vaginal stimulation were frequently tested. Rubbing her vulva always caused extension of the hindlegs with elevation of the pelvis, some stepping movements and dorsal arching of the back; the foreparts were never lowered. The posture thus assumed was quite unlike an estrual crouch and the stepping on extended hindlegs bore little resemblance to real treading. Vaginal stimulation by means of a rod or pipette was invariably accompanied by this same reaction and it caused angry vocalization. On five occasions a slight but definite after-reaction followed the vaginal stimulation: after licking her vulva the cat turned over onto her side, rubbed her cheek against the floor and stroked her head with a forepaw. These few responses were the only clear signs of estrual behavior she ever exhibited. Cat E85 appeared to be wholly indifferent to the presence of males and when mounted by one she merely stood or squatted occasionally attempting to free herself from his grasp.

The results obtained in E85 suggested to us that a lesion which destroys much of the hypothalamus and breaks all known descending connections of the remaining parts will seriously interfere with the capacity of a cat to exhibit estrual behavior. To our surprise another animal with equally large hypothalamic involvement showed all the essential features of estrual behavior. This was E84. An account of the condition of her brain stem must be read to appreciate fully the extent and character of the lesion (see also Fig. 248).

Cat E84. The bilateral lesion in this animal commences in the posterior part of the tuber (Fig. 248A) and reaches full size in the mammillary region of the hypothalamus (Fig. 248B), at the caudal end of which it extends from the base of the brain upward into the thalamus (Fig. 248C) and from one basis pedunculi to the other. The medial part of the left basis pedunculi is invaded slightly; on the right side the lesion extends to the basis without invading it appreciably and a few large myelinated fibers with globulated sheaths are present in the lateral margins of the glial scar. These appear to us to be the most medial fibers of the basis pedunculi. It therefore appears that all descending fibers from the hypothalamus which run in the medial forebrain bundle as well as all other known descending fibers (*e.g.*, those of the periventricular system) from this region were interrupted.

In addition to the hypothalamic destruction, the medial part of the subthalamus and the ventromedial part of the thalamus were destroyed. The lesion spares all of the thalamus except the midline group of nuclei and a little of the region adjacent to them. The medial lemniscus appears to be intact. The brachium conjunctivum is involved bilaterally.

Caudal to the main body of the lesion, a triangular area of anemic softening extended backward into the midbrain (Fig. 248D) to the anterior end of the red nuclei, destroying the ventral half of the central grey, all of the interpeduncular area and the adjacent portion of the medial tegmental region. A small continuation of this softening could be traced in the central grey as far back as the pons.

Cat E84 lived only two weeks after the lesions were made. Throughout her survival she was extremely somnolent. Unless disturbed she lay quietly on her side, apparently asleep. If a sufficiently strong stimulus were applied she righted herself without great effort. At the time of operation this animal was in nearly full estrus as a result of having received, six days before, an injection of 2,000 R.U. of estradiol benzoate. She was given another injection of the same amount while under the anesthetic, but received no further treatment. On the third postoperative day she responded to vulval stimulation by assuming a crouch and treading, but a tendency of her hind-feet to slip backward distorted the typical estrual picture. At this time vaginal stimulation intensified these responses and was followed by squirming and head rubbing; this was an unmistakable after-reaction. By the seventh day very excellent crouching and treading could be evoked by gently rubbing or tapping the vulva or perineum, and insertion of a rod in the

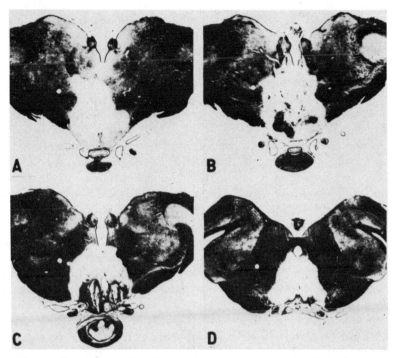

Fig. 248. Photomicrographs of four transverse sections through the brain stem of cat E84. Weil stain. A, B, C, D are at intervals of 1 mm. (Magoun and Bard).

vagina induced a delayed but really vigorous after-reaction, a prominent feature of which was sliding on her chin. Following this and before lapsing into somnolence she maintained a crouch and occasionally treaded.

It is no easy matter to reconcile the differences in estrual capacity shown by cats E85 and E84 on the basis of differences between the two lesions. In both the hypothalamus appeared to be completely destroyed at the mammillary level (Fig. 247 and 248) thus bilaterally interrupting all descending fibers of the medial forebrain bundle. Both lesions also obliterated that part of the hypothalamus which is most intimately related to the periventricular system of fibers. They also wholly destroyed the hypothalamic origins of the mammillothalamic and mammillotegmental tracts. The mammillary peduncle together with its hypothalamic projections was destroyed in both cats. In E84 the rostral hypothalamus including the entire supraoptic region was spared whereas in E85 much of this area was destroyed. It is, however, difficult to understand how this part of the brain could have been responsible for the much greater estrual performance of E84. The thalamus and subthalamus were invaded to a somewhat greater extent in E84. Mesencephalic damage was minimal in E85, considerable in E84. It appears that an explanation of the difference in estrual responses of these cats must await the results of further experimentation.

There are not yet available sufficient experimental facts to warrant any general statement about the relation of the hypothalamus to the excitation and execution of estrual behavior. The results which Magoun and I have already obtained appear, however, to demonstrate that after complete destruction of the caudal portion of the hypothalamus in cats typical estrual behavior can be evoked by administration of estrin. In the light of our observations on cat E84 the estrual behavior which Dempsey and Rioch were able to evoke after a pre-mammillary transection of the brain stem appears to have been dependent on the integrity of mesencephalic rather than hypothalamic mechanisms. On the other hand, we are faced with the fact that in some of our cats the capacity to exhibit estrual behavior appeared to be markedly reduced by lesions which involved much of the hypothalamus but scarcely invaded the midbrain. The possibility that the hypothalamus is involved in the central control of estrual behavior has not been excluded.

SUMMARY

1. Many forms of overt sexual behavior are reaction patterns which represent the welding together of individual responses to fashion a complex act directed toward a specific end. Such behavior is managed by central mechanisms which are activated either directly or indirectly by gonadal hormones. The question of the participation of the hypothalamus, or

of any other part of the brain, in this central control should be explored in relation to the rôles played by other levels in the mediation or integration of the items of behavior which, occurring together or in sequence, constitute the patterned response. Therefore considerable attention has been devoted to an examination of the management of sexual activities by the major levels of central integration.

2. In spinal, bulbospinal and low midbrain animals certain responses are evocable which appear to be fragments of the mating and copulatory patterns shown by intact individuals of the same species and sex. The most conspicuous examples are: Erection and a copulatory posture of the hindquarters in spinal dogs and cats; and a postural adjustment obtainable by vaginal stimulation in decerebrate female cats, whether estrual or anestrual, which results in an attitude suggestive of the peculiar feline estrual crouch. In spinal and decerebrate cats genital stimulation may produce alternating rhythmic movements which resemble the treading of estrual females, but again this response does not depend on a state of estrus. In fact the available evidence indicates that estrin exerts its specific effect on the central nervous system at some level above the lower portion of the mesencephalon. After truncation of the neuraxis anywhere below the caudal midbrain, cats and guinea pigs in the estrous state do not exhibit any reflex suggestive of estrual behavior which cannot also be evoked in anestrual preparations.

3. It has been demonstrated that in several species of mammals the full pattern of mating behavior can be elicited after removal of all cerebral cortex. When under the influence of ovarian hormones, wholly decorticate female cats, rabbits and guinea pigs exhibit typical estrual behavior. After removal of all cortex male rabbits are able to execute effective copulatory behavior. These patterns of sexual response must therefore be elaborated by subcortical mechanisms. There can be no doubt, however, that in any mammal the presence of the cerebral cortex greatly increases the number of circumstances which are able to modify the occurrence of sexual behavior. Evidence is lacking to show that the striatum plays a part in the control of sexual behavior and certain experimental results appear to exclude the possibility that the greater part of the thalamus is concerned in any essential way with estrual behavior.

4. The problem of the participation of hypothalamic mechanisms in the elaboration and execution of estrual behavior has received some experimental attention. The effects of brain stem lesions, which involve this division of the diencephalon, indicate that it may exert an influence on the capacity of female cats to display sexual excitement. It has been shown, however, that the full pattern of estrual behavior can be obtained in cats with complete destruction of either the middle or caudal portions of the hypothalamus. The estrual behavior which has been seen in guinea pigs and cats after premammillary transections of the brain stem probably depends on upper mesencephalic mechanisms. Further work must be done before any precise statement can be made concerning the relation of the hypothalamus to the central management of sexual behavior.

DISCUSSION

Dr. DAVID McK. RIOCH (St. Louis, Mo.): I have nothing very much to add to what Dr. Brooks and Dr. Bard had to say except to express my admiration for the care and exactness of their work. The results, I think, check with what has been done so far in terms of localization of sexual functions in brain stem. The last preparation of Dr. Bard's is one of the most interesting, I think, in the sense of ruling out the hypothalamus as the center for organization of at least the striated muscle pattern of sexual behavior and probably putting the central mechanism for such behavior in the reticular substance of the mesencephalon. That is another catch basket which will probably get overloaded very shortly.

I was wondering if Dr. Bard had any idea as to whether the pattern of behavior which he described in the decerebrate cat on vaginal stimulation—and I believe that this is the first time that a behavior really resembling sex behavior has been described in a decerebrate animal—has any relation to the flexor posture which Dusser de Barenne regarded as a sexual reflex in spinal cats. I have been inclined to doubt the sexual nature of the spinal posture, but I would like to hear if Dr. Bard has anything to add.

The experiments on cutting the pituitary stalk in the monkey that Dr. Brooks is now carrying out are also very interesting. Of course, it is known in the rat and also in the guinea pig that cutting the pituitary stalk does not result in interrupting the sexual cycle permanently. Apparently in these animals the cycle is controlled by some other mechanism. Dempsey has shown that adequately in the guinea pig and it has previously been shown in the rat. I think it would be very interesting if Dr. Brooks is able to show in the future that the monkey behaves in a different manner.

Dr. GEORGE B. WISLOCKI (Boston, Mass.): [At this point, apropos of the presentation of Dr. Chandler Brooks, Dr. Wislocki presented briefly a summary of the work of Dr. U. U. Uotila and of Dr. Edward Dempsey on the hypothalamic control of anterior pituitary function. Since their investigations are being presented in a separate chapter in the present volume, the reader is referred to those passages in considering the report of Dr. Brooks.]

Dr. LAWRENCE S. KUBIE (New York, N. Y.): I would like to ask Dr. Bard whether in his experimental work he has found any criteria for the occurrence of orgasm in the female cat. I ask this for two reasons. In the first place, the after-reaction which he describes suggests strongly a state of accumulated sexual excitement without discharge. That is interesting in itself. It is interesting if in animals the female has difficulties in achieving orgasm comparable to the difficulties one finds in women. This may be interesting both physiologically, and from an evolutionary point of view. In addition, however, the point may have a more important bearing upon Dr. Bard's experimental procedure. If we think of Dr. Bronk's report of yesterday and also of the fact that orgasm apparently involves a discharge of sympathetic activity, it becomes conceivable that the destruction of the hypothalamus, while leaving unaffected all of the preliminary sexual activities and the crouching pattern, could none the less by removing the supporting action of the hypothalamus on sympathetic discharges from the hindbrain, make the final orgasm itself impossible. In doing so the after-reaction would be unmodified or even exaggerated, exactly as Dr. Bard has described it; and would be an expression of the frustrated and unsatisfied sexual excitement of the female cat.

If it is impossible to judge of the occurrence of orgasm in the female cat, it might be illuminating to observe whether the male cat without a hypothalamus can achieve orgasm.

PRESIDENT FULTON: I wonder whether Dr. Rasmussen would care to comment on the innervation of the anterior lobe.

DR. RASMUSSEN: I am of the same opinion still—very few, probably a negligible number of nerve fibers from the infundibulum stem reaching the pars distalis of the hypophysis.

DR. JOSEPH HINSEY (New York, N. Y.): A number of years ago, we were interested in the phenomenon of ovulation. Working with Markee, we were able to show that pregnancy urine produced the phenomenon of ovulation when the hypothalamus was completely removed in the rabbit. Of course, it is known that ovulation can take place in the transplanted ovaries, and we came to the conclusion that certainly the action of pregnancy urine upon the hypophysis was one which did not involve any activity of the hypothalamus. The sparsity of nerve fibers in the pars distalis is something of some concern. The difficulty of staining these fibers is certainly real, but people who have been able to see them report there have been very few. Now it may be that there are few there, or their sparsity may be a reflection of the extreme difficulty of staining the nervous components of this region.

I have wondered whether or not the relationship of the hypothalamus to the activation of the pars distalis, might be one that would take place through an effect on the neurohypophysis. The blood supply, as Dr. Wislocki described that of the neurohypophysis and the sinusoids which communicated to the pars distalis, might give a method of transport of substances from the neurohypophysis to the pars distalis. I wonder what Dr. Brooks would have to say about that. We have had the same experience that he has had in finding an absence of ovulation after coitus in stalk-sectioned rabbits.

DR. R. LORENTE DE NÓ: The discussion has been very long, but I would like to ask Dr. Bard a question of a very general character. Dr. Bard has made a definition of center which appeals very much to me and that definition has no indication of inclusion of the localization of function. The center of the nervous system begins to be conceded as a group of neurons connected among themselves in many different ways and it appears that center means, so to say, a junction which is a transmission of a certain set of impulses, but there is no particular function localized in any particular center. Perhaps Dr. Bard will not have time to express his opinion on this point today, but I would like if he would make a statement for the publication of the book of this year.

DR. PERCIVAL BAILEY (Chicago, Ill.): I was going to say that one might suppose a priori the animal most likely to furnish an answer to this question would be the whale. In the whale, the anterior lobe is entirely separated from the posterior lobe. Unfortunately, Dr. Geiling has been unable to obtain the glands in sufficiently fresh state to make the preparations useful. It may, however, some day be possible to use this animal to answer that question.

DR. GEORGE B. WISLOCKI (Boston, Mass.): Getting into subsidiary discussion, having worked on the whale with Dr. Geiling, it isn't quite correct the way this has been stated by Dr. Bailey, in this sense, that in the whale, certainly the pars distalis, the anterior lobe, is separate from the neural lobe and quite distinctly so and separated by duras. There are no anatomical pathways or no vascular pathways that create a direct transfer from the neural process to the anterior lobe. Nevertheless, the anterior lobe has a tongue of pars tuberalis that goes up and embraces the stalk, just where the stalk or the median eminence connects with the hypothalamus. It is embraced there by a very fine and delicate tuberalis prolongation that goes up and encloses the stalk like a minute little ring. Hence, it is anatomically conceivable and likely that if the anterior lobe received innervation from above, it is by virtue of this very tiny prolongation or tongue of tissue that goes up, and there would be an anatomical pathway for innervation of the anterior lobe by the hypothalamus.

DR. BAILEY (Chicago, Ill.): Dr. Wislocki is not consistent with his own terminology. My remarks referred to the anterior lobe, which, by his own definition, does not include the pars tuberalis.

DR. CHANDLER McC. BROOKS (Baltimore, Md.): Dr. Rioch and Dr. Wislocki mentioned work of Uotila and Dempsey. I can say only that in so far as the present literature is concerned, there is still some uncertainty concerning the effect of stalk transection in the polyestrous forms. According to Schweizer, Charipper and Haterius, hypophysectomized guinea pigs with functional anterior lobe transplants come into permanent estrus but do not show cycles. It has been claimed also that in the rat after stalk transection cycles do reappear; but instead of being four-day cycles, they are long, twelve, sixteen to twenty-day cycles. Dr. Richter reported this. Unfortunately, that is the way my stalk-transected rats have thus far reacted. I had intended to study pseudopregnancy but the presence of long cycles rendered this difficult. I hope the group working at Harvard will study pseudopregnancy since they can obtain rats with stalks transected which show normal cycles.

In reply to Dr. Rasmussen, we are in perfect agreement; the fibers are rather few. I, however, do not know how to interpret the physiological evidence if there is not some way of getting impulses to the pituitary.

With respect to Dr. Hinsey's discussion and his suggestion that the pars neuralis may play some rôle in the excitation of pars distalis activity, that is a possibility. No one has been able to produce ovulation in the rabbit with pituitrin. That, however, does not settle the question. There is also the possibility of a hormone coming down from the hypothalamus. These possibilities, however, have less evidence to support them than does the hypothesis of a neural control of the pars distalis.

DR. PHILIP BARD (Baltimore, Md.): Dr. Rioch mentioned the reticular substance of the mesencephalon. We have our eye upon it and have at present one or two cats in which lesions have been made there. I think that before we get through, we will be able to tell you whether that particular region is of significance or not in relation to estrual behavior. There are other regions to be investigated, the subthalamus, for example. Although the results so far obtained are not conclusive, I am inclined to think that the subthalamus has little or nothing to do with this kind of behavior.

Dr. Kubie asked me a question concerning orgasm in the female cat. It is a point of some interest. The criterion, of course, in any work on animals must be based on the behavior of the animal. Faced with different kinds of behavior, one can, on the basis of that criterion, have all sorts of orgasms. In the cat the after-reaction may be the overt aspect of an orgasm. There is one point of interest which I did not mention in my talk. It is that in several of the cats with large hypothalamic lesions, the after-reaction was delayed. That, however, is also true of decorticate cats. Whether the accumulation of excitation which eventually leads to the discharge takes longer after removal of cortex or of hypothalamus, I don't know. I merely mention this as a suggestion in reply to your question, Dr. Kubie.

Dr. Lorente de Nó raised the question of a definition of a center. I purposely gave the definition which I did in order to clarify the material with which I have been working. He mentioned the matter of localization, the question of whether localization is involved in the concept of a center. I think that in my definition I did definitely refer to the matter of localization and one can quibble whether it is localization of function or functional localization and get, I think, nowhere. The important point from an experimental point of view is that if you can localize a process, you can then study it; but if you cannot localize it, attempts to study it are apt to be abortive. I draw your attention particularly to the development of our knowledge of the mammalian heart; there the establishment of localization of function led to an enormous advance and I think in studies on the central nervous system that point of view should be kept in mind. The function is what is represented in a center, a patterned response in this particular case. That is what is localized in the sense that I used the term.

PRESIDENT FULTON: It is a source of very great regret that Dr. Uotila cannot be here. This gifted young Finnish investigator has made a fundamental contribution that Dr. Wislocki, I am happy to say, has described to you. Owing to the fact that he was surgically trained, he was called back shortly after the outbreak of hostilities on the Finnish border, and he sailed a week ago last Saturday on the Stavangerfjord.

REFERENCES

BALL, JOSEPHINE. J. comp. Psychol., 1934, 18, 419–422.

BARD, P. Psychol. Rev., 1934, 41, 309–329; 424–449.

BARD, P. Amer. J. Physiol., 1935, 113, 5.

BARD, P. Amer. J. Physiol., 1936, 116, 4–5.

BARD, P. Res. Publ. Ass. nerv. ment. Dis., 1939, 19, 190–218.

BARD, P., AND RIOCH, D. McK. Johns Hopk. Hosp. Bull., 1937, 60, 73–147.

BEACH, F. A., JR. J. comp. Psychol., 1937, 24, 393–436.

BEACH, F. A., JR. J. genet. Psychol., 1938, 53, 109–148.

BROMILEY, R. B., AND BARD, P. (In preparation.)

BROOKS, C. McC. Amer. J. Physiol., 1937, 120, 544–553.

BROOKS, C. McC. Res. Publ. Ass. nerv. ment. Dis., 1940, 20, 525–550.

DAVIS, C. D. Amer. J. Physiol., 1939, 127, 374–380.

DEMPSEY, E. W., HERTZ, R., AND YOUNG, W. C. Amer. J. Physiol., 1936, 116, 201–209.

DEMPSEY, E. W., AND RIOCH, D. McK. J. Neurophysiol., 1939, 2, 9–18.

DUSSER DE BARENNE, J. G., AND KOSKOFF, Y. D. Amer. J. Physiol., 1932, 102, 75–86.

DUSSER DE BARENNE, J. G., AND KOSKOFF, Y. D. Amer. J. Physiol., 1934, 107, 441–446.

FULTON, J. F. Res. Publ. Ass. nerv. ment. Dis., 1939, 19, 219–236.

HERREN, R. Y., AND HATERIUS, H. O. Amer. J. Physiol., 1931, 96, 214–220.

MAES, J. P. Nature, Lond., 1939, 144, 598.

MAKEPEACE, A. W., WEINSTEIN, G. L., AND FRIEDMAN, M. H. Amer. J. Physiol., 1937, 119, 512–516.

MARSHALL, W. H. (In preparation.)

RANSON, S. W. Psychiat. neurol. Bl., Amst., 1934, 38, 534–543.

RANSON, S. W. Trans. Coll. Phys. Philad., 1934, 2, 222–242.

RANSON, S. W., AND MAGOUN, H. W. Ergebn. Physiol., 1939, 41, 56–163.

ROOT, W. S., AND BARD, P. Amer. J. Physiol., 1937, 119, 392–393.

SHERRINGTON, C. S. In: SCHÄFER: Text-book of physiology, 1900, 2, 782–883.

STONE, C. P. In: ALLEN, Sex and internal secretions, (pp. 1213–1262), Baltimore, Williams & Wilkins, 1939.

YOUNG, W. C., DEMPSEY, E. W. AND MYERS, H. I. J. comp. Psychol., 1935, 19, 313–335.

Reprinted from *Physiol. Rev.*, **47**, 289–316 (1967)

30

Cerebral and Hormonal Control of Reflexive Mechanisms Involved in Copulatory Behavior[1]

FRANK A. BEACH

Department of Psychology, University of California, Berkeley, California

I. FOUR PROPOSITIONS TO BE EVALUATED

The purpose of this review is to assemble and evaluate evidence bearing on the following propositions. *1*) The species-specific copulatory patterns of vertebrates consist in part of reflexes mediated by spinal and myelencephalic mechanisms that are capable of functioning after separation from more rostral parts of the central nervous system (CNS). *2*) In the intact adult organism these "lower centers" are under varying degrees of inhibitory control by more anteriorly placed neural tissue, and in some instances this control is inhibitory. *3*) Ovarian and testicular secretions may have relatively little or no direct effects on the centers located in the cord and the myelencephalon. *4*) One of the ways in which gonadal hormones influence the occurrence of copulatory behavior is by modifying cerebral control of the lower, reflexive mechanisms.

II. FEMALE MAMMALS

The literature contains a number of reports bearing directly or indirectly on several of the foregoing propositions as they apply to female rodents and carnivores. Unfortunately, there is very little relevant information concerning other orders of mammals.

[1] Financial assistance in the preparation of this article and support of the author's own researches cited herein were provided in part by USPHS Grant MH 04000.

289

A. Rodents

It is well established for a number of rodent species that with very few exceptions the female cooperates in coital relations with the male only when she is physiologically in estrus. Studies of intact female guinea pigs, rats, mice, and hamsters have revealed a clear-cut relation between the secretory cycle of the ovary and the periodic appearance of sexual receptivity (81). More direct proof that sexual acceptance of the male depends on ovarian hormones is provided by the results of a large number of investigations demonstrating that administration of exogenous estrogen or estrogen plus progesterone is followed by the appearance of sexually receptive behavior in the ovariectomized female of all species just mentioned (81).

In every order of mammals the details of the feminine copulatory pattern vary from species to species, but in all rodents that have been carefully studied one feature seems universally present. This is the adoption of a posture involving opisthotonos or lordosis with consequent exposure of the perineum, which insures maximal accessibility of the vaginal orifice and thus facilitates the mounted male's achievement of intromission. Very often the lordosis reflex is associated with dilation of the vaginal and anal orifices, and in species possessing a long tail this member usually is dorsiflexed or laterally deviated.

This pattern of simple reflexes cannot be evoked during anestrus or diestrus and is not exhibited in untreated, ovariectomized females; but if *propositions 1* and *2* are correct it should be possible to elicit lordosis and the associated responses in nonestrous animals after transection of the medulla or the spinal cord. Unfortunately the relevant critical experiments have not been conducted. Dempsey and Rioch (28) reported that female guinea pigs in estrus ceased showing coital responses when the brain was sectioned at the intercollicular level, but the tests for sexual activity were carried out promptly after transection, and since marked decerebrate rigidity developed immediately in all cases the failure to elicit specific reflexes is not particularly revealing.

The lack of definitive experiments to indicate the presence or absence of the lordosis response in female rodents after separation of the cord and myelencephalon from more rostral parts of the CNS deprives us of any direct evidence as to the possible existence of higher inhibitory mechanisms, but a number of studies involving injury to different areas in the forebrain have yielded results pertinent to this issue. Several investigators have reported the release of mating behavior from the usual hormonal controls after the destruction of circumscribed brain regions.

According to Zouhar and deGroot (84) female rats sustaining bilateral lesions in the medial habenular nuclei exhibited frequent matings during vaginal diestrus as well as during estrus. Lacking any detailed description of the behavior involved one cannot judge its completeness or normality, but since vaginal plugs were recovered from diestrous females it is certain that at the very least these animals displayed lordosis when mounted by males; otherwise intromission and the deposition of seminal plugs would have been impossible. DeGroot and Critchlow (27) reported frequent copulation at all stages of the vaginal cycle by female rats suffering destruction of the basomedial area of the amygdaloid nuclear complex.

Law and Meagher (54) observed directly the responses of female rats to sexually active males and found that females with lesions in the premammillary region of the posterior hypothalamus copulated during diestrus and exhibited a higher frequency of lordosis than normal females in full estrus. The same workers noted that comparable results could be produced by lesions in the preoptic area of the hypothalamus, and their conclusions were as follows: "That there is a system in the female rat which opposes or inhibits receptivity is apparent from the data for animals with posterior hypothalamic lesions and probably from the data for animals with anterior hypothalamic lesions as well. These influences may well be related to others originating in more remote areas, as reported for the amygdala and the cortex" (54, p. 1627).

A study of hypothalamic potentials evoked in female rats by vaginal stimulation during induced estrus and during diestrus led Law and Sackett (55, p. 43) to advance the following proposal: "One might visualize posterior hypothalamic areas as releasing the execution of copulatory behavior when signals from preoptic and supra chiasmatic critical areas relay the appropriate 'go' signal."

Goy and Phoenix (35) studied the sexually receptive behavior of ovariectomized guinea pigs before and after placement of lesions in the hypothalamus. Induction of physiological estrus was achieved by administration of estrogen and progesterone. Injury involving mammillary and premammillary structures was without effect. Destruction in the midventral region usually reduced or eliminated the mating reactions normally elicited after administration of the ovarian hormones. However, two spayed females with midventral lesions exhibited full sexual receptivity when they had not been given hormone treatment and when the vagina was in the typically anestrous condition. This finding led Goy and Phoenix (35, p. 37) to the following conclusion: "The possibility that the oestrogen-progesterone synergism common among laboratory rodents may suppress activity in a region which normally inhibits sexual activity deserves further attention."

It is well established that complete removal of the neocortex does not eliminate the copulatory behavior during estrus of female rats (26), guinea pigs (28), or rabbits (21). In fact a total loss of the neopallium sometimes results in the intensification of certain coitus-related responses. For example, Beach (8) found that 4 of 7 decorticate female rats in estrus showed lordosis when a pipette was inserted in the vagina, although before operation none of these individuals had reacted in this fashion to the same stimulus. A fifth decorticate "stiffened into the lordosis position" when picked up by the experimenter and maintained the same posture when she was replaced on the table. Abnormal strength of the lordosis reflex was evident when decorticate females were tested with potent males. When the male mounted, the female's lordosis posture was greatly exaggerated as illustrated in Figure 1.

After the copulating male dismounts, normal female rats either run away or sit up and clean the vaginal area. In either case the male must approach the female again and initiate the next mount, which in turn evokes another lordosis

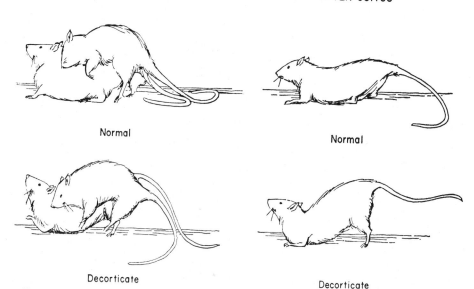

FEMALE MATING POSTURE

DURING COITUS   AFTER COITUS

Normal   Normal

Decorticate   Decorticate

FIG. 1. Tracings from 16-mm film of normal and cerebrally decorticated female rats.

response. Several of the decorticate females observed by Beach regularly adopted the lordosis pose on the male's first mount and then held this position for long periods while the male mounted, achieved insertion, dismounted, and then mounted again. Some of these females remained motionless at the same location while the male achieved 10 or more successive intromissions—an occurrence never observed before the females were decorticated. In a different study (7) one hemidecorticate female rat mated repeatedly and showed strong lordosis when she was in diestrus and her vaginal smear contained only leukocytes. Prior to brain injury this rat never accepted the male except during the period of vaginal estrus when the smear consisted of nucleated and cornified epithelium.

Since lordosis constitutes an essential element in the feminine copulatory patterns of several mammalian species it is properly designated as a "coitus-related" response. However, in some species the same reaction can occur under quite different circumstances. For example Boling et al. (20) observed that infant guinea pigs exhibit lordosis and pudendal dilation when the experimenter stimulates the animal's rump with his fingers. The lordosis response is normally present in both sexes at birth and can be experimentally elicited in the fetus 4 days before term (12). Reactivity to artificial stimulation dies out in normal young within a few hours postpartum, but nursing males and females continue to show lordosis and dilation when the mother licks the rump and perineal region.

Beach has suggested (12) that these reflexive responses constitute part of the infantile pattern of elimination in which maternal stimulation of the excretory orifices promotes the voiding of urine and feces. This hypothesis helps to explain the occurrence of lordosis and dilation in both sexes without the support of ovarian hormones. It has been proposed further that lordosis in response to tactile stimulation disappears when the control of excretory reflexes shifts to internal sensory receptors and when the spinal mechanisms involved come under the inhibitory control of more slowly maturing cerebral centers or systems. Lordosis and dilation in neonatal guinea pigs are thus conceived as resembling the forced grasp and Babinski reflexes of the infant primate. Finally, the hypothesis has been advanced that in the adult female, but not the male, the cerebral "lordosis-inhibiting mechanism" is itself inhibited by the synergistic action of estrogen and progesterone (12).

This analysis of an ontogenetic shift in the adaptive function and the physiological control of lordosis and pudendal dilation affords an explanation for certain otherwise contradictory findings reported by Young and his co-workers (82). Female guinea pigs given androgen during the appropriate stage of fetal development do not exhibit "heat behavior" or "sexual receptivity" in adulthood in response to the injection of estrogen and progesterone. The proferred explanation has been that early androgen treatment suppresses the development or "organization" of neural tissues destined to mediate "feminine behavior." Now, the feminine behavior referred to consists of displaying lordosis *in copula*, and for this reason it was initially somewhat perplexing to discover that androgen-treated females readily exhibit lordosis in infancy although they cannot be induced to do so in adulthood (62). The confusion is dispelled when it is recognized that the neonatal response is not "heat behavior," but part of the infantile excretory pattern that has no relation to reproduction and is independent of ovarian hormones. One effect of prenatal androgen treatment may be to reduce or eliminate the responsiveness of the cerebral lordosis-inhibiting mechanism to the action of estrogen and progesterone and thereby to prevent "heat behavior" in adulthood.

B. Carnivores

There is little doubt that the willingness of female carnivores to accept the male in coitus depends heavily on estrogenic stimulation. After observing 30 intact female cats in 340 mating tests and 8 spayed females in 300 tests Michael (64) concluded that the species-specific pattern of sexually receptive behavior can be elicited only when the vaginal smear indicates proestrus or estrus. In the felidae the female's behavior during copulation includes four distinct responses: *1*) pronounced flexion of the front limbs with consequent lowering of the foreparts of the body; *2*) moderate flexion of the hindlimbs that, combined with the more marked forelimb response, leaves the perineal area in an elevated position; *3*) lateral deviation of the tail; and *4*) "treading" or alternate stepping movements of the hindlegs.

Bromiley and Bard's studies of behavior evokable in spinal cats revealed that after cord transection at least the last three responses can be produced by gently tapping the perineal region (5). Maes (61) described the same phenomenon and reported that the reflexive reactions were evokable only when the spinal female

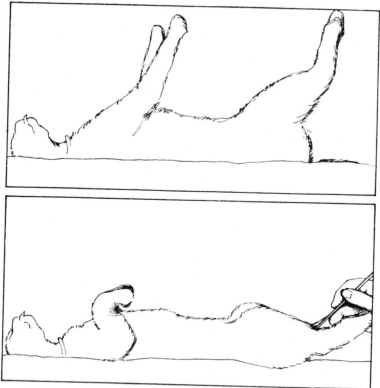

Fɪɢ. 2. Inhibition of decerebrate rigidity in a female cat by vaginal stimulation. [From Bard (5)].

was in estrus; but according to Bard, he and Bromiley elicited all three reactions repeatedly and reliably in a spayed female with a completely atrophic uterus. Because estrogen administration failed to alter the reflex excitability of his experimental female cats Bard (5, p. 559) reached the following conclusions: "On the basis of the experiments just described I believe it can be stated with some confidence that neither estrin nor any bodily alteration produced by estrin can cause a spinal cord, after separation from the brain, to mediate any reflex suggestive of normal estrual behavior *which cannot also be evoked in the anestrual spinal animal*" (italics added). In evaluating this type of generalization it is advisable to hold in mind the fact that no tests have been reported to indicate *1)* the intensity of stimulation necessary to evoke the reactions in question or *2)* the strength or frequency of the reactions themselves. The only fact established is that the responses were present postoperatively. The possibility that estrogen might lower the threshold of stimulation or exert quantitative effects on the surviving reactions cannot be excluded a priori.

In addition to studying spinal preparations, Bromiley and Bard examined the reflexive behavior of decerebrate cats. Decerebrate preparations invariably display the pattern of extensor rigidity illustrated in the top portion of Figure 2. However, one form of stimulation produces the response shown in the bottom portion of the figure and described by Bard (5, p. 560–561) as follows:

> ... all decerebrate female cats, whether wholly anestrual or fully in heat, regularly show a most striking collapse of the extensor rigidity of the forelegs when the distal vagina or its orifice is mechanically stimulated. The gentle insertion of the pipette or rod is the adequate stimulus ... The inhibition of the extensors of the foreleg is tonic in the sense that it is maintained as long as the vaginal stimulation is kept up ... but the moment the stimulus is withdrawn the member ... regains an attitude of rigid extension ... Moderate flexion of the hind legs almost invariably accompanies the foreleg collapse, but these limbs retain fair extensor tone. The response imposes on the standing decerebrate animal an attitude which is very suggestive of the estrual crouch of the normal female cat.

Three points connected with Bard's report deserve emphasis. First, these reflexes are elicitable in both estrous and anestrous females after decerebration and therefore are not dependent on ovarian hormones. Second, they are evoked specifically and exclusively by excitation of receptors in the interior of the vagina and cannot be called forth by stimulation of the external genitalia, anus, rectum, urogenital sinus, or urethra. They cannot, therefore, be produced in the male cat. Third, Bromiley and Bard examined a series of 10 decerebrate bitches and could not evoke, by stimulation of the vagina or any other area, either collapse of extensor rigidity in the forelimbs or hindlimb flexion. This failure is very significant because the copulatory posture of the estrous bitch, unlike that of the female cat, depends on maximal rigidity of all four limbs to support the weight of the mounted male. The reflexes elicited in the decerebrate female cat thus definitely constitute units in the species-specific feline mating pattern.

Bard (5) stated that female cats deprived of the cerebral cortex will accept the male during the period of estrus, but at all other times will respond antagonistically if he attempts to mount. However, Schreiner and Kling (72) found that anestrous females of this species sustaining bilateral invasion of the amygdaloid nuclei and adjacent structures often react to "handling" by arching the tail and assuming the typical coital posture. The same reactions can be elicited in some of these animals by inserting a glass rod in the vagina. Finally, according to these workers, amygdalectomized females not in estrus will tolerate mounting by the male and will display lordosis while mounted. This is behavior that Michael (64) asserts is never exhibited by female cats with an intact nervous system.

In partial confirmation of Schreiner and Kling's findings, Green et al. (36, p. 519) stated that three female cats with lesions in the pyriform cortex (and involving the amygdala in 2 of the 3 cases) showed reactions judged to be "suggestive of the abnormal estrous behavior as described by Schreiner and Kling."

Gastaut (34) described the responses of amygdalectomized cats to rhythmic manual stimulation of the perineal region. These included elevation of the pelvis and lateral deviation of the tail. This behavior could be evoked in anestrous females after but not before brain injury.

There appears to be some justification for concluding that in female carnivores, as in female rodents, certain types of injury to the CNS can result in the appearance of coitus-related reflexes in the absence of those ovarian hormones that are essential for the occurrence of the same responses in animals with an intact nervous system.

C. *Primates*

There is very little to report in connection with the neural basis for reflexive reactions associated with coital behavior in female primates. According to Klüver (49) the female *Macacus rhesus* exhibits "intensification of sexual responses" after extirpation of the temporal lobes; and this postoperative change has been observed in individuals previously deprived of uteri and ovaries. Unfortunately the descriptions of behavior are idiosyncratic and qualitative, and no preoperative base line is presented. Furthermore, it is known that ovariectomized monkeys without brain lesions will at least occasionally copulate with males. For these reasons Klüver's report is exceedingly difficult to interpret.

No usable descriptions of sexual functions in women suffering neurological disease, injury, or surgical insult have been found in the medical literature.[2] This lacuna probably reflects in part the failure of male clinicians to accord feminine psychosexual problems the same attention and interest that is devoted to understanding their counterparts in the male. Although as noted later there is an appreciable body of evidence relating to sexual functions in paraplegic men, a search for comparable reports describing women with spinal injury has been unsuccessful.

III. MALE ANURANS

Some of the earliest experiments yielding evidence relevant to the propositions set forth in the introduction to this review were carried out on frogs and toads. The copulatory pattern of most anurans is relatively simple. The male clasps the female from the rear and maintains his embrace throughout the period of oviposition, fertilizing the eggs as they emerge from the female's cloaca. Breeding is strictly seasonal but in several species can be induced at different times of the year by gonadotrophic stimulation.

In 1910 Steinach (76) described a series of experiments revealing that although male frogs (*Rana fusca* and *R. esculenta*) do not normally exhibit the mating clasp outside of the breeding season they readily do so after destruction of a brain area "*in den distalen Teilen der Corpora bigemina und in Kleinhirn.*" Steinach (76, p. 555) was convinced that the anterior medulla contains a center that inhibits the clasp reflex: "... *der Umklammerungsmechanismus der Froschmännchens ausserhalb der Brunzeit unter der Herrschaft eines Hemmungstonus steht, und ... die Grundbedingung für das Zustandekommen der natürlichen Brunst auf Herabsetzung, beziehungweise Sistierung dieses Hemmungstonus beruht.*"

Steinach further proposed that during the breeding season the inhibitory center is itself inhibited by the "*innersecretion*" of the testis. The action of this secre-

---

[2] The few clinical accounts I have encountered were scientifically valueless.

tion is as follows: "*Die Wirkung is eine elektive. Das Sekret greift die den Reflex beherr-schenden Hemmungszentren an, schäct oder vernichtet den Hemmungstonus und schafft auf diese Weise die Disposition zur Umklammerung.*" (76, p. 560).

Writing at the same time as Steinach, Busquet (22) reported similar results obtained in experiments on *Rana esculenta* and *R. temporaria.* He found that the clasp reflex appeared in out-of-season males after the brain had been transected, "*immédiatement au-dessous du bulbe.*" Busquet demonstrated that the same operation is not followed by clasping in females or in immature males.

Control of the anuran clasp reflex was studied much later by Smith (75), working with *Rana temporaria* and *Bufo vulgaris.* This investigator discovered that decapitated, nonbreeding male frogs and toads exhibited a powerful clasping reaction in response to electrical stimulation of the exposed end of the severed cord. Decapitated females in contrast responded to the same stimulus with tonic exten-sion of the forelimbs. Smith applied mechanical stimuli to the exposed brains of males while they were in amplexus during the breeding season. When the mid-brain was stimulated with the point of a needle the male relaxed his clasp, re-leased the female in normal fashion, and resumed copulation after a delay of 15–30 min. The investigator (75, p. 1) was led to postulate the existence of " . . . an inhibitory centre in the midbrain, stimulation of which caused cessation of the embrace and the disappearance of sexual activity for a short time."

Aronson and Noble (3) investigated the central neural control of mating responses in the leopard frog *Rana pipiens* and found that the clasp survived in breeding males after complete transection through the caudal medulla. In fact the response appears to have been intensified after this operation. Qualitative differ-ences between the reactions of transected males and of intact specimens led Aron-son and Noble to postulate a "spinal clasp reflex" that normally is modified by midbrain mechanisms. They observed that males with extensive lesions in the preoptic area and in the anterior cerebrum did not approach the female and initiate amplexus. Animals suffering destruction of the preoptic region and the dorsal hypothalamus failed to release their clasp after the female had oviposited. Finally, although clasping occurred after medullary transection, males operated on in this fashion often failed to achieve the correct orientation to the clasped female.

The South African clawed toad *Xenopus laevis* never copulates spontaneously in the laboratory, but Shapiro (73) demonstrated that males and females of this species become reproductively active after treatment with extracts of human pregnancy urine. With reference to the appearance of the mating clasp in his ex-perimental males Shapiro proposed the following interpretation: " . . . it is sug-gested that the anterior pituitary-like substance extracted from pregnancy urine acts by inhibiting the center in the corpora bigemina and the hindbrain, thus re-leasing the clasping reflex in the male toad. . . . It is possible, however, that the extract does not act directly on the clasping reflex inhibitory center, but does so only through the intermediation of the gonads" (73, p. 112).

Weisman and Coates (80) obtained results leading them to agree that the male *Xenopus* possesses "an inhibitory center in the brain which prevents mating during the year." The release from inhibition during a breeding season they re-

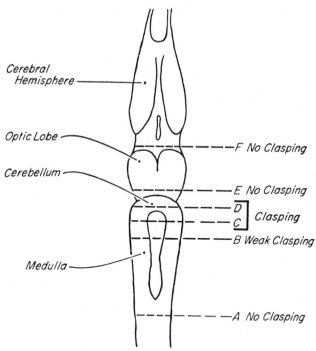

Cerebral
Hemisphere

Optic Lobe

Cerebellum

—F No Clasping

—E No Clasping

—D
—C   Clasping

—B Weak Clasping

Medulla

—A No Clasping

FIG. 3. Effects of brain transection on the mating clasp of male *Xenopus*. [From Hutchison and Poynton (46)].

ferred to an endocrine secretion: " . . . perhaps from the testes or the pituitary gland in turn [which] inhibits the inhibitory center of the brain and allows clasping to proceed. . . . This release of the inhibitory center in turn stimulates the male to activity. In the female, the reverse is found, and she is physiologically depressed while the male is stimulated" (180, p. 34).[3]

Results reported more recently by Hutchison and Poynton (46) are congruent with the suggestions of Shapiro and with those of Weisman and Coates as they apply to the male *Xenopus*. Hutchison and Poynton found that although the sexually active male cannot be induced to clasp an artificial stimulus such as the experimenter's finger, this response is easily elicited after certain types of brain injury.

Some of the results of transecting the CNS at different levels are shown in Figure 3, taken from the original publication. Spinal males (transection A) showed no coordinated behavior. Midbrain transection at level B permitted survival of a

------

[3] The relative quiesence of the gonadotrophically treated female may be due at least in part to the fact that her abdomen is distended by recently ovulated eggs. Aronson and Noble (3) found that immediately after ovipositing, the female *Rana pipiens* becomes active and attempts to dislodge the male if he does not promptly release his clasp. However, if a female is allowed to deposit her eggs, and large amounts of physiological saline are then injected into her abdominal cavity, she will permit clasping, remain relatively quiet, and go through the entire pattern of oviposition for a second time.

weak clasp in response to the insertion of two fingers between the toad's forelimbs. Transections in the anterior medulla (C and D) were followed by the appearance of a strong and persistent clasp response to the experimenter's fingers. The response was never displayed spontaneously, but when a male sustaining one of these transections was placed on the back of a female she was clasped convulsively as soon as contact had been established. In such a case amplexus usually was maintained for the entire 30-min test and the male resisted any attempt to dislodge him. However, the mating pattern was not completely normal, as is indicated by the following description of animals with operation at levels C and D (46, p. 48).

> These observations indicate that males transected in the anterior medulla: (a) can initiate a clasp in response to the appropriate stimuli; these clasp patterns are never fired off spontaneously, (b) are capable of displaying coordinated clasp patterns which serve to prevent the transected male from being dislodged from the female; these clasp patterns are identical to the emergency patterns displayed by an intact male in amplexus, (c) are unable to re-orient themselves into the correct clasp position if they are clasping in the wrong position, (d) cannot terminate a clasping spell.

In several respects these defects described by Hutchison and Poynton are remarkably similar to abnormalities observed in brain-operated *Rana pipiens* by Aronson and Noble (3).

Hutchison and Poynton found that male *Xenopus* subjected to transection in the midbrain (E) or forebrain (F) could never be stimulated to clasp fingers inserted between their forelimbs.

It seems clear that in this species the mating embrace depends on neural mechanisms located just caudal to the most rostral portions of the medulla. Hutchison and Poynton (46, p. 58) drew the following conclusions; "Transections in the posterior medulla evidently eliminated the centres responsible for the regulation of the clasp patterns, whereas transections anterior to the medulla left intact centres which normally prevent the clasp patterns from being elicited."

The female *Xenopus*, like females of most other anuran species, does not normally display the clasp reflex. However, she apparently differs from females of several species of *Rana* (22), because Hutchison and Poynton (46, p. 48) found that after transection of the anterior medulla females exhibited "clasping patterns which were basically similar to those of transection C males."

As already noted, intact captive *Xenopus* will not mate until they have received a series of daily injections of gonadotrophic hormone. Of major importance, therefore, is Hutchison and Poynton's discovery that clasping of an artificial stimulus or of a female can be readily and reliably elicited in untreated males after the brain has been sectioned at the level of the anterior medulla. The authors (46, p. 59) interpret their findings as follows: "These results indicate that the sex hormones exert their influence on sensitive regulatory regions in the anterior brain but do not directly affect the clasp control mechanisms in the spine and medulla."

One general observation worthy of mention at this juncture has to do with certain differences in the details of amplexus. Each anuran species possesses a characteristic method by which the male clasps and holds the female. Male ranidae and bufonidae have different clasp patterns, and superimposed on these familial differences are additional intrageneric variations. The *Xenopus* male, for example, displays an extremely posterior clasp so that his body tends to trail behind that of the female. In contrast male *Rana esculanta* and *temporaria* employ a postaxillary clasp that places them directly on top of the receptive female. Now, when the crucial myelencephalic or spinal mechanisms are released from inhibition by transection at a higher level of the CNS, the clasp pattern that appears is the one that would normally occur in males of that species during the breeding season. The effective brain operation does not simply produce a generalized, tonic flexor rigidity of the forelimbs. Instead it obviously frees for action the precise mechanisms that control the normal species-specific copulatory response.

IV. MALE MAMMALS

It would be highly desirable to include at this point information about those reflexive responses involved in the mating activities of reptiles and birds, but no relevant accounts appear in the literature. In contrast, pertinent data concerning several species of mammals are available, and it is clear that spinal mechanisms are capable of mediating several of the reflexes that normally occur during copulation.

A. Rodents

Direct spinal action is obviously involved when erection and seminal emission are produced in rats and rabbits by low-voltage shock to the sacral cord (29).

In 1897 Spina (cited in ref. 4) noted that when the spinal cord of a male guinea pig is cut between T_{12} and L_1 the animal promptly exhibits erection and expulsion of semen. Bacq (4) performed the same operation on several guinea pigs and described the following sequence of events that began from 1 to 7 min after the cord was transected: "... rhythmic movements occur in the anogenital region; the penis becomes longer and thicker; the glans has the aspect of a fan divided in two lobes; the two horny appendices are protruded. It is a complete erection which ends by the discharge of a vitreous mass, the semen" (4, p. 322).

Many writers fail to distinguish between the emission of seminal fluid and the occurrence of the total ejaculatory pattern, but Beach et al. (16) have recently presented evidence to show that important differences are involved. It appears that in rodents, emission and the formation of a seminal plug can occur without involving all of the neural elements that are activated when the male ejaculates *in copula*. Herberg (43) found that seminal emission can be induced in the male rat by electrical stimulation of the ventromedial fibers of the median forebrain bundle in the hypothalamus. However, Clemens [Beach et al. (16)] discovered that emission produced in this manner is not equivalent to an ejaculation produced by coitus. The latter is always followed by specific changes in the behavior when mating is resumed and these changes are not apparent when a male copulates with an estrous female shortly after experiencing an electrically induced emission.

By the use of indwelling electrodes Vaughan and Fisher (79) stimulated the dorsolateral hypothalamus of male rats while these animals were caged with receptive females. In 3 of 30 cases the male began mounting when the current was turned on and stopped when stimulation was discontinued. Erection was constant throughout the period of stimulation. The frequency of emission was greatly increased, but there is some reason to suspect that this result represented the same phenomenon produced by Herberg, because although large numbers of seminal plugs were emitted, the normal behavioral sequelae to ejaculation did not appear. It may be that in both studies spinal mechanisms mediating emission were released from inhibition by higher centers and that the entire "ejaculatory mechanism" (16) was never called into play.

Other indications that parts of the ejaculatory mechanism may discharge when cerebral control is relaxed are found in descriptions of "spontaneous emission" in various species. Seminal plugs are frequently produced by male rats that are caged individually and prevented from licking their own genitalia (67). Emission has never been directly observed in the rat, but this same phenomenon in other species almost always occurs when the male is asleep—a state in which cerebral activity is modified and inhibitory control over more caudal regions of the CNS may be partially relaxed. This hypothesized state of cerebral disinhibition might also help to explain the occurrence of certain reflexive actions Pearson (68, p. 41) has observed in captive shrews (*Blarina brevicauda*): "These occur while the animal is lying asleep on its side or back. First, the penis becomes erect; then the animal stirs restlessly and there are twitching movements of the hind legs . . . after several seconds a number of thrusts are made, followed by one vigorous, deep thrust which is accompanied by repeated extension of the hind limbs as in actual copulation." These behavioral manifestations may occur in sleeping males simply because other, prepotent reflexes are reduced at this time, but there is also the aforementioned possibility that during sleep the critical spinal or bulbospinal mechanisms that mediate the responses in question are temporarily released from the inhibitory control of higher centers. In a report not available until this review was in preparation, Hart (Hart, B. L. Sexual reflexes and mating behavior in the male rat. *J. Comp. Physiol. Psychol.* in press, 1967) described sexual reflexes evoked in spinal male rats by genital stimulation. The patterning of reflexive responses in the spinal preparation bore several striking resemblances to the copulatory performance of the intact male of this species. It was tentatively concluded that sexual reflexes are released from suprasegmental inhibition by midthoracic transection. The results suggest not only that the final ejaculatory reflex is dependent on release from inhibition of more rostral origin, but that the sequential pattern of successive preejaculatory intromissions and their (presumed) additive effects may be "programed" at a spinal level; and furthermore that the temporary postejaculatory refractory period is a reflection of transitory reduction of excitability in spinal mechanisms.

In some rodents the sexually active male mounts the receptive female, achieves intromission, ejaculates, and then withdraws. In other species the normal coital pattern involves a number of successive mounts with brief insertion, prompt withdrawal and dismount, a short rest interval, and then another mount and another insertion; ejaculation takes place during the final intromission of a series. This latter pattern is characteristic of the rat, mouse, and hamster, and one interpretation includes the assumption that the stimuli produced by successive intromissions have additive effects that cumulate at some locus in the CNS, eventually reaching a threshold level and resulting in discharge of the neural mechanisms responsible for ejaculation (9).

The concept of an additive function of successive intromissions is supported by experimental evidence (17), but the eventual occurrence of ejaculation could be interpreted as involving the release of lower circuits instead of a more direct stimulative effect. The release hypothesis involves the assumption that in the resting, nonaroused male the ejaculatory mechanism is under restraint from other parts of the CNS and that one important function of penile stimulation is to overcome this inhibition. If this were indeed the case variations in the amount or frequency of penis stimulation necessary to elicit ejaculation might reflect differences in the degree or intensity of the hypothetical inhibition. According to this view any procedure or circumstance resulting in the occurrence of ejaculation after less than the normal amount of genital stimulation could be viewed as having reduced the effectiveness of the mechanisms tending to inhibit this response.

In the case of the male rat a reduced inhibition of the ejaculatory mechanism might be inferred when evidence is produced to show that ejaculation occurs *in copula* after fewer intromissions than normally are needed to elicit it. This effect has been reliably produced by separating the male and female after each intromission and keeping them apart for a longer period of time than would normally elapse before the male mounted again. If the enforced interval is too short it has no effect and if it is too long ejaculation does not occur (52, 69).

Electrical stimulation of the hypothalamus in copulating male rats appears to permit "ejaculation" after relatively few mounts (79), although as noted earlier, this may reflect the occurrence of emission without activation of the full ejaculatory pattern. Lisk (58) placed small lesions in the brains of male rats near the diencephalic-mesencephalic junction and then housed each male individually with an estrous female. Trays beneath the living cages were inspected daily for the presence of seminal plugs, and the experimental rats appeared to produce many more plugs than unoperated controls or operates with lesions in other brain regions. This may have been due to increased copulatory activity, but the absence of direct behavioral observations makes it impossible to reject various alternative interpretations (e.g., an increase in spontaneous emissions). In any case Lisk's findings are in harmony with the concept of decreased inhibition resulting from localized brain injury. Lesions similar to those described by Lisk were also placed by Heimer and Larsson (41) in the brains of male rats. These investigators observed the behavior of their subjects in standardized mating tests before and after operation. Postoperatively 7 of 16 males "tended to ejaculate after relatively few intromissions and with a short latency" (41, p. 3).

Reduction in the frequency of pre-ejaculatory intromissions occurs in male rats after they have undergone a series of electroconvulsive shocks (13). A series of preliminary studies carried out on male rats in the author's laboratory has yielded results that tentatively suggest that temporary reduction in the number of intromissions preceding ejaculation may occur after the inhalation of fumes of ethyl alcohol, injection of small doses of Nembutal, or a brief period of etherization. None of these experiments yielded statistically significant results, but in all cases the differences were in the direction predicted by the present hypothesis. All these treatments might be expected to interfere with the function of cerebral mechanisms to a greater degree than with that of neural mechanisms situated in the brain stem and spinal cord, and thus might reduce the efficiency of "higher" inhibitory centers while sparing the more caudally located mechanisms responsible for the mediation of the ejaculatory response.

Stone and Ferguson (77) observed that when sexually rested male rats were allowed uninterrupted access to receptive females copulatory activity continued until several ejaculations had occurred. The behavior of males was modified during the prolonged mating tests; one consistent change was a reduction in the number of penile insertions preceding the second and all subsequent ejaculations. This effect of an initial ejaculation was confirmed by Beach and Jordan (14) and has since been observed in several additional laboratories (33, 51). This same phenomenon characterizes the coital behavior of the male hamster, *Cricetus auratus* (15). One possible interpretation congruent with the notion of cerebral inhibition is that in males that have not recently copulated the level of activity in the hypothesized "ejaculation-inhibiting center" is high, and that after this mechanism has once been suppressed or inactivated sufficiently to permit the occurrence of an initial ejaculation it does not regain its full strength before copulation is resumed, with the result that subsequent disinhibition and release of the lower centers implicated in ejaculation demand less peripheral stimulation.

B. Carnivores

Von Bechterew (19) observed that penile erection and seminal emission can be evoked in dogs by electrical stimulation of the appropriate areas of the cerebral cortex; but it is clear that both erection and emission, as well as other "coitus-related" reflexes, survive after severance of the cord. Sherrington's (74, p. 852–853) description is classic.

> In the spinal mammal erection can be reflexly elicited. In the dog, after spinal transection above the lumbar region, movements due to the skeletal musculature are easily excited from certain genital regions. Touching the preputial skin evokes bilateral extension of the knees and ankles, and to a less extent at the hip-joints; accompanying this there is depression of the tail. If through the skin more posteriorly the glans penis be pressed, the posterior end of the body is curved downwards, pushing the penal bone forward. These reflex movements suggest themselves as belonging to the act of copulation, and this suggestion is strengthened by failure to obtain similar movement on touching the analogous parts of the spinal bitch.

Quite recently Hart and Kitchell (40) verified and extended the observations reported by Sherrington more than 60 years earlier. These investigators examined the responses of chronic spinal dogs to various types of genital stimulation. They found that males maintained for 60–185 days after transection between T_6 and T_8 react to appropriate manipulation of the penis with the display of erection and seminal expulsion. These investigators compared operated and normal males and found that the reflexive responses to penile stimulation were more consistent and intense in the spinal animal. It was tentatively suggested that the difference reflected a release of spinal mechanisms from supraspinal inhibition.

In a series of unpublished experiments the present author found that adult male dogs that have been castrated either before or after puberty react to penile manipulation with the same erectile, ejaculatory and postural reflexes that are characteristic of intact individuals. It would appear, therefore, that the spinal circuits mediating these reflexes are capable of functioning in the complete absence of testicular hormone. These particular animals were neurologically intact, and the statement that the reflexes were spinally mediated is an inference. However, it is clear that testicular hormone is not essential, and a report that became available while this review was in preparation strengthens the conclusion implied above (Hart, B. J. Sexual reflexes and mating behavior in the male dog. *J. Comp. Physiol. Psychol.* in press, 1967). Hart studied the effects of penile manipulation in male beagles after complete transection between T_7 and T_{10}. Spinal animals displayed several of the reflexes that characterize the copulatory pattern of the unoperated male. These included the shallow pelvic thrusting that normally precedes insertion, the very rapid thrusting and accompanying hind leg movements associated with initial penetration and ejaculation, and the protracted maintenance of full erection and tumescence that causes the male to "lock" or "tie" with the bitch in normal coitus. Evidence was also presented to support the hypothesis that reflexes directly associated with the "intense ejaculatory reaction" normally are inhibited by supraspinal mechanisms and facilitated by neural disinhibition. Hart's spinal dogs were supplied with exogenous androgen, whereas the writer's castrates had an intact nervous system. However, the two sets of findings combined are interpreted as supporting the present thesis that the reflexes in question are mediated by spinal mechanisms capable of functioning in the absence of gonadal androgen. At the same time it should be stressed that these reflexes definitely are influenced in some manner by testosterone when this hormone is present in sufficient quantities. For example, observations by the writer have shown that in some dogs castration tends to shorten the length of time the male remains tied to the bitch during coitus. Whether or not this represents direct facilitation of spinal elements remains to be determined.

Bromiley and Bard experimented with decerebrate dogs, and Bard reported that it was "extremely difficult to evoke any marked degree of reflex penile vasodilation in decerebrate dogs" (5, p. 563). Inasmuch as Hart and Kitchell had no difficulty in eliciting full erection and emission after thoracic transection, it seems that neural mechanisms that are spared in the decerebrate male must exert an inhibitory influence upon centers lying below T_8.

The male cat, like the dog, exhibits several elements of the normal mating pattern after transection of the spinal cord. Dusser de Barenne and Koskoff (30) described a combination of reflexes elicitable in spinal male cats after complete section of the medulla slightly caudal to the calamus scriptorium after previous removal of all brain tissue anterior to the intercollicular level. The spinal cats showed: *1*) loss of the decerebrate rigidity produced by the first operation; *2*) replacement of extensor tone in the hind legs by a marked, sustained flexor posture (flexor rigidity of the rear legs); *3*) marked priapism.[4] In the spinal preparation a dorsiflexion of the tail by the experimenter induced augmentation of the penile erection and "quick twitchings in the anal and perineal musculature as well as an increased flexion of the hind legs" (30, p. 80). All these reactions are shown by the normal male *in copula* in association with "spontaneous" dorsiflexion of the tail (present author's observations). When priapism was maximal in the spinal animal, dorsiflexion of the tail or tapping the male's back sometimes resulted in the ejaculation of semen.

The spinal cat and dog thus are similar in their retention of several aspects of the normal mating posture plus the capacity for erection and emission. Dusser de Barenne and Koskoff pointed out that the cat's flexor rigidity of the hind legs, which may be retained for as long as 24 hr, is clearly reflexive since it was abolished by sectioning the lumbosacral dorsal roots. These workers interpreted their findings as follows (30, p. 83–84):

> The abduction and strong flexion of the hind legs is part of the normal copulation posture of the male cat. This abduction of the legs, together with the sensory stimulation of the skin of the abdomen and the medial aspect of the thighs, are probably the imitation in these experiments of the adequate stimuli which are active in the male cat during normal copulation by the contact of these regions with the hindquarters of the female. It is plausible to interpret the adduction of the shoulders and the flexion of the elbows often present in these animals as part of a grasping reflex which may be seen in the front legs of the male during coitus. Furthermore the augmentation of the FR [flexor rigidity] and priapism together with ejaculation which results from dorsiflexion of the tail is in agreement with the fact that during copulation the tail of the male cat assumes a similar posture. *There can be little doubt, therefore, that this syndrome is the manifestation of a sexual reflex pattern, released in the lumbo-sacral cord after secondary spinal transection* (italics added).

Several experiments have yielded evidence implicating more rostral parts of the CNS in the control of various portions of the feline coital pattern. For example, MacLean (59) related hippocampal activity to "certain aspects of courtship" behavior in male cats. Among other responses, penile erection was noted to occur in conjunction with chemical excitation of the dorsal hippocampus and with after-discharges induced by electrical stimulation.

Earlier it was pointed out that male rodents of some species exhibit erection and seminal emission while they appear to be sleeping, and the point was made that during sleep higher centers in the forebrain may partially relax any control

---

[4] Dusser de Barenne and Koskoff noted that the manifestation of priapism in their spinal males was subject to seasonal variations, appearing most marked in January and February and frequently being weak or absent during March and April. This finding is of particular interest because Aronson and Cooper (2) have recently reported annual or semiannual cyclicity in the reactions to estrous females by male cats in which the glans penis has been desensitized by section of the dorsal nerve.

they exert over reflexive responses mediated by more caudally situated mechanisms. In this connection it should be added that "spontaneous" emission sometimes occurs in sleeping or somnolent carnivores.

Aronson's observations of a male cat illustrate this phenomenon (1). A 2-year-old male, immediately on awakening from a sound sleep, exhibited "characteristic pelvic movements" and emitted fluid "crowded with motile sperm." "For 2½ weeks thereafter, one or more spontaneous emissions became an almost daily occurrence. During several of the observations of this behavior, a series of marked movements of the genital region was observed. From their localization, it is believed that these involved contractions of the bulbocavernosus, ischiocavernosus, and other perineal muscles, and therefore constituted the external appearance of an ejaculation" (1, p. 226). The movements described may have been the same as the "quick twitchings in the anal and perineal musculature" seen in the spinal cats studied by Dusser de Barenne and Koskoff.

In considering evidence pertaining to the behavior of female cats sustaining certain types of brain injury it was recorded that coitus sometimes occurs during diestrus. The term "hypersexual," which has unfortunately been applied to this kind of behavior in females, is also encountered in accounts of certain responses shown by male animals after rhinencephalic ablations. For example, Schreiner and Kling (72) reported the display of "hypersexual" behavior in male cats with lesions in the amygdala. The behavior in question consisted primarily of indiscriminate mounting reactions directed toward conspecific males, anestrous females, animals of other species, and even inanimate objects. The investigators apparently were unaware of the fact that similar behavior is sometimes exhibited by cats without neural operations (37, 65), but it is nevertheless possible that the frequency of such responses was increased in the brain-operated males. In fact this seems likely in view of Green et al.'s description (36) of mounting reactions shown by male cats in response to a wide variety of biologically inappropriate stimulus objects after placement of lesions in the pyriform cortex. Green and his co-workers advanced the following interpretation: ". . . the most likely explanations seem to be that either the pyriform cortex normally exerts some restraining action on sexual behavior or that destruction of this localized area leads to certain perceptual changes whereby every furry object is interpreted as an estrous female" (36, p. 529).

Evidence relating to male rodents and carnivores clearly indicates that a number of the reflexive responses that normally are incorporated in the species-specific mating pattern can be mediated by spinal mechanisms. This statement applies not only to the reflexes involved in erection and emission, but also to various postural adjustments of the extremities and hindquarters. The hormonal contribution to these various reflexes remains to be determined for most species, but in the male dog they seem to survive in the absence of testicular secretions.

C. Primates

Knowledge of the neural control of coitus-related reflexes in nonhuman primates is disappointingly limited, but it is known that penile erection occurs in infant monkeys, chimpanzees, and humans and can therefore be mediated by an immature CNS functioning under very low androgen levels.

In an exhaustive study of the development of reflexes in *Macacus rhesus* Hines (44) noted that erection was present by the third postnatal week. It occurred "spontaneously" as well as in response to palpation of the adductor muscles of the thigh. This reflex appears well before maturation of the pyramidal tract or of the motor and premotor cortex. It contrasts with the cremasteric reflex, which is not seen until the testes are mature. One therefore is inclined to speculate that the baby monkey's erection is independent of either higher cerebral control or androgenic support.

Another response Hines observed in a 17-day-old male consisted of "rhythmic extension and flexion of the legs so that the glans penis rubbed against an object (his mother's back)" (44, p. 198). Various coitus-related activities have been evoked in male squirrel monkeys (*Saimiri sciureus*) by electrical stimulation of the brain (60). Excitation of the spinothalamic tract and of associated caudal thalamic areas was sometimes accompanied by seminal emission and "other signs of orgasm." In some instances discharge of semen preceded the appearance of a "throbbing erection."

The most widely quoted experiment touching on the problem of neural control of complex sexual reactions in monkeys is undoubtedly Klüver and Bucy's description of various effects of temporal lobe removal in the rhesus macaque (50).[5] Several weeks after operation frequent erections were noted. These were associated with prolonged licking and sucking of the phallus and with persistent pulling at the penis and scrotum—reactions that might with some justification be interpreted as pruritic instead of "hypersexual." Finally, the report states that temporal lobectomized males engaged in homosexual mounting as well as in "protracted bouts" of heterosexual copulation; but without any knowledge of the behavior shown by the same monkeys before brain injury it is extremely difficult to identify, let alone interpret, possible postoperative changes.

It is impossible to exaggerate the importance of establishing "preoperative norms" in every study that involves manipulation of the nervous system and description of postoperative behavior. Hamilton (38) and Kempf (47) observed frequent homosexual mounting in unoperated captive monkeys, and according to Carpenter (23) the same behavior is shown by free-living male macaques. In the same report Carpenter stated that during a single copulatory sequence feral males were seen to mount and penetrate an estrous female more than 100 times. Inexperienced observers (such as Klüver and Bucy) might be forgiven for interpreting such a performance as "protracted," but to describe it as abnormal or "hypersexual" without any knowledge of the animal's normal preoperative capacity is less excusable.

Reports of "hypersexuality" in men after bilateral removal of the temporal

---

[5] This now-classic account of so-called "hypersexuality" occupies only three-fourths of a page in the original article, and it is obvious that this particular result of the operation was totally unexpected and of marginal interest to the investigators, who were principally concerned with the intellectual deficits produced by the experimental lesions.

lobes (78) include no evidence of any relevance to the present discussion, but accounts of reflexive responses elicitable after total spinal separation are illuminating. One of the earliest accounts was that of Riddoch (70), who described one patient with a complete spinal transection as being capable of erections and "ejaculation." This man reacted to stimulation of the genital region with pelvic movements and violent kicking of the legs. Viewing all these responses in concert, Riddoch proposed the existence of a "spinal coital reflex" in man.

Clinical descriptions published since Riddoch's day support his contention that the capacity for erection and seminal emission is retained in some paraplegic men. One patient with injury between T_6 and T_7, which produced total motor and sensory paralysis below the xiphoid, was able to have intercourse although he experienced no sensation of orgasm (66). This man displayed "an increase in the spasms of his legs and abdomen during ejaculation" (66, p. 905). According to Zeitlin et al. (83), who examined a number of paraplegic men, when tactile stimulation was effective in eliciting what their patients regarded as some sort of sexual climax, the reaction characteristically included "an uninhibited spastic contraction of the legs and perineal region" (83, p. 343).

It is thought provoking to compare such accounts of activity in spinal men with Humphrey and Hooker's observations on the human fetus (45, p. 4–5).

> At $18\frac{1}{2}$ weeks ... one stimulus, passing downward over the penis, was followed by tonic flexion of both thighs, tonic extension of both knees, sharp dorsiflexion of both ankles and marked fanning of all toes, accompanied by some pelvic flexion on the trunk ... Another stimulus, over the scrotum, to the penis on the right side was followed by a quick double kick of the right leg followed by flexion of the right and left thighs perpendicular to the trunk, ... flexion of the pelvis on the trunk and pelvic rotation to the right. ... The character of the reflexes clearly indicates their basic relationship to sexual activity. Postnatally the activity of these reflex arcs is varied, modified, facilitated or suppressed by other nervous system centers. Essentials of the reflex activity pattern are retained, however.

V. DISCUSSION AND CONCLUSION

A. *Problems of Definition and Classification*

There is a natural inclination to refer to most of the responses dealt with in this review as "sexual reflexes," but there are very important objections to such careless diction in scientific discussion. Assigning a name to any phenomenon constitutes an act of implicit classification, and the casual or intuitive use of terms such as "sexual," "abnormal," "hypersexual," etc., has already created serious and unnecessary confusion in the taxonomy of behavior. In actual fact it has been demonstrated that different investigators regularly employ exceedingly disparate working definitions of "sexual behavior" (11). With respect to the classification of separate reactions that appear in the adult copulatory pattern, the present author has previously expressed the following point of view (11, p. 546):

> Many of the elements eventually involved in the coitus of adult mammals can occur independently long before puberty. ... mounting behavior is not uncommonly shown by

very young males, and penile erection is often possible at birth; but to identify these as "sexual" responses is most strongly inadvisable. Throughout the individual's life these reactions may occur in several different contexts with different outcomes. To assign them arbitrarily and exclusively to a single functional pattern can lead only to confusion. Defining erection in a week-old puppy as "sexual behavior" simply because the adult male has an erection while copulating is no more logical than affirming that the voiding of urine by the infant male is "territorial behavior" because the adult male urinates on rocks and trees in connection with the establishment of territorial boundaries.

To these examples may be added the reflexes of lordosis and pudendal dilation in guinea pigs that appear as elements in the pattern of excretory behavior during infancy and function as part of the coital pattern in the adult female (12).

In the literature dealing with mammalian copulation the term "estrus" is often given different meanings. Derived from the Latin word for gadfly it originally signified the mating frenzy that male zoologists imputed to female mammals of other species. As a result of advances in the knowledge of reproductive physiology estrus came to be identified with a specific vaginal condition; and today when most endocrinologists or neurophysiologists state that a rat, mouse, cat, or monkey is in estrus, they mean simply that the vaginal smear contains primarily nucleated and cornified epithelium. Further confusion arose when some writers implicitly defined "estrual" behavior as that presumed to be dependent on estrogenic hormone. The undesirable outcome of this type of classification is clearly revealed in some of the studies mentioned in the present review. In his summary of experiments on decerebrate and spinal female cats Bard wrote as follows: "It is not easy to assay the significance of those reflexes of the spinal animal which resemble certain components of the estrual behavior of normal individuals. They cannot properly be called estrual, for their occurrence is not dependent upon a state of estrus. But as non-specific items of response they may be used by prespinal levels of integration in fashioning the estrual pattern in the normal animal." (15, p. 559).

Their discussion of the lordosis and dilation shown by neonatal guinea pigs reveals that Boling et al. (20) were caught in the same semantic *cul de sac* that confused Bard. "Whether the response in young animals is true heat behavior stimulated by a hormonal action to which newborn young are subjected *in utero* is not known. . . . Until the 'heat response' can be stimulated in newborn young with the regularity which is shown normally, we shall not know how to regard this postpartum behavior" (20, p. 129–130).

The "solution" to this pseudoproblem is simply to define behavior in purely descriptive terms. Lordosis is no more than "a curvature of the spine with a forward convexity" (*Gould's Medical Dictionary*, 5th ed.). Opisthotonos is opisthotonos and vaginal dilation is vaginal dilation, whether they occur in an adult female after injections of estrogen and progesterone or in a newborn animal in response to maternal licking of the rump. Considered in isolation there is no reflexive response that can properly be defined as "sexual." In the final analysis it is probably advisable to avoid applying this term even to the complex patterns that have been examined in the foregoing pages, but to substitute the wording "copulatory behavior" with the understanding that the referent is an integrated pattern of separate responses functionally coordinated in space and time.

B. Re-examination of Original Propositions

Proposition 1 as set forth in the introduction to this review was that the species-specific copulatory patterns of vertebrates are in part comprised of reflexes mediated by spinal and myelencephalic mechanisms that are capable of functioning after separation from more rostral parts of the CNS.

There is evidence to support this thesis with respect to male anurans, rodents, carnivores, and primates. The "mating clasp" persists in several species of frogs and toads after brain transection in the anterior medulla (22, 46, 76). Survival of "copulatory" postures and movements plus the occurrence of penile erection and seminal emission have been observed in male guinea pigs (4), dogs (40), cats (30), and humans (66) after complete severance of the spinal cord.

Confirmatory evidence with respect to female vertebrates is less plentiful, and this may be due in part to the fact that for most species the female's coital role appears to be primarily a passive one. One female mammal that actively and obviously cooperates with the male during intercourse is the cat; and in this species most if not all of the reflexive elements in the feminine coital pattern can be elicited after high spinal section (5).

It is to be emphasized that those portions of the normal copulatory pattern that can be evoked in the spinal or bulbospinal preparation are, as stated in the original proposition, species-specific responses—i.e. decerebrate male frogs of different species still clasp the female in characteristically different ways, and the spinal female cat exhibits the feline mating posture, which cannot be evoked in the spinal bitch.

The proposal that neural tissue in the hindbrain and cord is capable of mediating various coitus-related responses is in harmony with the finding that some of these reflexes are elicitable in fetal or neonatal organisms before the higher brain regions are fully functional. Lordosis and pudendal dilation can be evoked prenatally in guinea pigs (12), and the human fetus reacts to penile stimulation with leg and pelvic movements defined as basically related to sexual activity (45). Penile erection is common in infant rhesus monkeys and humans, and stimulation of the phallus in both species evokes thrusting movements of the leg and pelvic musculature similar to those associated with coitus in adulthood (44, 48). In male puppies less than 1 month old, penile massage produces erection plus urethral and anal contractions resembling those accompanying ejaculation in the adult male (author's observations).

The foregoing observations should not be allowed to result in underestimation of the importance of the diencephalon and telencephalon in the organization and control of copulatory activity throughout the vertebrate class. As suggested earlier, these contributions may involve inhibition of more caudal mechanisms, but they comprise much more than this.

Specific positive functions are easily demonstrated. For example, although the decerebrate male frog will clasp the female when placed on her back, he will not swim toward her and initiate amplexus (3, 46). Furthermore, if the bulbospinal animal's clasp is inappropriately oriented (e.g., if he grasps the female's head), he does not reorient to the normal posterior position as does the intact male (46). Similarly, the male cat deprived of both frontal lobes is incapable of effective coition because, although he repeatedly grips and mounts the receptive female, his activities are so uncoordinated that they never result in intromission (18). The female rat lacking a neocortex readily displays lordosis, but this reaction is so poorly integrated with or "phased into" other elements in her copulatory pattern that male rats offered a choice of sexual partners show a clear-cut preference for normal in contrast to brain-operated females (8).

It is appropriate here to mention Bard's (5) definition of a "center" for sexual activity as "a collection of neurones possessing the capacity to weld together individual responses to form a complex reaction pattern." Bard believed that such a mechanism was situated in the hypothalamus, and in connection with our fourth proposition it is instructive to note his next statement: "The center may also serve as a focus when afferent impulses or changes in the blood act to evoke or repress the total response" (5, p. 552).

An additional type of forebrain function is suggested by the fact that the decorticate female rat does not orient her behavior to the male. Normal females in estrus approach an inactive male, investigating and incidentally stimulating him in such a way that he begins to respond with reciprocal investigation, which may in turn lead to copulation. This type of behavior, which has been termed "appetitive" in contradistinction to "consummatory," is not seen in decorticated female rats (8). Nor is it present in decorticated males of the same species, and for this reason they never copulate (6). Females lacking appetitive behavior will copulate when a male brings the appropriate stimuli to bear; but males with the same behavioral deficiency never take the initiative and, in fact, show no sign of interest or excitement in the presence of receptive females. The present author once proposed that sexual intercourse in male animals depends on one hypothetical mechanism for arousal (AM) and another for the execution of the copulatory pattern (CM) (9). In the intact male if the AM is not activated, the CM is never called into play. AM functions would seem to depend on the forebrain, whereas at least some parts of the CM are spinal or myelencephalic.

The anestrous spinal or decerebrate female cat or dog may react to local stimulation by assuming the copulatory posture, deviating the tail, etc., but these reactions are purely reflexive and stand in marked contrast to the behavior of intact, sexually receptive females. During estrus some female cats exhibit the mating crouch when the male merely stands beside them without making any physical contact, and some receptive bitches react to a male 10 ft away by aiming their hindquarters in his direction and pulling their tails sharply to one side.

Two points seem clear. *1)* Neural circuits situated in the medulla and cord are capable of independently mediating various reflexive responses which are normally incorporated into the species-specific coital pattern. *2)* Nervous tissue lying rostral to the brain stem organizes these reflexes into a biologically effective spatiotemporal pattern and brings them under the influence of exteroceptive stimuli. Anterior parts of the brain also contribute an "appetitive" element to the total pattern of precoital interaction.

Proposition 2 held that in the intact, adult animal the spinal and bulbospinal mechanisms mediating coitus-related reflexes are under varying degrees of inhibitory control by more rostrally located neural circuits or centers.

This seems to be reasonably well established in the case of male anurans, but evidence relating to mammals is not plentiful. To test the hypothesis thoroughly would necessitate accurate measurement of the intensity of stimulation sufficient to evoke the reflexes in intact as well as in brain-transected subjects. Also essential would be quantitative data concerning the strength, duration, and latency of each type of response under both conditions.

There are some reports that at least indirectly support the proposition. After removal of the neocortex some female rats display lordosis in response to mechanical stimulation of the vagina—a reaction that was not present preoperatively (7). If decorticated females are mounted by males the lordosis posture is markedly exaggerated and may be maintained for an abnormally long time after the male has dismounted (8).

When female cats in a state of decerebrate rigidity are stimulated by inserting a glass rod in the vagina, extensor tonus in the limbs disappears and they assume the position normally exhibited *in copula* (5).

During infancy male and female guinea pigs react to licking by the mother with the display of lordosis and perineal dilation, but this response dies out toward the end of the nursing period. It has been suggested that this change reflects the intrusion of inhibitory control by brain mechanisms that are not functional in early postnatal life (12).

Both intact and spinal dogs react to manipulation of the penis with erection and seminal emission, but in the chronic spinal preparation these responses are "more consistent and intense" (40).

Males of several anuran species that never exhibit the mating clasp outside of the annual breeding season will readily do so if the brain is transected near the rostral border of the medulla (22, 46, 75, 76).

Proposition 3 stated that gonadal hormones may have little or no effect on myelencephalic and spinal neural tissues responsible for various coitus-related reflexes. Since this is essentially a negative conclusion supporting evidence must be evaluated with great care. It seems to be fairly well proven that these mechanisms are capable of functioning at some level in the absence of testicular and ovarian hormones, but it cannot be asserted categorically that the nerve cells involved are totally unreactive to gonadal secretions. Furthermore, the possibility has not been excluded that androgenic or estrogenic secretions from the adrenals may exert organizational effects on the embryonic nervous system. Nevertheless, evidence in favor of the third proposition comes from various sources.

Lordosis and pudendal dilation in infant guinea pigs prematurely delivered by caesarean section and reared on an artificial diet can safely be characterized as "estrogen-free" responses (12). Amplexus in out-of-season, brain-operated male frogs and toads is probably independent of testis hormone (46, 76). The occurrence of copulatory responses in diestrous female rats (54) and guinea pigs (35) with various types of brain injury shows that lordosis can be evoked in the absence of the normal amounts of ovarian hormone. Equally pertinent is the fact that decerebrate and spinal cats exhibit coitus-related reflexes in anestrus and even after ovariectomy (5).

Castrated male dogs display erection, normal ejaculatory reflexes, and coital movements (40) even if the operation is performed before puberty (author's observations). Under such circumstances the presence of testosterone can safely be discounted.

Proposition 4 was that gonadal hormones might promote the occurrence of coital behavior in intact animals by modifying the inhibitory effects exerted by higher neural mechanisms. This is a suggestion of considerable antiquity, having been advanced by Steinach in 1910 as an outgrowth of his experiments on the clasp reflex in male frogs (76). Since that time several additional investigations of anuran behavior have yielded results that could be interpreted in the same way (46, 73, 75, 80).

As far as mammals are concerned, evidence is rapidly accumulating to support the thesis that the effects of gonadal hormones on copulatory behavior involve some kind of change in the brain. Direct application of crystalline estrogen to the hypothalamus is followed by the appearance of coital responses in ovariectomized cats (39) and rats (56). Injection (31) or implantation (24) of androgenic material in the brain of the castrated male rat induces the resumption of copulatory behavior. Findings of this nature might reasonably be interpreted as evidence for an excitatory or stimulative action of hormones on localized brain regions. This view may be correct, but there are some reasons for considering the possibility that the total effect includes an element of disinhibition or release.

A strong argument for this suggestion is found in the results of various experiments showing that after particular types of brain injury some female rats, guinea pigs, and cats will exhibit their species-specific coital responses during diestrus or even in the total absence of ovarian hormones. Several investigators reporting such findings have suggested that inhibitory mechanisms were incapacitated by the operation (35, 54).

Precisely how hormones might reduce the inhibitory activity of a cerebral mechanism is a matter for speculation. Fisher (32) increased copulatory activity in castrated male rats by injecting testosterone into the lateral preoptic area and was able to obtain similar results when he substituted a chelating agent, versine. He suggested that instead of exciting or firing nerve cells, hormones remove blocking agents in the neural target area.

Meyerson (63) induced "coital behavior" (lordosis) in spayed rats by systemic injections of estrogen and progesterone and then showed that the behavioral response was suppressed by the administration of α-methyl-DOPA. This pharmacological agent inhibits monoamine oxidase and increases the level of monoamine. His findings led him to the following conclusion: "The results of the present investigation indicate a possible connection between cerebral monoamines and the estrogen-progesterone activated heat behavior, namely that there is a heat inhibiting system in the brain which is dependent on serotonin or catecholamines or both. Thus sexual behavior could possibly be activated by a decreased activity in central nervous pathways mediating heat inhibition" (63, p. 30–31).

In concluding this review it must be acknowledged that evidence directly relevant to some of the original propositions has proven to be disappointingly limited. In addition it should be admitted that some of the data cited are obviously amenable to alternative interpretations. In spite of these criticisms it has seemed worthwhile to assemble the facts and ideas presented here in the hope of *1*) evoking more adequate interpretations of the available evidence and *2*) stimulating research that will fill some of the lacunae such a review inevitably reveals and renders prominent.

REFERENCES

1. ARONSON, L. R. Behavior resembling spontaneous emissions in the domestic cat. *J. Comp. Physiol. Psychol.* 42: 226–227, 1949.

2. ARONSON, L. R., AND M. L. COOPER. Seasonal variations in mating behavior in cats after desensitization of glans penis. *Science* 152: 226–230, 1966.

3. ARONSON, L. R., AND G. K. NOBLE. The sexual behavior of Anura. 2. Neural mechanisms controlling mating in the male leopard frog, *Rana pipiens. Bull. Am. Mus. Nat. Hist.* 86: 87–139, 1945.

4. BACQ, Z. M. Impotence of the male rodent after sympathetic denervation of the genital organs. *Am. J. Physiol.* 96: 321–330, 1931.

5. BARD, P. The hypothalamus and sexual behavior. *Res. Publ. Assoc. Res. Nervous Mental Disease* 20: 551–579, 1940.

6. BEACH, F. A. Effects of cortical lesions upon the copulatory behavior of male rats. *J. Comp. Psychol.* 29: 193–244, 1940.

7. BEACH, F. A. Effects of injury to the cerebral cortex upon the display of masculine and feminine mating behavior by female rats. *J. Comp. Psychol.* 36: 169–198, 1943.

8. BEACH, F. A. Effects of injury to the cerebral cortex upon sexually-receptive behavior in the female rat *Psychosomat. Med.* 6: 40–55, 1944.

9. BEACH, F. A. Characteristics of masculine "sex drive." In: *Nebraska Symposium on Motivation*, edited by M. R. Jones. Lincoln: Univ. of Nebraska Press, 1956, p. 1–32.

10. BEACH, F. A. Biological bases for reproductive behavior. In: *Social Behavior and Organization among Vertebrates*, edited by W. Etkin. Chicago: Univ. of Chicago Press, 1964, 117–142.

11. BEACH, F. A. Retrospect and prospect. In: *Sex and Behavior*, edited by F. A. Beach. New York: Wiley, 1965, p. 535–569.

12. BEACH, F. A. Ontogeny of "coitus-related" reflexes in the female guinea pig. *Proc. Natl. Acad. Sci. U. S.* 56: 526–532, 1966.

13. BEACH, F. A., A. C. GOLDSTEIN, AND G. JACOBY. Effects of electroconvulsive shock on sexual behavior in male rats. *J. Comp. Physiol. Psychol.* 48: 173–179, 1955.

14. BEACH, F. A., AND L. JORDAN. Sexual exhaustion and recovery in the male rat. *Quart J. Exptl. Psychol.* 8: 121–133, 1956.

15. BEACH, F. A., AND R. RABEDEAU. Sexual exhaustion and recovery in the male hamster. *J. Comp. Physiol. Psychol.* 52: 56–61, 1959.

16. BEACH, F. A., W. H. WESTBROOK, AND L. G. CLEMENS. Comparisons of the ejaculatory response in men and animals. *Psychosomat. Med.* 28: 749–763, 1966.

17. BEACH, F. A., AND R. WHALEN. Effects of intromission without ejaculation upon sexual behavior in male rats. *J. Comp. Physiol. Psychol.* 52: 476–481, 1959.

18. BEACH, F. A., A. ZITRIN, AND J. JAYNES. Neural mediation of mating in male cats. II. Contribution of the frontal cortex. *J. Exptl. Zool.* 130: 381–402, 1955.

19. BECHTEREW, W. VON. *Die Functionen der Nervencentra* Jena: Fischer, 1911, vol. 3, p. 1692–1697.

20. BOLING, J. L., R. J. BLANDAU, J. G. WILSON, AND W. C. YOUNG. Postparturitional heat responses of new born and adult guinea pigs. Data on parturition. *Proc. Soc. Exptl. Biol. Med.* 42: 128–132, 1939.

21. BROOKS, C. McC. The role of the cerebral cortex and of various sense organs in the excitation and execution of mating activity in the rabbit. *Am. J. Physiol.* 120: 544, 1937.

22. BUSQUET, H. Existence chez la Grenouille male d'un centre medullaire permanent presidant à la copulation. *Compt. Rend.* 68: 880–881, 1910.

23. CARPENTER, C. F. Sexual behavior of free ranging Rhesus monkeys (*Macaca mulatta*). *J. Comp. Psychol.* 33: 113–162, 1942.

24. DAVIDSON, J. M. Activation of the male rat's sexual behavior by intra-cerebral implantation of androgen. *Endocrinology* 79: 783–794, 1966.

25. DAVIDSON, J. M., AND C. H. SAWYER. Evidence for an hypothalamic focus of inhibition of gonadotrophin by androgen in the male. *Proc. Soc. Exptl. Biol. Med.* 107: 4–7, 1961.

26. DAVIS, C. D. The effect of ablations of neocortex on mating, maternal behavior and the production of pseudopregnancy in the female rat and on copulatory activity in the male. *Am. J. Physiol.* 127: 374, 1939.

27. DEGROOT, J., AND V. CRITCHLOW. Effects of "limbic system" on reproductive functions of female rats (abstr.). *Physiologist* 3: 49, 1960.

28. DEMPSEY, E. W., AND D. McK. RIOCH. The localization in the brain stem of the oestrus responses of the female guinea pig. *J. Neurophysiol.* 2: 9–18, 1939.

29. DURFEE, T., M. W. LERNER, AND N. KAPLAN. The artificial production of seminal ejaculation. *Anat. Record* 76: 65, 1940.

30. DUSSER DE BARENNE, J. G., AND Y. O. KOSKOFF. Flexor rigidity of the hind legs and priapism in the "secondary" spinal preparation of the male cat. *Am. J. Physiol.* 102: 75–86, 1932.

31. FISHER, A. E. Maternal and sexual behavior induced by intracranial chemical stimulation. *Science* 124: 228–229, 1956.

32. FISHER, A. E. Behavior as a function of certain neurobiochemical events. In: *Current Trends in Psychological Theory.* Pittsburgh: Univ. of Pittsburgh Press, 1961, p. 70–86.

33. FISHER, A. E. Effects of stimulus variation on sexual

satiation in the male rat. *J. Comp. Physiol. Psychol.* 55: 614-620, 1962.

34. GASTAUT, H. Correlations entre le système nerveux vegetatif et le système de la vie de relation dans le rhinencephal. *J. Physiol.* (Paris) 44: 431-470, 1952.

35. GOY, R. W., AND C. H. PHOENIX. Hypothalamic regulation of female sexual behavior; establishment of behavioral oestrus in spayed guinea-pigs following hypothalamic lesions. *J. Reprod. Fertil.* 5: 23-40, 1963.

36. GREEN, J. D., C. D. CLEMENTE, AND J. DE-GROOT. Rhinencephalic lesions and behavior in cats. An analysis of the Klüver-Bucy syndrome with particular reference to normal and abnormal sexual behavior. *J. Comp. Neurol.* 108: 505-545, 1957.

37. HAGAMEN, W. D., AND F. ZITMANN. "Hypersexual" activity in normal male cats. *Anat. Record* 133: 388, 1959.

38. HAMILTON, G. V. A study of sexual tendencies in monkeys and baboons. *J. Animal Behavior* 4: 295-318, 1941.

39. HARRIS, G. W., R. P. MICHAEL, AND P. P. SCOTT. Neurological site of stilboestrol in eliciting sexual behavior. In: *Ciba Found. Symp., Neurol. Basis Behaviour*. Boston: Little, Brown, 1958, p. 236-254.

40. HART, B. L., AND R. L. KITCHELL. Penile erection and contraction of penile muscles in the spinal and intact dog. *Am. J. Physiol.* 210: 257-262, 1966.

41. HEIMER, L., AND K. LARSSON. Drastic changes in the mating behaviour of male rats following lesions in the junction of diencephalon and mesencephalon. *Experientia* xx: 1-4, 1964.

42. HEIMER, L., AND K. LARSSON. Mating behaviour in male rats after destruction of the mamillary bodies. *Acta Neurol. Scand.* 40: 353-360, 1964.

43. HERBERG, L. J. Seminal ejaculation following positively reinforcing electrical stimulation of the rat hypothalamus. *J. Comp. Physiol. Psychol.* 56: 679-685, 1963.

44. HINES, M. The development and regression of reflexes, postures, and progression in the young macaque. *Carnegie Inst. Wash. Publ.* 541: 153-209, 1942.

45. HUMPHREY, T., AND D. HOOKER. Human fetal reflexes elicited by genital stimulation. *Proc. Intern. Congr. Neurol., 7th*, Rome. 1961, p. 3-6.

46. HUTCHISON, J. B., AND J. C. POYNTON. A neurological study of the clasp reflex in *Xenopus laevis* (Daudin). *Behaviour* xxii: 41-63, 1963.

47. KEMPF, E. J. The social and sexual behavior of infra-human primates with some comparable effects in human behavior. *Psychoanal. Rev.* 4: 127, 1917.

48. KINSEY, A. C., W. B. POMEROY, AND C. E. MARTIN. *Sexual Behavior in the Human Male*. Philadelphia: Saunders, 1948.

49. KLÜVER, H. Brain mechanisms and behavior with special reference to the rhinencephalon. *Lancet* 72: 567-574, 1952.

50. KLÜVER, H., AND P. C. BUCY. Preliminary analysis of functions of the temporal lobes in monkeys. *Arch. Neurol. Psychiat.* 42: 979-1000, 1939.

51. LARSSON, K. Conditioning and sexual behavior in the male albino rat. *Acta Psychol.* Gothoburgensia, Stockholm: Almqvist & Wiksell, 1956, vol. 1, p. 1-269.

52. LARSSON, K. Excitatory effects of intromission in mating behaviour of the rat. *Behaviour* xvi: 65-73, 1960.

53. LAW, O. T. *Neural Mechanisms of Sexual Behavior*. Paper presented at 1965 meeting of the Western Psychological Association, 1965.

54. LAW, T., AND W. MEAGHER. Hypothalamic lesions and sexual behavior in the female rat. *Science* 128: 1626-1627, 1958.

55. LAW, O. T., AND G. P. SACKETT. Hypothalamic potentials in the female rat evoked by hormones and by vaginal stimulation. *Neuroendocrinology* 1: 31-44, 1965-1966.

56. LISK, R. D. Diencephalic placement of estradiol and sexual receptivity in the female rat. *Am. J. Physiol.* 203: 493, 1962.

57. LISK, R. D. The hypothalamus and hormone feedback in the rat. *Trans. N. Y. Acad. Sci.* 27: 35-38, 1964.

58. LISK, R. D. Inhibitory centers in sexual behavior in the male rat. *Science* 152: 669-670, 1966.

59. MacLEAN, P. D. Chemical and electrical stimulation of hippocampus in unrestrained animals: Part II. Behavioral findings. *Arch. Neurol. Psychiat.* 78: 128-142, 1957.

60. MacLEAN, P. D., S. DUA, AND R. H. DENNISTON. Cerebral localization for scratching and seminal discharge. *Arch. Neurol.* 9: 485-497, 1963.

61. MAES, J. P. Neural mechanism of sexual behavior in the female cat. *Nature* 144: 598-599, 1939.

62. MEIDINGER, R. Differential effect of testosterone propionate given prenatally on sexually dimorphic and sexually isomorphic measures of behavior in the guinea pig. *Proc. Am. Soc. Zool.* 2: 540, 1962.

63. MEYERSON, B. J. The effect of neuropharmacological agents on hormone-activated estrus behaviour in ovariectomized rats. *Arch. Intern. Pharmacodyn.* 150: 4-33, 1964.

64. MICHAEL, R. P. Sexual behaviour and the vaginal cycle in the cat. *Nature* 181: 567-568, 1958.

65. MICHAEL, R. P. Observations on the sexual behaviour of the domestic cat (*Felis catus* L.) under laboratory conditions. *Behaviour* xviii: 1-24, 1961.

66. MUNRO, D. H., H. W. HORNE, JR., AND D. P. PAULL. The effect of injury to the spinal cord and cauda equine on the sexual potency of men. *New Engl. J. Med.* 239: 903-911, 1948.

67. ORBACH, J. Spontaneous ejaculation in rat. *Science* 134: 1072-1073, 1961.

68. PEARSON, O. P. Reproduction in the shrew (*Blarina brevicauda* Say). *Am. J. Anat.* 75: 39-93, 1944.

69. RASMUSSEN, E. W. The effect of an enforced pause between each coitus on the number of copulations necessary to achieve ejaculation in the albino rat. Unpublished manuscript cited by author in: Experimental homosexual behavior in male albino rats. *Acta Psychol.* 11: 303-334, 1955.

70. RIDDOCH, G. The reflex functions of the completely divided spinal cord in man, compared with those associated with less severe lesions. *Brain* 40: 264-402, 1917.

71. SAWYER, C. H. Blockade of the release of gonadotrophic hormones by pharmacologic agents. *Proc. Soc. Intern. Congr. Endocrinol., No. 83.* Amsterdam: Excerpta Med. Found., 1964, p. 629-634.

72. SCHREINER, L., AND A. KLING. Behavioral changes following rhinencephalic injury in the cat. *J. Neurophysiol.* 16: 643-649, 1953.

73. SHAPIRO, H. A. The influence of the pituitary-like substance in human pregnancy urine on the motor components of sexual behavior in the South African clawed toad (*Xenopus laevis*). *S. African J. Med. Sci.* 1: 107-113, 1936.

74. SHERRINGTON, C. S. The spinal cord. In: *Text-*

Book of Physiology, edited by E. A. Shafer. New York·
Macmillan, 1900, vol. 2, p. 782–883.

75. SMITH, C. L. The clasping reflex in frogs and toads
and the seasonal variation in the development of the
brachial musculature. *J. Exptl. Biol.* 15: 1–9, 1938.

76. STEINACH, E. Geschlechtstrieb und echt sekundäre
Geschlechtsmerkale als Folge der innersekretorischen
Funktion der Keimdrüsen. *Zentr. Physiol.* 24: 551–566,
1910.

77. STONE, C. P., AND L. W. FERGUSON. Temporal
relationships in the copulatory acts of adult male
rats. *J. Comp. Psychol.* 30: 419–433, 1940.

78. TERZIAN, H., AND G. D. ORE. Syndrome of Klüver
and Bucy reproduced in man by bilateral removal
of the temporal lobes. *Neurology* 5: 373–380, 1955.

79. VAUGHAN, E., AND A. E. FISHER. Male sexual

behavior induced by intracranial electrical stimulation. *Science* 137: 758–760, 1962.

80. WEISMAN, A. I., AND C. W. COATES. The African frog (*Xenopus laevis*) in pregnancy diagnosis.
Res. Bull. N. Y. Biol. Res. Found. 1–134, 1944.

81. YOUNG, W. C. The hormones and mating behavior.
In: *Sex and Internal Secretions*, edited by W. C. Young.
Baltimore: Williams & Wilkins, 1961, p. 1173–1239.

82. YOUNG, W. C., R. W. GOY, AND C. H. PHOENIX.
Hormones and sexual behavior. *Science* 143: 212–218,
1964.

83. ZEITLIN, A. B., T. L. COTTRELL, AND F. A.
LLOYD. Sexology of the paraplegic male. *Fertility
Sterility* 8: 337, 1957.

84. ZOUHAR, R. L., AND J. DeGROOT. Effects of
limbic brain lesions on aspects of reproduction in
female rats (abstr.). *Anat. Record* 145: 358, 1963.

Reprinted from *Science*, **161**, 1355–1356 (1968)

Autoradiographic Localization of Radioactivity in Rat Brain after Injection of Tritiated Sex Hormones

Abstract. *Radioactivity was found in cell bodies of neurons and glial cells throughout brains of male and female rats that had been injected with either testosterone-H³ or estradiol-H³. Uptake by limbic and hypothalamic structures was higher and longer lasting than that in nonlimbic structures. In all brains, the preoptic area, prepiriform cortex, olfactory tubercle, and septum had particularly high, long-lasting uptake of both hormones.*

Steroid sex hormones affect release of gonadotropic hormones from the pituitary (*1*) and elicit mating behavior (*1, 2*) in many species, including rats. Testosterone is very effective in males, and estradiol is very effective in females. However, testosterone can also affect sexual behavior when injected into female rats (*3*), and estrogens can affect male rats (*4*).

Testosterone and estradiol seem to affect pituitary function and mating behavior by acting on the brain, particularly in the preoptic area and hypothalamus (*1, 2*). In males, the brain takes up testosterone-H³ from the bloodstream (*5*); in females, the brain takes up estradiol-H³ (*6*). These findings were demonstrated by scintillation counting of extracts of homogenized brain tissue from rats given previous intravenous or subcutaneous injections. However, this technique does not afford high spatial resolution; different brain structures, different cell types, and different parts of cells are grouped together. The spatial resolution given by

autoradiography is high enough that these problems can be solved, and this method has been useful in the description of hexoestrol-H³ uptake by the brain of the female cat (*7*). I have used autoradiography to determine the uptake of testosterone-H³ and estradiol-H³ in the brains of male and female rats.

Nineteen rats, about 90 days old, were used. All rats were gonadectomized about 2 weeks before the experiment to reduce endogenous amounts of sex hormones. Each rat was anesthetized briefly with ether, and a radioactive hormone dissolved in 0.05 ml of 25 percent ethanol was injected into the femoral vein. Six males and two females were injected with testosterone-1, 2-H³ (150 or 200 μc; specific activity, 46.5 c/mmole) and were killed ½, 2, or 3 hours later. Six females and two males were injected with estradiol-17β-6, 7-H³ (110, 150 or 200 μc; specific activity, 42.4 c/mmole) and were killed ½ or 2 hours later. Brains from three uninjected rats were prepared as above so that the presence of autoradiographic artifacts could be determined.

At the end of the designated period after injection, the rats were decapitated, and the brains were removed quickly and frozen in dry ice. Frozen sections were cut at 8 μm in a cryostat at −15° to −20°C and mounted directly from the microtome knife onto microscope slides, which were then placed immediately on a 60°C hot plate to dry. The mounted sections were stored over calcium chloride in a 60°C oven. The tissue was then fixed by immersion in 1 percent osmium tetroxide for 10 minutes, and then in 10 percent neutral formalin for 10 minutes. After drying overnight in a well-ventilated, 80°C oven, the slides were dipped in Kodak NTB-3 nuclear emulsion, dried in the air, and packed in lightproof boxes with drying agent. After exposure for 9 months at about 5°C, the autoradiograms were developed for 7 minutes in Kodak D19 at 19°C and fixed in Kodak fixer. Alternate autoradiograms were stained in cresyl violet or Mayer's hematoxylin just dark enough so that cell bodies could be located in the sections; prolonged staining often stained or loosened the emulsion.

Reduced grains over cell bodies were counted in 20 regions in each brain. Only those grains in the same plane of focus as the top of the section were counted. For each brain region in each animal, about 100 cells were counted.

Table 1. Distribution of radioactivity over cell bodies in the brains of castrated male and ovariectomized female rats after injection of testosterone-H³ or estradiol-H³. Amount of uptake is expressed in average number of reduced grains per cell with the standard error of the mean.

Brain structures	Testosterone-H³				Estradiol-H³			
	Male		Female		Male		Female	
	½ hour	3 hours	½ hour	2 hours	½ hour	2 hours	½ hour	2 hours
Olfactory bulb								
Granule cell layer	8.1 ± 0.6	8.0 ± 0.9	22.9 ± 1.8	12.0 ± 1.0	14.0 ± 1.2	3.0 ± 1.9	21.4 ± 2.8	9.8 ± 0.9
Mitral cell layer	7.9 ± 1.0	6.1 ± 0.7	15.0 ± 1.1	7.2 ± 1.2	25.1 ± 3.9	8.8 ± 0.9	22.9 ± 2.4	10.5 ± 1.7
Limbic and hypothalamic structures								
Preoptic area	14.9 ± 1.3	8.4 ± 1.0	12.3 ± 1.4	15.5 ± 1.7	9.7 ± 2.1	15.8 ± 1.8	21.6 ± 2.7	20.5 ± 2.6
Prepiriform cortex	8.8 ± 0.8	6.7 ± 0.9	18.0 ± 1.3	15.6 ± 2.3	19.0 ± 1.0	12.7 ± 2.0	22.7 ± 1.7	19.0 ± 3.0
Septum	11.9 ± 0.8	8.3 ± 0.9	7. ± 1.4	11.3 ± 1.6	7.0 ± 0.9	10.0 ± 2.1	21.8 ± 1.6	20.1 ± 2.1
Olfactory tubercle	14.6 ± 1.6	7.5 ± 1.2	12.9 ± 1.6	13.2 ± 2.1	9.8 ± 1.5	12.8 ± 2.0	21.9 ± 2.1	20.2 ± 2.5
Amygdala	12.0 ± 0.8	0.4 ± 0.3	29.4 ± 5.1	7.2 ± 0.9	12.2 ± 2.1	10.2 ± 1.5	20.7 ± 1.6	18.5 ± 2.1
Hippocampus, dentate gyrus	11.3 ± 1.0	0.4 ± 0.2	16.1 ± 1.4	8.4 ± 1.1	8.9 ± 1.7	3.4 ± 1.2	19.8 ± 1.5	17.8 ± 2.4
Hippocampus, Ammon's horn	15.5 ± 0.7	0.5 ± 0.3	18.3 ± 1.3	8.7 ± 1.0	9.5 ± 1.8	3.5 ± 0.9	22.2 ± 1.9	22.5 ± 2.4
Cingulate gyrus	10.0 ± 0.7	2.8 ± 0.9	13.8 ± 2.3	11.6 ± 1.1	21.0 ± 1.4	15.1 ± 2.1	21.3 ± 2.6	20.3 ± 2.0
Ventromedial hypothalamus	13.1 ± 0.9	0.8 ± 0.4	18.5 ± 3.0	8.0 ± 0.9	21.3 ± 2.5	9.8 ± 1.1	26.7 ± 2.2	20.7 ± 2.6
Nonlimbic structures								
Sensory-motor cortex	7.2 ± 0.6	1.0 ± 0.5	15.1 ± 2.9	6.8 ± 1.2	16.3 ± 2.1	5.0 ± 2.3	17.9 ± 3.0	11.0 ± 1.4
Prefrontal cortex	7.8 ± 0.8	3.1 ± 1.2	15.3 ± 2.1	7.7 ± 1.4	8.1 ± 1.9	4.2 ± 1.0	18.7 ± 2.1	13.9 ± 2.3
Cerebellar granule cells	10.8 ± 0.6	0.3 ± 0.2	10.6 ± 1.7	6.1 ± 0.8	4.9 ± 0.5	2.8 ± 0.9	13.1 ± 1.1	3.2 ± 0.9
Caudate	9.1 ± 1.3	2.5 ± 0.7	10.0 ± 1.5	8.6 ± 1.2	4.8 ± 1.1	4.5 ± 0.9	11.8 ± 1.8	9.4 ± 1.1
Ventrobasal thalamus	4.1 ± 0.7	0.1 ± 0.2	6.3 ± 0.8	4.9 ± 1.1	7.7 ± 1.2	4.2 ± 1.2	11.8 ± 1.2	7.8 ± 1.3
Superior colliculus	6.2 ± 0.5	0.1 ± 0.2	7.8 ± 1.2	3.9 ± 0.8	8.9 ± 1.3	2.4 ± 1.2	19.9 ± 2.0	9.8 ± 1.5
Superior central nucleus (Bechterew)	6.3 ± 0.9	0.3 ± 0.2	7.5 ± 1.1	4.2 ± 1.1	8.6 ± 1.4	3.8 ± 1.3	20.3 ± 2.8	10.5 ± 1.8
Medial habenula	8.2 ± 0.7	0.7 ± 0.4	7.4 ± 1.3	6.0 ± 1.2	7.9 ± 1.1	4.7 ± 1.0	16.5 ± 1.4	11.4 ± 1.2
Spinal cord	5.9 ± 0.8	0.1 ± 0.1	6.9 ± 0.9	7.9 ± 1.0	10.4 ± 1.6	1.8 ± 1.6	16.0 ± 2.2	5.1 ± 1.0

Care was taken to insure a uniform sampling method in all animals. In many other structures throughout the brain, amount of uptake was estimated qualitatively.

Evidence of both testosterone and estradiol uptake throughout brains from both males and females was seen at short (½ hour) and at longer (2 or 3 hour) times after injection. In all animals injected with radioactive hormone, reduced grains were found regularly over diverse types of neurons and glial cells throughout the brain. No evidence was seen for exclusive uptake by or absence of uptake from any particular type of nerve cell or glial cell. Almost all grain reduction was located over the cell body. No concentrations of reduced grains were seen in the brains of the control animals.

One-half hour after injection of

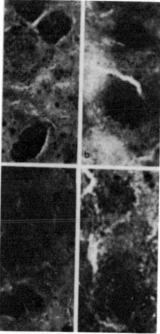

Fig. 1. Representative autoradiographs of brain tissue from male rats injected with testosterone-H³. Uptake by cells in the preoptic area is high ½ and 3 hours after injection; cortical cells show very little uptake 3 hours after injection. Cell bodies are lightly stained with cresyl violet: (a) three preoptic area cells ½ hour after injection (×1300); (b) two cortical cells ½ hour after injection (×1200); (c) three preoptic area cells 3 hours after injection (×1300); (d) two cortical cells 3 hours after injection, showing virtually no grain reduction (×1300).

radioactive hormone, the average number of reduced grains over cell bodies in most limbic and hypothalamic structures was greater than that in most nonlimbic structures. For testosterone, the ratio of the number of reduced grains over cell bodies in limbic and hypothalamic structures to that in nonlimbic structures was 1.75 in male brains and 1.73 in female brains. For estradiol, the ratio was 1.42 in male brains and 1.37 in female brains. Uptake was highest throughout the brains of females injected with estradiol.

Radioactive testosterone and estradiol disappeared at a faster rate from nonlimbic structures than from limbic and hypothalamic structures (Fig. 1). By 2 or 3 hours after injection, the ratio of limbic and hypothalamic uptake to nonlimbic uptake had increased from the value at ½ hour. For testosterone, the ratio was 4.15 in male brains and 1.81 in female brains. For estradiol, the ratio was 3.00 in male brains and 2.03 in female brains. This difference between limbic and hypothalamic, and nonlimbic uptake was statistically significant in all four groups of animals.

Among limbic and hypothalamic structures 2 or 3 hours after injection, uptake of testosterone and estradiol in both male and female brains was highest in anteriorly placed structures in the ventral forebrain, such as the preoptic area, prepiriform cortex, septum, and olfactory tubercle (Table 1). The pattern of long-lasting (2 or 3 hour) uptake among other limbic and hypothalamic structures and in the olfactory bulb depended on the hormone injected and the sex of the recipient rat. Estradiol uptake in female brains was relatively high in posteriorly placed structures such as the hippocampus, ventromedial hypothalamus, and amygdala, but was low in the olfactory bulb compared to limbic and hypothalamic uptake. Testosterone uptake in male brains was low in posterior limbic structures, but was as high in the olfactory bulb as in anterior limbic structures. That is, the ratio of uptake in the olfactory bulb to that in posterior limbic structures was high for testosterone in males and low for estradiol in females. It was intermediate for testosterone in females and for estradiol in males.

A major problem in the preparation of autoradiograms is diffusion of the radioactive chemical from its original site of uptake. In this experiment, fixation of tissue with osmium tetroxide and formalin was used to retain radio-

active testosterone in the tissue sections. There was apparently little or no diffusion, for the ratio of the number of grains concentrated over cell bodies to the numbers spread between cell bodies was very high in all brain structures examined (Fig. 1). The method of fixation did not cause artifactual grain reduction, for the autoradiograms prepared from the brain sections of the uninjected control animal were free of grain accumulations. Finally, the fixation technique did not lead to excess variability in measurements of uptake, for the standard errors of the means in Table 1 are acceptably low.

Thus, anteriorly placed limbic and hypothalamic structures in the ventral forebrain took up and retained either sex hormone in brains from either sex of rat. The time course of uptake in these structures resembled that of estradiol uptake in target organs such as the uterus (8). High uptake in these structures may be specific for sex steroids, for the uptake of cholesterol (9), certain drugs (10), and norepinephrine (11), by these structures is not particularly high. The type of physiological effects of sex hormones on these structures may differ from the type of effects elsewhere in the brain. However, the functional significance of uptake of sex steroids in these structures must be investigated with other experimental techniques.

DONALD W. PFAFF
Rockefeller University, New York 10021

References and Notes

1. G. W. Harris, in *The Pituitary Gland*, G. W. Harris and B. T. Donovan, Eds. (Univ. of California Press, Berkeley, 1966), vol. 2, p. 99; B. T. Donovan, *ibid*, p. 49.
2. F. A. Beach, *Hormones and Behavior* (Hoeber, Harper, New York, 1948); C. H. Sawyer, in *Handbook of Physiology*, J. Field, Ed. (American Physiological Society, Washington, D.C., 1960), vol. 2, p. 1225; W. C. Young, in *Sex and Internal Secretions*, W. C. Young, Ed. (Williams and Wilkins, Baltimore, 1961), vol. 2, p. 1173.
3. F. A. Beach, *Endocrinology* 31, 673 (1942); J. Ball, *J. Comp. Psychol.* 29, 151 (1940).
4. F. A. Beach, *Endocrinology* 31, 679 (1942); J. Ball. *J. Comp. Psychol.* 28, 273 (1939).
5. J. Resko, R. Goy, C. Phoenix, *Endocrinology* 80, 490 (1967).
6. A. Eisenfeld and J. Axelrod, *J. Pharmacol. Exp. Therap.* 150, 469 (1965); J. Kato and C. Villee, *Endocrinology* 80, 567 (1967).
7. R. Michael, *Brit. Med. Bull.* 21, 87 (1965).
8. E. Jensen and J. Jacobson, *Rec. Progr. Hormone Res.* 18, 387 (1962).
9. F. Chevallier and L. Petit, *Exp. Neurol.* 16, 250 (1966).
10. L. Roth and C. Barlow, *Science* 134, 22 (1961).
11. M. Reivich and J. Glowinski, *Brain* 90, 629 (1967).
12. This research was performed while the author was at the Massachusetts Institute of Technology. Supported by an AEC grant to Dr. J. Altman and by grants from NASA (NsG 496) and the John A. Hartford Foundation to Dr. H. L. Teuber. I thank Drs. Altman and Teuber for their support and encouragement.

28 June 1968

Author's Note: The illustrations in this paper have suffered from multiple reproductions. For better resolution the reader is, therefore, referred to the original article.

Reprinted with permission from *J. Nervous Mental Disease*, **132**, 239–248 (1961)

COMPONENTS OF EROTICISM IN MAN: I. THE HORMONES IN RELATION TO SEXUAL MORPHOLOGY AND SEXUAL DESIRE[1]

32

JOHN MONEY, Ph.D.[2]

When systematic experimentation is ethically forbidden, the study of rare clinical cases can help to elucidate the components of eroticism in man. This paper is the first of a series of three[3] reporting the study of rare clinical cases of several varieties (Table 1), interspersed with references to relevant published findings of other investigators.

THE MORPHOLOGIC COMPONENT

MORPHOLOGY AND SOCIAL MATURITY

Morphologic maturity of the body, with respect to sexual characteristics, is hormonally controlled. Estrogens are specific for female characteristics, androgens for male, regardless of genetic sex. Pubic and axillary hair are not sex specific. These characteristics of maturity can develop in gonadless boys or girls, apparently under the stimulus of androgens secreted from the adrenal cortex.

Morphologic maturity is a prerequisite of adult eroticism insofar as fully matured copulatory organs are a prerequisite of completely adult erotic experience. Morphologic maturity also plays another, more indirect role in adult eroticism, namely as a prerequisite of normal social maturation in the teens. Social maturity is significantly

dependent on social interaction with persons of one's own age, especially during the teen-age and early adult years. The person who looks sexually infantile is not fully acceptable to his group, since he looks like someone much younger. His physiognomy and physique do not declare him as a sexually attractive person desirable for flirting, dating or marriage. It is difficult for the boy who reaches the middle or late teens looking juvenile, to behave in a socially mature and grown-up way, for other people, even close friends and family who know his age, habitually and unwittingly react to him on the basis of his physical appearance. They expect him to behave like a juvenile and treat him accordingly. Undeveloped boys are more handicapped than girls: it is more feasible for a girl to disguise, with facial make-up and brassieres padded with falsies, some of the stigmata of sexual infantilism than it is for a beardless, high-voiced boy. Girls as much as boys, however, are adversely affected by the handicap of physical infantilism. Among the ten girls and twelve boys of Table 1 who were over sixteen and physically unmatured, only one girl and two boys had a history of teenage social dating life that approached the norm for their age group.

The obverse of sexual infantilism with respect to the influence of morphologic development on social maturity is found in sexual precocity (15, 26). In this instance, the child looks physically mature while chronologically still a child. The mental age, by contrast, remains concordant with the birthday age. Other people, unwittingly influenced by appearance, regard the child as older than his years and frame their expectancies accordingly. It is therefore possible for the child to become accelerated in

[1] The research program from which this paper derives was supported, 1951 through 1957, by a grant from the Josiah Macy, Jr. Foundation. As of 1957, the United States Public Health Service assumed financial responsibility under the terms of grant #M1557. The endocrine clinic of Lawson Wilkins, Professor of Pediatrics, on whose unfailing cooperation the program has been dependent, has provided an indispensable wealth of clinical material.

[2] Departments of Pediatrics and Psychiatry, The Johns Hopkins University School of Medicine, Baltimore, Maryland.

[3] The second paper will appear in a subsequent issue of this journal; the third is already published (24).

social maturation—indeed desirably so, if social, recreational and educational experience are suitably planned and coordinated to favor this acceleration.

The sexually precocious child who fails to receive an opportunity for accelerated social maturation may thereby suffer a certain amount of psychologic trauma, though not inevitably so. By contrast, the sexually infantile teenager seems exceptionally prone to psychologic trauma relative at least in part to his impaired opportunities for social development. The longer the lag in his social maturation persists, the more difficult becomes the problem of social and psychologic adjustment as a maturing and mature adult, after hormonal substitution therapy has been instituted. A teenager with pubertal failure not corrected until approaching the age of twenty or later, has time in which the patterns of social infantilism may become cemented. If ancillary psychopathologic adjustment to the handicap also develops, that too may become stabilized.

A parallel state of affairs exists in the case of a virilized hyperadrenocortical girl not

TABLE I
*Various Sex-Hormone Anomalies and
Numbers of Patients*

Unmatured males aged 16 or older, not dwarfed, with hypogonadism and sex-hormone deficiency	11
Unmatured male, aged 21, with anorchia	1
Unmatured females over 16, dwarfed in stature, with gonadal aplasia and sex-hormone deficiency:	
(a) sex chromatin negative (XO)	7
(b) sex chromatin positive (XX)	2
Female, aged 35, sex-hormone deficiency secondary to pituitary deficiency	1
Female hyperadrenocortical hermaphrodites precociously virilized, studied when feminized on cortisone therapy:	
(a) between ages 8 and 14 years	11
(b) between ages 15 and 47 years	17
Boys with precocious puberty of hyperadrenocortical onset, not arrested by treatment	8
Boys with idiopathic sexual precocity	9
Girls with idiopathic sexual precocity	16
Youths with gynecomastia	10
Klinefelter's syndrome	9

correctively treated with cortisone until her twenties or older. When treatment finally is given in such cases, the patients abruptly face all the social adjustments of adolescence anachronistically and alone. They have no social support from age mates who are in the same boat. Some of them never really succeed: their patterns of social infantilism or psychopathological adjustment never completely yield to the new way of life open to them.

The early teen years appear to be the critical period for optimal, all-round adolescent development. It is then that adolescent physical growth coincides with the presence of age-mates in the same growth phase, and this coinciding is basic to healthy adolescent psychologic development.

SOCIAL HANDICAP OF DEFORMITIES

Not only failure or acceleration but also deviation of morphologic development may have an adverse effect on social maturation and adult eroticism. Dwarfism or extreme obesity, for example, makes a person sexually *persona non grata* to some extent among his age mates. Among anomalies, specifically sexual, the breast-swelling of adolescent gynecomastia in boys is particularly morale-crushing. All ten of the patients in Table 1 were profoundly embarrassed by the girlish-looking swelling of their breasts and avoided exposing their chests to the view of others. Though there were extensive personal variations according to age and social background, all of the boys were more inhibited in their dating and girlfriend relationships than would otherwise have been likely. One older patient had the incipient delusion that his breasts caused other people to suspect homosexuality and three others, younger, needed reassurance that gynecomastia was not the first hideous sign of an unwanted sex transformation.

Defects of the genital organs are usually a worse social-erotic handicap than gynecomastia. To look normal is a great satisfaction to patients with such a deformity. For this reason, complete therapy for boys with total testicular aplasia consists not only of androgen substitution, but also of surgical implantation of prosthetic testes. For hermaphrodites with ambiguous-looking external genitals, the sooner the incongruity can be corrected surgically, the greater the benefit for their social development, indeed for the very basis of their psychosexual identity.

SEX HORMONES AND SEX DESIRE

In real life, sexual desire occurs in relation to a predicate—desire for someone or something. In the abstract, sexual desire is a concept representing what is left when the predicates are hypothetically removed from specific manifestations of sexual desire. Sexual desire is manifested in erotic imagery, sensation, report and action. Sexual drive and libido may be regarded as synonymous with sexual desire.

In man the predicates, that is the goals or objects of sexual desire appear much more variably determined than by hormones or other biochemical substances. Sexual desire in the abstract, however, appears to be intimately related to the body's metabolism and especially to its hormonal functioning after puberty.

STARVATION AND DESIRE

The relationship between general body metabolism and sexual desire was shown in the case of starvation by Keys, Brožek, Henschel, Mickelson and Taylor (20). Sexual desire then disappears and is completely supplanted by desire for food. Sex imagery and conversation is replaced by fantasy and talk of food. In extreme starvation there is actual failure of sex-hormone secretion. Jacobs (19) wrote of the castration syndrome in emaciated, starving male prisoners of war. Body hair became soft and sparse. Sebaceous glands became unfunctional. Erections and nocturnal emissions disappeared, along with reporting and joking of sexual desire. The effects of the syndrome were readily reversed on two occasions when food packages arrived. Then some of the men grew breasts (gynecomastia), temporarily, presumably because the liver, damaged by starvation, was unable to metabolize estrogens circulating anew in the usual way. Impotence and sterility did not turn out to be permanent sequelae as the men had feared they might be.

HORMONAL MECHANISM AND DESIRE

The mechanism whereby the sex hormones may activate sexual desire in man is not understood. It is most likely that, once an optimal level of circulating hormone has been reached, the control of sexual desire is strictly nonhormonal. Fluctuations of sexual desire might then be explained as dependent on factors like cognitional stimulation and the time interval since the last sexual release.

If the control of sexual desire is attributed directly to the hormones, then fluctuations of desire would need to be attributed to fluctuations in the quantity or potency of circulating hormone, for there is always some hormone circulating in the bloodstream. On the basis of evidence from patients being treated with hormonal substitution therapy, the quantity of hormone given, once an optimal blood level has been attained, bears no direct relationship to sexual desire or sexual activity. Apparently there is a certain amount of hormone that the body can use; anything in excess of that amount is superfluous waste to be excreted. Thus, in the untreated adrenogenital syndrome, urinary androgen excretion is excessively high without excessive, parallel elevation of the patient's sexual desire or activity.

It has sometimes been assumed that sexual desire in women fluctuates with the hormonally controlled menstrual cycle. The trouble with such an assumption is that different authorities attribute the peak of desire to different phases of the cycle. The evidence is far from convincing. Predictable cycles of sexual desire in men have seldom been taken seriously.

Quite apart from the fluctuation issue, there are two rival propositions to support an argument in favor of hormonal influence on sexual desire in man. One is that the sex hormones have a primary activating effect on the reproductive organs from which messages are relayed to the brain and consciousness. The other is that the sex hormones directly stimulate a sex center in the central nervous system, probably in the hypothalamus, from which stimuli are relayed to the genitals. Harris, Michael and Scott (16), in an admirably controlled experiment, showed that stilbestrol microimplanted in the posterior hypothalamus of spayed cats led to the full development of mating behavior, despite a genital tract that remained anestrous. The significance of this finding in the cat for the nonestrous human species cannot yet be conjectured.

So much for the unsolved problem of the mechanism and locale of the influence of sex hormones on sexual desire. It nonetheless remains clear, from the study of hormone-deficient patients, that the sex hormones do exert an influence on sexual desire. This influence is effected *via* the sex organs which are brought to maturity by the sex hormones and kept erotically functional by them. Consider first androgen. The penis and clitoris both respond to initial doses of androgen with extensive dilation of the vasculature and growth in size. Thereafter maintenance of an erection by complete engorgement of the organ with blood is facilitated by androgen. Tumescence of the penis can occur in the absence of post-pubertal levels of androgen, but the erection is generally not complete and not long-lasting.

The genital effects of androgen are over and above the more generalized correlation between androgen and nitrogen metabolism. Correction of an androgen deficiency is accompanied by an increase in nitrogen metabolism and by increased general muscular size and strength. Androgen-deficient men on substitution therapy (Table 1) be-

came sexually more interested, bold and participative, but not because they had formerly been invalids too weak to participate. They did not become more violent, assaultive or aggressive, but more desirous of sexual activity and more competent in sexual performance.

ANDROGEN DEFICIENCY: CASTRATION

The relationship between androgen and sexual performance in males can be studied in hypogonadism, agonadism and castration with and without hormonal substitution therapy. The literature on castration, which is deficient on findings in female castration, has been confusing and imprecise with regard to the erotic functioning and psychology of castrates. Opinions conflict and there is a regrettable lack of well-documented psychologic data. Beach in 1958 (2) and Kinsey, Pomeroy, Martin and Gebhard in 1953 (22) made comprehensive reviews of the literature, incorporating and bringing up to date earlier surveys. Notable for good psychologic data are case reports by, Foss (10) and Hamilton (14). Notable for its breadth of scope as well as detail is a recent followup study by Bremer (4) of 244 legal castrates in Norway.

The obvious and most significant generalization to be drawn from the literature on castration is that the effect of the condition on sexual desire and erotic functioning in man is extremely variable. In some instances, loss of both potency and desire follow castration. The loss may be rapid or insidious over a period of months. In some cases the castrate loses potency and the capacity for genital gratification, but continues to experience sexual desire. In a few instances, both potency and desire remain and the castrate continues to copulate successfully for months or even years, though with lessened frequency. It is almost certain that he has no emission, but reports are uninformative as to whether some sort of orgastic climax is experienced or not. Postcastration impairment of desire, potency, emission or orgasm is readily reversed by androgen substitution therapy.

There are no statistics from which to draw comparisons between the phenomenology of postpubertal and prepubertal castration. The one anorchid patient in the present series, though physically prepubertal when first interviewed at the age of twenty-one, experienced sexual desire, was married and had satisfactory sexual intercourse about three times a week according both to his own testimony and that of his wife. He had no emission or orgasm, as he did after androgen therapy, but simply an unclimactic feeling of pleasure. A prepubertal sexual history of this type is probably rare in eunuchs and the hypogonadal, who appear more often to have a prepubertal sex history limited to fantasy, with little sexual action. The twenty-two gonad-deficient men and women of Table 1 showed (or retrospectively reported) a wide variability of erotic interest and activity before hormonal substitution therapy was begun and also in response to induced puberty. Some of them remained erotically relatively indifferent after treatment. But in no instance did erotic imagery and activity decrease after treatment. The general trend was quite clear to see: hormonal medication rehabilitated the erotic life and genital function. The effect was more clearly evident in the men than the women of this particular series.

SUBSTITUTION THERAPY LAPSED: MEN

Sex hormone deficiency can be investigated in eunuchism and hypogonadism not only in studies of sexual function before and after treatment is begun, but also in studies of the effects of lapses of treatment once these same patients have commenced taking hormones.

Of the eleven hypogonadal men entered in Table 1, there were five who discontinued androgen medication for a period of three months or longer. Three of them went through periods of discontinued treatment while being adjusted to a medicational routine; one discontinued treatment to avoid the problems of activated sex drive— he was physically deformed and morally very straight-laced; and one became periodically lax about getting an injection of long-acting testosterone each month.

In each case, the absence of androgen from the tissues made a decided difference to sexual desire and sexual performance. The most sensitive indicator was the ejaculate from the seminal vesicles and prostate. The men reported that it gradually diminished in volume until no fluid was emitted at all. The feeling of orgasm either was lost through disuse of the penis erotically, or was changed in quality. There were fewer erections. There was a lessened initiative to masturbate or to make coital advances. With loss of ejaculation, the men also lost whatever erotic, ejaculatory dreams (wet dreams) they had been having. They considered that waking erotic imagery and daydreams diminished in frequency of appearance.

One may generalize and say that these men did not lose completely their erotic imagery, their erotic sensations, or their erotic actions and behavior. What did happen was that eroticism, whether in imagery, sensation or activity, did not get initiated with the same frequency as before. This failure of initiation showed up in the involuntary failure of the penis to erect or hold an erection, and in the failure of other, more voluntary erotic action and coordinated endeavor as well.

The man who was married when he discontinued treatment had a good barometer of failure of his erotic initiative, namely his wife's comments and complaints. In fact, this man, and the two others who also got married, found that it paid them not to be lax about their injections. They reported a slackening of erotic initiative, including erectile potency, if they delayed even a week in getting their monthly injection of long-acting testosterone.

The conclusion to be drawn from periods of interrupted treatment in men with sex-hormone failure is that androgen is necessary not only to induce morphologic maturity of the sex organs, but also for the maintenance of sexual desire and active eroticism.

SUBSTITUTION THERAPY LAPSED: WOMEN

There were four hypogonadal women who discontinued estrogen medication for three months or longer. One of them was off treatment for three months by her physician's request. She happened to get married during the first month. Two others were single and celibate. One of these two discontinued treatment with stilbestrol for eighteen months, before resuming on premarin, as stilbestrol produced unpleasant gastric symptoms. The other single woman, a graduate nurse, had been off treatment for two years, having discovered that the only sequel of significance in her rather puritanical life was cessation of the menses. The fourth woman was divorced. She discontinued treatment after a doctor scared her about the carcinogenic dangers of estrogen, and found no ill effects from not taking estrogen. She continued to have sexual liaisons, casually, with men she met professionally in show business during the five years off treatment, though eventually she discovered that vaginal tightness and dryness due to lack of estrogen-stimulated secretions was undeniably a handicap.

These four women had been on cyclic estrogen therapy so that they menstruated upon cyclic withdrawal of estrogen for a week each month. Following total withdrawal of estrogen, they ceased to menstruate altogether. Vaginal smears showed that the vaginal mucosa underwent involutional changes as in postmenopausal women. There were no definitive reports of hot flashes or malaise typical of the climacteric, however. Vaginal secretions concomitant with sexual arousal no longer appeared.

The women reported nothing to indicate any change in their erotic imagery, sensations or actions. The two who were having intercourse claimed definitely that they reached the climax of orgasm, as they did when taking estrogen.

The evidence from these four women is consistent with common knowledge concerning postmenopausal disappearance of estrogen in ordinary women. Though there are exceptions, erotic imagery, sensations and actions are not abolished and sexual desire is not necessarily even lessened in the usual course of diminished estrogen production at the menopause.

ANDROGEN: THE LIBIDO HORMONE

The relationship of estrogen to eroticism in the adult female seems, after the pubertal estrogenic functions of maturing the reproductive tract and feminizing the body morphology in general, to be restricted to maintaining the lubricant secretions of the vagina preparatory to copulation (31). The primary estrogenic function would appear to be monitoring endometrial growth in close coordination with the gestagenic function of monitoring nidation and gestation.

Since maintenance of sexual desire and erotic functioning in men appears to be significantly dependent on androgen, it would be an odd biological discontinuity that ordained erotic imagery, sensations and actions in the female of the human species to be, by contrast with estrous behavior of lower mammals, independent of hormonal functioning.

It being mechanically possible for a flaccid, unconscious female body to receive the male in coitus, though not *vice versa*, one may use the logic of extreme absurdity to argue that passivity and female eroticism are synonymous and that there is indeed a biological discontinuity between human male and female so far as erotic desire and hormones are concerned. Such a position is not to be taken seriously.

A tenable hypothesis is that sexual desire with its attendant imagery, sensations and actions is maintained well-functioning in both men and women by androgens. In women the androgens of eroticism may conceivably be of an origin as yet unknown, or they may be metabolically derived from gestagens. The most likely explanation, however, is that these androgens are of adrenal origin. Waxenberg, Drellich and Sutherland (33) reported an excellent study of eroticism in twenty-nine women who had had both ovaries and both adrenals removed in the treatment of breast cancer. Loss of only the ovaries and ovarian hormones had no definite adverse effect on sexual drive, activity and response, but all three were diminished or abolished in most of the women after their adrenals also had been removed. The adrenals secrete some estrogen, but larger amounts of androgen. The authors concluded that the loss of adrenal androgens was the responsible factor in the women's lessened or abolished eroticism.

There are other lines of evidence that can be marshaled to support the hypothesis that androgen is the libido hormone in both men and women. Salmon and Geist (31) assembled some very impressive evidence from groups of women patients who were given variously androgen, estrogen or both. Androgen was found to increase the sensitivity of the genitals, especially the clitoris, as well as to heighten desire and increase sexual gratification; but estrogen was necessary to abolish vaginal dryness and tenderness and so expedite coitus. It is a common part of the clinical lore of gynecology that many women for whom androgen therapy is prescribed report a definite increase of sexual desire as a side effect: this has been commented on by Salmon (30); Carter, Cohen and Shorr (5); Foss (11); Kupperman and Studdiford (23); and Dorfman and Shipley (6). Conversely, in the clinical lore of urology, many men for whom estrogen is prescribed report great diminution or total abolition of sexual desire and activity: among others Foote (9) and Paschkis and Rakoff (29) have written of this.

Gestagens seem to have a similar libido-inhibitory function in men, as reported by Greenblatt, Mortara and Torpin (13) and Heller (17).

The androgen-libido relationship is far from being a simple linear correlation. There are irregularities that cannot at present be explained and that will require controlled pharmaceutical experiments, preferably on human subjects, for their proper resolution. One needs to know, for example, more about quantitative relationships. Apparently there is an optimal point beyond which added androgen does not increase frequency or degree of eroticism. Heller and Maddock (18) reported no increase in sex drive when hormonally normal men were given testosterone. In women it seems likely that a large enough quantity of androgen will suppress estrogenic function while maintaining libido, whereas more equal amounts of the two hormones may have a coexistent or even synergistic effect.

Yet another type of information is needed concerning specificity and failure of target-organ response. To illustrate: there are some fertile females, and some simulant females with feminizing testes and male sex-chromatin pattern, who fail to grow sexual hair, and in one case of Wilkins (34) attempted treatment with testosterone by mouth and locally applied in an ointment failed to have any effect. There is no doubt that one occasionally encounters instances of target-organ failure of erotic sensitivity and response, irrespective of androgen titers in the bloodstream. One patient, a woman with gonadal aplasia and male sex-chromatin pattern, was given 10 mg. of methyl testosterone a day for a month. There was no change in her grossly inert eroticism, nor in her very placid behavior. Another patient, a man with Klinefelter's syndrome and female sex-chromatin pattern, was unique among others of his type in that he showed neither morphologic nor libidinal response to large doses of long-acting testosterone (400 mg. of testosterone enanthate a month for three months).

Further data are needed also on the time factor in relation to hormonal effects. It seems likely that the body may develop some degree of hormonal tolerance, perhaps by way of conditioning and learning. Thus, it was commonly reported by hypogonadal men patients that erections were triggered off with greater ease and frequency, by even the mildest friction of the clothing, during the first month of androgen therapy, than was subsequently the case. Salmon and Geist (31) reported a similar transient exaggeration of libido in some of their women patients treated with testosterone for various endocrine disorders.

Finally, one needs to know more of the fine points concerning the biological effects of the many biochemical forms of the sex hormones that can be differentiated in fractionation analysis. It is known, for instance, that some of the structural variants of testosterone are biologically inert. It is quite possible that others have potencies and functions that are somewhat specific and different from one another.

Irrespective of problems to be solved, the sex drive appears from the hormonal point of view to be neither male nor female, but undifferentiated—an urge for the warmth and sensation of close body contact and genital proximity.

HORMONES AND MALE-FEMALE DIFFERENCES OF DESIRE

It has long been a popular theory that the sexual inclinations, thoughts and dreams of men and women are different in content because the hormones of the two sexes are different. From this doctrine derives the assumption that homosexuality is a hormonal disorder—an assumption held in discredit by the majority of endocrine authorities today. Heller and Maddock (18) reviewed studies made in the 1940's by Glass, Deuel and Wright (12), Neustadt and Myerson

(27), and Sevringhaus and Chornyak (32), to test the hypothesis that homosexuals (male) differ from control subjects in their urinary output of androgens and estrogens. The studies failed to give an affirmative result. Urinary output of both androgens and estrogens was found to have the same variability in homosexual men as in men in the control group. The estrogen-androgen ratio was higher than expected more often in the homosexual than in the control group of one study. The significance attached to this ratio was criticized by Kinsey (21) and is even more open to criticism today in the light of new techniques of estrogen assay. The rather crude bioassay methods used in the 1940's have been largely supplanted by the more accurate and refined fractionation methods of contemporary research.

Failure to detect an hormonal anomaly in homosexuals has been paralleled by a failure to effect a cure with hormonal treatment. Androgen does not masculinize the erotic inclination of a male homosexual. It either leaves his eroticism unchanged or else intensifies it in the homosexual direction.

The obverse of a lack of abnormal hormonal findings in homosexuality is the lack of increased incidence of homosexuality in sex-endocrine dysfunction. Estrogen levels are elevated in males with gynecomastia sufficiently to cause prominent breast feminization. There is no corresponding feminization of the personality. Androgen levels are elevated in females with hyperactive adrenals or, more rarely, a virilizing tumor. There is no corresponding lesbian virilization of the personality.

There is no special affinity between homosexuality and lack of hormones to bring about complete maturation of secondary sexual characteristics in men and women, respectively. In men, Klinefelter's syndrome is an instructive example of weakened androgen action (plus seminiferous tubule dysgenesis, sterility and possible gynecomastia), because it is also accompanied usually by a supernumerary chromosome. The sex-chromosome constitution is XXY, whereas the normal male is XY and female XX, and the total chromosome count is 47, not 46—Ferguson-Smith (7). There is no corresponding psychosexual epicenism as part of the syndrome though, of course, homosexuality may occur in a man with the condition and was so observed in one of the nine cases of Table 1. In two other cases, both boys in their teens, there was a possible homosexual tendency disguised by the sexual indifference and low-powered sexual activity which was so commonly evident in the syndrome.

Homosexuality may occur among those with a chromosomal abnormality, but chromosomal aberration is not a demonstrable characteristic of homosexuals, as was shown in a properly controlled study by Pare (28) and another by Bleuler and Wiedemann (3). Barr and Hobbs (1) found typical male chromatin pattern in the cell nuclei of five cases of genuine male transvestism.

HORMONES AND SEX-CHROMOSOME STATUS

In the ordinary course of events, hormonal status as male or female correlates perfectly with chromosomal status as male (XY) or female (XX), and both these variables correlate perfectly with the reproductive anatomy as male or female. These correlations are not invariant, however. They go awry in certain endocrine disorders (see above) and in hermaphroditism (25). In the syndrome of testicular feminization, for instance, gonads that are histologically testes produce powerful estrogens. They are found in the groins of a person who is morphologically entirely womanly (and psychologically womanly also), but who has no uterus and tubes. The sex chromosomes are XY and the total chromosome count is 46.

Normal female hormonal functioning has been observed in women who have 47 chromosomes and an XXX sex-chromosome count. Male hormonal functioning has been found in men who have supernumerary sex chromosomes, namely XXXY as well as XXXY (8).

The relationship between sex chromosomes and sex hormones is a very loose and indirect one. The sex hormones exert their influence relatively independently of the cytogenic structure of the body's cells. Just as the sex hormones are nonspecific for the direction of erotic inclination, correspondingly the sex chromosomes are not specific determinants of the direction of erotic orientation.

SUMMARY

It is not in a rigid, inalterable way that the male or female status of the sex hormones parallels the male or female status of the sex chromosomes, and deviations do occur. Similarly, there are deviations from the normally expected correlation between gonadal structure and hormonal status. Embryonic and fetal sexual differentiation is very sensitive to male or female hormonal balance, irrespective of chromosomal and gonadal structure. Hormonal balance at puberty may or may not be congruous with the sex of the reproductive organs, but estrogens are specific for female and androgens for male secondary sexual characteristics, irrespective of genetic or gonadal sex. Anomalies and incongruities of secondary sexual characteristics have repercussions on social and psychosexual maturation. The level of sex drive or libido is hormonally influenced, and androgen is probably the libido hormone in both men and women. The male and female sex hormones have a direct male and female effect, respectively, on the genitalia, maintaining them erotically functional. The direction or content of erotic inclination in the human species is not controlled by the sex hormones. Hormonally speaking, the sex drive is neither male nor female but undifferentiated—an urge for the

warmth and sensation of close body contact
and genital proximity.

REFERENCES

1. BARR, M. L. AND HOBBS, G. E. Chromosomal
   sex in transvestites. Lancet, 1: 1109–1110,
   1954.
2. BEACH, F. A. *Hormones and Behavior: A survey
   of Interrelationships Between Endocrine
   Secretion and Patterns of Overt Response.*
   Hoeber, New York, 1948.
3. BLEULER, M. AND WIEDEMANN, H. R. Chromo-
   somengeschlecht und psychosexualität.
   Arch. Psychiat. Nervenkr., 195: 14–19, 1956.
4. BREMER, J. *Asexualization, A Follow-up Study
   of 244 Cases.* Macmillan, New York, 1959.
5. CARTER, A. C., COHEN, E. J. AND SHORR, E.
   The use of androgens in women. Vitamins
   Hormones, 5: 317–391, 1947.
6. DORFMAN, R. I. AND SHIPLEY, T. A. *Androgens: Biochemistry, Physiology and Clinical
   Significance.* Wiley, New York, 1956.
7. FERGUSON-SMITH, M. A. Cytogenetics in man.
   Arch. Intern. Med., 105: 627–639, 1960.
8. FERGUSON-SMITH, M. A. AND JOHNSON, A. W.
   Chromosome abnormalities in certain diseases of man. Ann. Intern. Med., 53: 359–
   371, 1960.
9. FOOTE, R. M. Diethylstilbestrol in the management of psychopathological states in
   males. J. Nerv. Ment. Dis., 99: 928–935, 1944.
10. FOSS, G. L. Effect of testosterone propionate
    on a postpubertal eunuch. Lancet, 2: 1307–
    1309, 1937.
11. FOSS, G. L. The influence of androgens on
    sexuality in women. Lancet, 1: 667–669, 1951.
12. GLASS, S. J., DEUEL, H. J. AND WRIGHT, C. A.
    Sex hormone studies in male homosexuality.
    Endocrinology, 26: 590–598, 1940.
13. GREENBLATT, R. B., MORTARA, F. AND TOR-
    PIN, R. Sexual libido in the female. Amer. J.
    Obstet. Gynec., 44: 658–663, 1942.
14. HAMILTON, J. B. Demonstrated ability of
    penile erection in castrate men with mark-
    edly low titers of urinary androgens. Proc.
    Soc. Exp. Biol. Med., 54: 309–312, 1943.
15. HAMPSON, J. G. AND MONEY, J. Idiopathic
    sexual precocity in the female: Report of
    three cases. Psychosom. Med., 17: 16–35, 1955.
16. HARRIS, G. W., MICHAEL, R. P. AND SCOTT,
    P. P. Neurological site of action of stilboes-
    trol in eliciting sexual behaviour. In *Ciba
    Foundation Symposium on the Neurological
    Basis of Behaviour.* Little, Brown, Boston,
    1958.
17. HELLER, C. G. Personal communication. In
    Best, W. and Jaffe, F. S., eds. *Simple Meth-
    ods of Contraception*, p. 37. Planned Parent-
    hood Federation of America, New York,
    1958.
18. HELLER, C. G. AND MADDOCK, W. O. The
    clinical uses of testosterone in the male.
    Vitamins Hormones, 5: 393–423, 1947.
19. JACOBS, E. C. Effects of starvation on sex
    hormones in the male. J. Clin. Endocr., 8:
    227–232, 1950.
20. KEYS, A. B., BROŽEK, J., HENSCHEL, A., MIC-
    KELSON, O. AND TAYLOR, H. L. *The Biology
    of Human Starvation.* Univ. of Minnesota
    Press, Minneapolis, 1950.
21. KINSEY, A. C. Homosexuality. Criteria for a
    hormonal explanation of the homosexual.
    J. Clin. Endocr., 1: 424–428, 1941.
22. KINSEY, A. C., POMEROY, W. B., MARTIN, C.
    F. AND GEBHARD, P. H. *Sexual Behavior in
    the Human Female.* Saunders, Philadelphia,
    1953.
23. KUPPERMAN, H. S. AND STUDDIFORD, W. E.
    Endocrine therapy in gynecologic disorders.
    Postgrad. Med., 14: 410–425, 1953.
24. MONEY, J. Components of eroticism in man:
    Cognitional rehearsals. In Wortis, J., ed.
    Recent Advances in Biological Psychiatry, pp.
    210–225. Grune & Stratton, New York, 1960.
25. MONEY, J. Hermaphroditism. In Ellis, A. and
    Abarbanel, A., eds. *The Encyclopedia of
    Sexual Behavior.* Hawthorn Books, New
    York, 1961.
26. MONEY, J. AND HAMPSON, J. G. Idiopathic
    sexual precocity in the male: Management,
    report of a case. Psychosom. Med., 17: 1–15,
    1955.
27. NEUSTADT, R. AND MYERSON, A. Quantitative
    sex hormone studies in homosexuality,
    childhood, and various neuropsychiatric
    disturbances. Amer. J. Psychiat., 97: 524–
    551, 1940.
28. PARE, C. M. B. Homosexuality and chromo-
    somal sex. J. Psychosom. Res., 1: 247–251,
    1956.
29. PASCHKIS, K. E. AND RAKOFF, A. E. Some
    aspects of the physiology of estrogenic hor-
    mones. Recent Progr. Hormone Res., 5:
    115–149, 1950.
30. SALMON, U. J. Rationale for androgen therapy
    in gynecology. J. Clin. Endocr., 1: 162–179,
    1941.
31. SALMON, U. J. AND GEIST, S. H. Effect of an-
    drogens upon libido in women. J. Clin.
    Endocr., 3: 235–238, 1943.
32. SEVRINGHAUS, E. J. AND CHORNYAK, J. A
    study of homosexual adult males. Psycho-
    som. Med., 7: 302–305, 1945.
33. WAXENBERG, S. E., DRELLICH, M. G. AND
    SUTHERLAND, A. M. Changes in female sex-
    uality after adrenalectomy. J. Clin. Endocr.,
    19: 193–202, 1959.
34. WILKINS, L. *The Diagnosis and Treatment of
    Endocrine Disorders in Childhood and Ado-
    lescence*, 2nd ed. Thomas, Springfield, Ill.,
    1957.

Hormones and Sexual Differentiation

IV

Editor's Comments on Papers 33 Through 36

Using cross-sexual gonadal transplants in rats, Steinach in 1912 offered one of the first experimental demonstrations of a dependence of mammalian sexual development on gonadal hormones. The "natural" experiments reported by Frank Lillie (Paper 33) further revealed the ability of the male gonadal secretion in cattle to at least partially alter the morphological development of the genetic female. In the late 1930s a variety of experiments such as those of Green and Ivy (Paper 34) appeared, demonstrating more conclusively that the male hormone testosterone is capable of anatomically masculinizing the female. Further, it became apparent as a result of the work of Phoenix, Goy, Gerall, and Young (Paper 35) with the guinea pig that sex differences in behavior could also be influenced by perinatal gonadal secretions.

Subsequent research with a variety of other species has confirmed the importance of androgenic secretions during critical periods in development in the suppression of gonadal cyclicity and in the establishment of typically male sexual behavior in mammals. In addition, a potential site of action of masculinizing hormones in the hypothalamus has been recently suggested by the results of the work of Barraclough and Gorski (Paper 36).

Reprinted from *Science*, **43**, 611–613 (1916)

33

uals, in one of them the sexual organs remain in the undifferentiated stage, so that the animal superficially resembles a female and ordinarily is recorded as such, although it is barren. The records for monozygotic twins accordingly go to increase the homosexual female and the heterosexual classes, while the homosexual male class in which part of them really belong, does not receive any increment.

Cole thus tentatively adopts the theory, which has been worked out most elaborately by D. Berry Hart, stated also by Bateson, and implied in Spiegelberg's analysis (1861), that the sterile free-martin is really a male co-zygotic with its mate.

Cole's figures represent the only statistical evidence that we have on this subject. Let us follow his suggestion and take from the heterosexual class enough cases to make the homosexual male twins equal in number to the homosexual female pairs; this will be approximately one fourth of the class, leaving the ratio 2:3:2 instead of 1:4:2. Which one of these is the more satisfactory sex ratio I leave others to determine; I wish only to point out the fatal objection, that, according to the hypothesis, the females remaining in the heterosexual class are normal; in other words, on this hypothesis the ratio of normal free-martins (females co-twin with a bull) to sterile ones is 3:1; and the ratio would not be very different on any basis of division of the heterosexual class that would help out the sex ratio. Hitherto there have been no data from which the ratio of normal to sterile free-martins could be computed, and Cole furnishes none. I have records of 21 cases statistically homogeneous, 3 of which are normal and 18 abnormal. That is, the ratio of normal to sterile free-martins is 1:6 instead of 3:1.

This ratio is not more adverse to the normals than might be anticipated, for breeders' associations will not register free-martins until they are proved capable of breeding, and some breeders hardly believe in the existence of fertile free-martins, so rare are they.

My own records of 41 cases of bovine twins (to date, February 25, 1916), all examined *in utero*, and their classification determined anatomically without the possibility of error,

SPECIAL ARTICLES

THE THEORY OF THE FREE-MARTIN

THE term free-martin is applied to the female of heterosexual twins of cattle. The recorded experience of breeders from ancient times to the present has been that such females are usually barren, though cases of normal fertility are recorded. This presents an unconformable case in twinning and sex-determination, and it has consequently been the cause of much speculation.

The appearance of an abstract in SCIENCE[1] of Leon J. Cole's paper before the American Society of Zoologists on "Twinning in Cattle with Special Reference to the Free-Martin," is the immediate cause of this preliminary report of my embryological investigation of the subject. Cole finds in a study of records of 303 multiple births in cattle that there were 43 cases homosexual male twins, 165 cases heterosexual twins (male and female), and 88 cases homosexual female, and 7 cases of triplets. This gives a ratio of about 1♂♂:4♂♀:2♀♀, for the twins instead of the expected ratio of 1:2:1. Cole then states:

The expectation may be brought more nearly into harmony with the facts if it is assumed that in addition to ordinary fraternal (dizygotic) twins, there are numbers of "identical" (monozygotic) twins of both sexes, and that while in the case of females these are both normal, in the case of a dividing male zygote, to form two individ-

[1] Vol. XLIII., p. 177, February 4, 1916.

SCIENCE

give $14\male\male : 21\male\female : 6\female\female$. It will be observed that this agrees with expectation to the extent that the sum of the homosexual classes is (almost) equal to the heterosexual class; and it differs from expectation inasmuch as the $\male\male$ class is over twice the $\female\female$ class instead of being equal to it, as it should be if males and females are produced in equal numbers in cattle. The material can not be weighted statistically because every uterus containing twins below a certain size from a certain slaughter house is sent to me for examination without being opened. Cole's material shows twice as many female as male pairs, and the heterosexual class is about one third greater than the sum of the two homosexual classes. I strongly suspect that it is weighted statistically; the possibility of this must be admitted, for the records are assembled from a great number of breeders. But, whether this is so or not, if we add the sterile free-martin pairs of my collection to the male side in accordance with Cole's suggestion, we get the ratio $32\male\male : 3\male\female : 6\female\female$, which is absurd. And if we take Cole's figures, divide his heterosexual class into pairs containing sterile females and pairs containing normal females according to the expectation, 6 of the former to 1 of the latter, and add the former to his male class, we get an almost equally absurd result ($184\male\male : 23\male\female : 88\female\female$). On the main question our statistical results are sufficiently alike to show that the free-martin can not possibly be interpreted as a male. The theory of Spiegelberg, D. Berry Hart, Bateson and Cole falls on the statistical side alone.

But the real test of the theory must come from the embryological side. If the sterile free-martin and its bull-mate are monozygotic, they should be included within a single chorion, and there should be but a single corpus luteum present. If they are dizygotic, we might expect two separate chorions and two corpora lutea. The monochorial condition would not, however, be a conclusive test of monozygotic origin, for two chorions originally independent might fuse secondarily. The facts as determined from examination of 41 cases are that about 97.5 per cent. of bovine twins are monochorial, but in spite of this nearly all are dizygotic; for in all cases in which the ovaries were present with the uterus a corpus luteum was present in each ovary; in normal single pregnancies in cattle there is never more than one corpus luteum present. There was one homosexual case (males) in which only one ovary was present with the uterus when received, and it contained no corpus luteum. This case was probably monozygotic.

There is space only for a statement of the conclusions drawn from a study of these cases, and of normal pregnancies. In cattle a twin pregnancy is almost always a result of the fertilization of an ovum from each ovary; development begins separately in each horn of the uterus. The rapidly elongating ova meet and fuse in the small body of the uterus at some time between the 10 mm. and the 20 mm. stage. The blood vessels from each side then anastomose in the connecting part of the chorion; a particularly wide arterial anastomosis develops, so that either fetus can be injected from the other. The arterial circulation of each also overlaps the venous territory of the other, so that a constant interchange of blood takes place. If both are males or both are females no harm results from this; but *if one is male and the other female, the reproductive system of the female is largely suppressed, and certain male organs even develop in the female. This is unquestionably to be interpreted as a case of hormone action.* It is not yet determined whether the invariable result of sterilization of the female at the expense of the male is due to more precocious development of the male hormones, or to a certain natural dominance of male over female hormones.

The results are analogous to Steinach's feminization of male rats and masculinization of females by heterosexual transplantation of gonads into castrated infantile specimens. But they are more extensive in many respects on account of the incomparably earlier onset of the hormone action. In the case of the free-martin, nature has performed an experiment of surpassing interest.

Bateson states that sterile free-martins are found also in sheep, but rarely. In the four

Reprinted from *Science*, **86**, 200–201 (1937)

34

THE EXPERIMENTAL PRODUCTION OF IN-
TERSEXUALITY IN THE FEMALE RAT
WITH TESTOSTERONE

IN a previous report[1] the observations of Hain[2] on the production of hypospadias in the female off-spring of the rat by the injection of the-mother with estrone,[3] before or immediately after birth, have been confirmed. We have now found that estradiol[3] injected into the mother (2.0–3.0 mg) antepartum or into the *new-born* female (0.2–0.4 mg) also produces hypospadias. The male offspring were apparently not influenced. On the basis of embryological facts, it was suggested[1] that the hypospadias was due to an hypotrophic defect. This immediately suggested the idea that testosterone[3] when given to the pregnant rat might (a) cause hypospadias in the male offspring, or (b) produce an arrest of the development of the vagina in the female, or (c) produce intersexuality (free-martin) in the female. These latter two possibilities have now been shown to be true.

Testosterone and testosterone propionate in varying doses have been administered to rats at varying periods of pregnancy. A large percentage of resorptions or still births have resulted. To date, however, seven litters have been obtained. In these litters there were twenty-seven normal males and nineteen females with varying degrees of intersexuality. Three of these litters are now completely mature (67 to 70 days, with weight 150 to 170 gms) and display no evidence of further "feminization."

A rudimentary but patent vagina is present in the female adult offspring of one litter, whose mother received a small dose of testosterone late in pregnancy (3.3 mgm on nineteenth day). In all other animals the pelvic vagina is absent (distal portion). In all animals dissected to date, mature or new born, attached to the proximal urethra at the base of the bladder, are paired glandular structures that histologically resemble prostate. In addition other glandular tissue that histologically resembles seminal vesicles is found.

In adult animals that have been killed apparently normal ovaries (corpora lutea present in one case) have been found with seemingly normal oviducts and uteri that end in a dilated, bulbous structure that represents the proximal portion of the vagina.

Other animals, whose mothers received the male sex hormone earlier in pregnancy, show varying degrees of inhibition of Mullerian duct, and stimulation of Wolffian duct derivatives. In one animal, whose mother received 2.5 mgm testosterone propionate on the twelfth day of pregnancy, the oviducts are seemingly absent and the uteri are represented by a very rudimentary structure immediately posterior to the bladder. Prostate, seminal vesicle and a rudimentary vas deferens are present.

The external genitalia of affected animals vary. The offspring of mothers receiving the male sex hormone late in pregnancy have a crescentic fold of skin that represents the vaginal orifice surrounding the caudal base of an organ that resembles a "hypospadic" clitoris. With larger doses given earlier in pregnancy the organ resembles a penis, but is smaller than the penis of a litter brother.

It is rational to expect that male sex hormone may influence the development of the primordia of the female genitalia as late as the twelfth to fourteenth day of gestation, because the early development of the rat embryo is very slow, *e.g.*, the mesodermal layer does not appear until the ninth day.[4] It is known that if in cattle[5] or pigs[6] the circulation of twin male and female embryos is interconnected, the female becomes modified in the male direction. Lillie[5, 6] postulates that this is due to the effect of a male sex hormone upon the anlage of the genitalia of the female embryo. Further, a report by Dantchakoff,[7] which was found during the preparation of this communication, shows that on injecting testosterone into the amniotic sack of embryo guinea pigs, agenesis of the vagina and other evidences of intersexuality result. Thus our results on the rat are quite analogous to those on the guinea pig.

R. R. GREENE
A. C. IVY

NORTHWESTERN UNIVERSITY
MEDICAL SCHOOL

[1] R. R. Greene, *Proceedings Soc. Exp. Biol. and Med.*, 36: 503, 1937.
[2] A. M. Hain, *Quart. Jour. Exp. Physiol.*, 25: 131, 303, 1935; *ibid.*, 26: 290, 293, 1936.
[3] We desire to thank Dr. Oliver Kamm, of Parke, Davis and Company for the estrone, and Dr. E. Schwenk, of the Schering Corporation, for the estradiol and testosterone, used in this work.
[4] G. C. Huber, *Jour. Morph.*, 25: 247, 1915.
[5] F. R. Lillie, *Jour. Exp. Zool.*, 23: 371, 1917.
[6] W. Hughes, *Anat. Rec.*, 41: 213, 1928–29.
[7] V. Dantchakoff, *Compt. rend. Soc. d. Biol.*, 174: 516 (March), 1937.

twin pregnancies of sheep that I have so far had the opportunity to examine, a monochorial condition was found, though the fetuses were dizygotic; but the circulation of each fetus was closed. This appears to be the normal condition in sheep; but if the two circulations should anastomose, we should have the conditions that produce a sterile free-martin in cattle. The possibility of their occurrence in sheep is therefore given.

The fertile free-martin in cattle may be due to cases similar to those normal for sheep. Unfortunately when the first two cases of normal cattle free-martins that I have recorded, came under observation I was not yet aware of the significance of the membrane relations, and the circulation was not studied. But I recorded in my notebook in each case that the connecting part of the two halves of the chorion was narrow, and this is significant. In the third case the two chorions were entirely unfused; this case, therefore, constitutes an *experimentum crucis*. The male was 10.4 cm. long; the female 10.2 cm. The reproductive organs of both were entirely normal. The occurrence of the fertile free-martin is therefore satisfactorily explained.

The sterile free-martin enables us to distinguish between the effects of the zygotic sex-determining factor in mammals, and the hormonic sex-differentiating factors. The female is sterilized at the very beginning of sex-differentiation, or before any morphological evidences are apparent, and male hormones circulate in its blood for a long period thereafter. But in spite of this the reproductive system is for the most part of the female type, though greatly reduced. The gonad is the part most affected; so much so that most authors have interpreted it as testis; a gubernaculum of the male type also develops, but no scrotal sacs. The ducts are distinctly of the female type much reduced, and the phallus and mammary glands are definitely female. The general somatic habitus inclines distinctly toward the male side. Male hormones circulating in the blood of an individual zygotically female have a definitely limited influence, even though the action exists

from the beginning of morphological sex-differentiation. A detailed study of this problem will be published at a later date.

FRANK R. LILLIE

UNIVERSITY OF CHICAGO

Copyright © 1959 by J. B. Lippincott Co.

Reprinted from *Endocrinology*, **65**, 369–382 (1959)

ORGANIZING ACTION OF PRENATALLY ADMINISTERED TESTOSTERONE PROPIONATE ON THE TISSUES MEDIATING MATING BEHAVIOR IN THE FEMALE GUINEA PIG[1]

CHARLES H. PHOENIX, ROBERT W. GOY, ARNOLD A. GERALL
AND WILLIAM C. YOUNG

Department of Anatomy, University of Kansas, Lawrence, Kansas

35

ABSTRACT

The sexual behavior of male and female guinea pigs from mothers receiving testosterone propionate during most of pregnancy was studied after the attainment of adulthood. As a part of the investigation, the responsiveness of the females to estradiol benzoate and progesterone and to testosterone propionate was determined.

The larger quantities of testosterone propionate produced hermaphrodites having external genitalia indistinguishable macroscopically from those of newborn males. Gonadectomized animals of this type were used for tests of their responsiveness to estradiol benzoate and progesterone and to testosterone propionate. The capacity to display lordosis following administration of estrogen and progesterone was greatly reduced. Male-like mounting behavior, on the other hand, was displayed by many of these animals even when lordosis could not be elicited. Suppression of the capacity for displaying lordosis was achieved with a quantity of androgen less than that required for masculinization of the external genitalia.

The hermaphrodites receiving testosterone propionate as adults displayed an amount of mounting behavior which approached that displayed by the castrated injected males receiving the same hormone.

The data are uniform in demonstrating that an androgen administered prenatally has an organizing action on the tissues mediating mating behavior in the sense of producing a responsiveness to exogenous hormones which differs from that of normal adult females.

No structural abnormalities were apparent in the male siblings and their behavior was essentially normal.

The results are believed to justify the conclusion that the prenatal period is a time when fetal morphogenic substances have an organizing or "differentiating" action on the neural tissues mediating mating behavior. During adulthood the hormones are activational.

Attention is directed to the parallel nature of the relationship, on the one hand, between androgens and the differentiation of the genital tracts, and on the other, between androgens and the organization of the neural tissues destined to mediate mating behavior in the adult.

Received February 9, 1959.

[1] This investigation was supported by research grant M-504 (C6) from the National Institute of Mental Health, Public Health Service.

INVESTIGATORS interested in reproductive behavior have demonstrated that one role of the gonadal hormones in adult male and female mammals is to bring to expression the patterns of behavior previously organized or determined by genetical and experiential factors (1, 2, 3, 4, 5). The hypothesis that these hormones have an organizing action in the sense of patterning the responses an individual gives to such substances has long been rejected (6, 7, 8, 9, 10). As far as the adult is concerned, this conclusion seems well founded. Female hormone, instead of feminizing castrated male rats as Kun (11) claimed, increased their activity as males (6). Male and female guinea pigs gonadectomized the day of birth, and a female rat with a congenital absence of the ovaries, displayed normal patterns of behavior when injected with the appropriate hormones as adults (3, 9, 12).

Unexplored since the studies of Dantchakoff (13, 14, 15), Raynaud (16) and Wilson, Young and Hamilton (17), is the possibility that androgens or estrogens reaching animals during the prenatal period might have an organizing action that would be reflected by the character of adult sexual behavior. If the existence of such an action were revealed, it would 1) extend our knowledge of the role of the gonadal hormones in the regulation of sexual behavior by providing information bearing on the action of these hormones or related substances during the prenatal period, 2) be suggestive evidence that the relationship between the neural tissues mediating mating behavior and the morphogenic fetal hormones parallels that between the genital tissues and the same hormones, and 3) direct attention to a possible origin of behavioral differences between the sexes which is *ipso facto* important for psychologic and psychiatric theory (18). Although comprehensive experiments have not yet been performed, initial investigations with an androgen have yielded effects which are so much more in line with current thought in the area of gonadal hormones and sexual differentiation (19, 20, 21, 22) than the earlier experiments on behavior, that the results are summarized here.

MATERIALS AND METHODS

(the production of hermaphrodites)

Most of the experimental animals were born to mothers which had received intramuscular injections of testosterone propionate[2,3] during much of gestation. One group was composed of females in which there were no visible abnormalities of the external

---

[2] Testosterone propionate (Perandren propionate) was supplied by Ciba Pharmaceutical Products, Inc.

[3] The injections were made by Mr. Myron D. Tedford, a Public Health Service Predoctoral Fellow, who is using these and other animals treated similarly for a study of the structural changes in the gonads, genital tracts, and external genitalia, and the course of gestation. We are indebted to him for supplying us with the animals whose behavior was investigated.

genitalia. These are referred to as the *unmodified females*. Their mothers were given an initial injection of 1 mg. of testosterone propionate some time between day 10 and day 27 after conception and 1 mg. every third or fourth day thereafter until the end of pregnancy.

The larger group was composed of females in which the external genitalia at the time of birth were indistinguishable macroscopically from those of their male siblings and untreated males. These animals are designated *hermaphrodites*. Laparotomy was necessary in order to distinguish these genetical females from males; it was performed within the first week after birth. Their mothers received an initial injection of 5 mg. of testosterone propionate on day 10, 15, 18, or 24 of the gestation period and 1 mg. daily thereafter until day 68.

Control animals were females and males from untreated mothers from the same stock as the experimental animals.

All these animals, i.e. the unmodified females, the hermaphrodites, their male siblings, and the control females and males, were used in four experiments designed to test the effects of testosterone propionate received prenatally on the responsiveness of the animals as adults to male and female hormones.

EXPERIMENTAL

Experiment I. *The behavior of gonadectomized adult unmodified females and hermaphrodites injected with estradiol benzoate and progesterone.*

Subjects

Fourteen females from untreated mothers.
Fourteen unmodified females.
Nine hermaphrodites.
Eight males from untreated mothers.

Except for four unmodified females gonadectomized when they were 45 days old, all the unmodified females and hermaphrodites were gonadectomized at 80 to 150 days of age. No data from the laboratory indicate that the response to exogenously administered sex hormones is influenced by age at the time of gonadectomy. The eight males were castrated before they were 21 days old.

Tests

After gonadectomy, when the animals were 90 to 160 days old, tests were made of the responsiveness to 1.66, 3.32, and 6.64 µg. of subcutaneously injected estradiol benzoate followed 36 hours later by 0.2 mg. of progesterone.[4] Observations were continuous for 12 hours, beginning immediately after the injection of progesterone. Following the procedure of Goy and Young (4) hourly checks were made for the occurrence of the lordosis reflex in response to fingering. Individual records were kept of this measure of behavior and of the frequency of male-like mounting.

In three tests the control females, hermaphrodites, and males were ob-

---

[4] Estradiol benzoate (Progynon-B) and progesterone (Proluton) were supplied by the Schering Corporation.

served for the occurrence of mounting in the absence of exogenous hormone. The unmodified females were given one such test.

The means and medians of the measures of behavior for which data were obtained were calculated from the individual averages and they are based on the data from the animals which responded to the hormones. For purposes of statistical analysis, maximum values (12 hours) for latency and 0 values for all other measures were arbitrarily assigned to the individuals failing to respond.

Results

The data bearing on all the measures of the estrous response except mounting are summarized in Table 1. The lower values for the per cent of

TABLE 1. DURATION OF HEAT AND LORDOSIS IN GONADECTOMIZED GUINEA PIGS GIVEN DIFFERENT AMOUNTS OF ESTRADIOL AND 0.2 MG. OF PROGESTERONE

Subjects	Tests* N	Per cent of tests positive for estrus	Mean latency in hours	Mean duration of heat in hours	Median duration of max. lord. in seconds
		1.66 μg.			
Control females	19	89	5.7	5.7	11.5
Unmodified females	20	65	6.5	2.8	8.5
Hermaphrodites	9	22	8.5	2.5	2.0
Castrated males	8	38	6.0	1.2	2.0
		3.32 μg.			
Control females	33	94	4.4	7.3	12.3
Unmodified females	38	68	5.6	2.8	5.1
Hermaphrodites	18	22	8.0	2.0	3.0
Castrated males	16	31	4.5	3.2	2.7
		6.64 μg.			
Control females	28	96	3.7	7.2	9.3
Unmodified females	22	77	5.8	3.3	6.0
Hermaphrodites	18	22	9.2	2.0	2.0
Castrated males	16	0	—	—	—

* All the animals were given one or more tests at each level of hormone.

tests positive for estrus, the mean duration of heat, and the median duration of the maximum lordosis were conspicuous effects of the treatment given prenatally and the differences among the groups are highly significant (P <.001). Among the two groups of experimental females and the castrated males, the low gutteral growl which is so characteristically a part of the pattern of lordosis in normal females, was commonly, and in some individuals always, lacking. Had the estimation of the duration of maximum lordosis been based only on complete responses, the differences among the groups would have been even greater.

Variations in medians for the duration of maximum lordosis were not systematically related to quantity of estradiol given prior to the tests. The analysis, therefore, was based on the medians of individual averages over all dosages. These medians were 11.3, 6.5, 2.3, and 2.5 seconds for control females, unmodified females, hermaphrodites, and castrated males, respectively. The median of the unmodified females, which most closely resembles that of the control females, is significantly different ($U = 22$, $P < .002$) from that of the controls.

Other differences also are indicative of the changes that were induced. Per cent response and duration of heat tended to increase in the control groups as the quantity of injected estradiol was increased. Latency which is related inversely to the duration of heat (4) decreased. Among the experimental groups (unmodified females, hermaphrodites, and castrated males), similar relationships were seen only in the unmodified females.

In general the suppression of the capacity to display lordosis was proportional to the quantity of androgen injected prenatally. Amounts in-

TABLE 2. THE QUANTITY OF MOUNTING WITH AND WITHOUT
ESTRADIOL AND PROGESTERONE

Subjects	Without hormone	With hormone*
	Mean number of mounts	Mean number of mounts
Control females	0	10.7
Unmodified females	0	8.8
Hermaphrodites	4.4	5.6
Castrated males	11.8	16.7

* Variation in the amount of mounting was not related to the quantity of estradiol. The means therefore are based on the averages for each individual whether the dosage was 1.66, 3.32, or 6.64 μg. of the hormone.

sufficient to alter external genital structures resulted in disturbances in the lordosis in only 50% of the animals, but the larger amounts that produced the hermaphrodites affected the lordosis in all. Within each group the effect on lordosis was not related to the quantity of androgen received prenatally. Among unmodified females, even siblings differed, one showing complete suppression of the lordosis and the other responding normally. The findings demonstrate that suppression of the capacity for displaying lordosis does not depend on masculinization of the external genitalia; clearly less androgen was required for the former than for the latter.

Additional evidence for the masculinizing effect of the prenatally administered androgen is provided by the data on the male-like mounting displayed by each group (Table 2). When estradiol and progesterone were injected all groups displayed mounting, and the differences among the groups are not statistically significant. In contrast, on tests when no hormones were given, the hermaphrodites and castrated males were the only animals that mounted.

The interval from the beginning of the test to the display of mounting differed among the groups. Of the males which mounted, all did so at least once during the first hour. Of the 7 hermaphrodites which mounted, 5 or

71% mounted at least once during the first hour, but only 1 normal female (7%) and 1 unmodified female (7%) mounted this early in the test. The modal time for the onset of mounting was the 1st hour for the castrated males and hermaphrodites and the 6th and 7th hours for the control females and unmodified females, respectively. In this respect the hermaphrodites closely resembled the castrated males and seem to have been masculinized. The latency of mounting in the unmodified females was not different from that in the control females.

In one way the mounting performance of the unmodified females did differ from that of the controls. More unmodified than control females displayed mounting on tests after injections when the lordosis reflex could not be obtained. Of 8 unmodified females which failed to show lordosis, 6 or 75% mounted. Because of the small number of control females which failed to display lordosis after injection, older data on normal females from the same genetical stock are used for comparison. These data combined with those from the present study reveal that of 38 normal females failing to display lordosis after injection with comparable amounts of estradiol and progesterone only 4 or 10.5% mounted. The difference between the proportions of control females and unmodified females displaying mounting in the absence of lordosis is significant (C.R. = 4.02, P < .001). Inasmuch as mounting was displayed spontaneously by the hermaphrodites, it was not possible with the animals available to determine the extent to which this behavior was being shown in response to the estradiol and progesterone.

Conclusions

1. Prenatally administered testosterone propionate suppressed the capacity for displaying lordosis following gonadectomy and the injection of estradiol and progesterone. The effect was manifested either by an absence of lordosis or by a marked abnormality in its character when it was displayed.

2. Suppression of the capacity for displaying lordosis was achieved with a smaller quantity of the androgen than was necessary for the gross modification of the external genitalia.

3. The capacity to display male-like mounting was not suppressed.

4. Quantities of testosterone propionate sufficient to suppress lordosis and masculinize the genitalia also reduced the interval before mounting behavior was displayed.

Experiment II. *Permanence of the effects of prenatally administered androgen.*

Subjects

Group 1. Three hermaphrodites used in the previous experiment.

Group 2. Seven unmodified females used in the previous experiment.

Group 3. Eight control females used in the previous experiment.

Group 4. Six hermaphrodites injected with 500 μg. of testosterone propionate per 100 gm. body weight per day from birth to 80 days of age.

Group 5. Six normal females injected with the same amount of testosterone propionate from birth to 80 days of age.

Group 6. Five mothers of hermaphrodites injected with testosterone propionate during pregnancy as described in Materials and Methods.

Group 7. Eight untreated females comparable in age with those injected with testosterone propionate during pregnancy.

The animals in Groups 1 through 5 were gonadectomized when they were 80 to 150 days of age, those in Groups 6 and 7 when they were 1.5 to 3 years old. The operations on the animals in Group 6 were performed approximately 10 months after the last injection of testosterone propionate.

Tests

All the animals received 3.32 μg. of estradiol benzoate followed 36 hours later with 0.2 mg. of progesterone. The tests were similar to those given the hermaphrodites, unmodified females, and controls in Experiment I. The number, however, differed for each group and is shown in the description of the results. The values reported in the tables and the statistical treatment of the data were determined by the methods described in Experiment I.

Results

The behavior of the 3 hermaphrodites, the 7 unmodified females, and the 8 control females is summarized in Table 3 and compared with that dis-

TABLE 3. BEHAVIORAL RESPONSES TO 3.32 μG. OF ESTRADIOL AND
0.2 MG. OF PROGESTERONE

		Tests at 6–9 months of age	Tests at 11–12 months of age
Hermaphrodites (Group 1)	Per cent response	33.0	0
	Latency to heat in hours	7.5	—
	Duration of heat in hours	2.5	—
	Median maximum lordosis in seconds	2.0	—
	Mean number of mounts	3.0	45.2
Unmodified females (Group 2)	Per cent response	55.0	71.0
	Latency to heat in hours	6.3	7.5
	Duration of heat in hours	2.2	2.3
	Median maximum lordosis in seconds	4.0	5.8
	Mean number of mounts	8.7	17.5
Normal females (Group 3)	Per cent response	95.0	94.0
	Latency to heat in hours	4.4	6.1
	Duration of heat in hours	7.2	4.5
	Median maximum lordosis in seconds	10.0	10.2
	Mean number of mounts	9.9	9.6

TABLE 4. PER CENT RESPONSE, DURATION OF HEAT, AND MAXIMUM LORDOSIS AFTER CESSATION OF TREATMENT WITH TESTOSTERONE PROPIONATE FROM BIRTH TO 80 DAYS OF AGE

| | | Approximate age in days at time of test | | | |
		90	140	160	175
Hermaphrodites (Group 4)	Per cent response	0	0	0	0
	Mean duration of heat in hours	—	—	—	—
	Median maximum lordosis in seconds	—	—	—	—
Females (Group 5)	Per cent response	0	84	66	66
	Mean duration of heat in hours	0	4.6	1.7	3.7
	Median maximum lordosis in seconds	0	9.0	5.5	9.5

played during the earlier tests when the animals were 6 months old. The results reported in Table 3 are based on at least 2 tests of each individual at each age level. No significant change occurred in the hermaphrodites and unmodified females for per cent response, latency to heat, duration of heat, and the duration of maximum lordosis. The normal females, however, showed a significant decrease in the duration of heat (T = 0, P = .01), reflecting perhaps a decrease in responsiveness to the hormones as the animals aged. The increase in mounting is significant for the unmodified females (T = 0, P = .02). The 3 hermaphrodites displayed increased mounting behavior, but the increase could not be evaluated statistically. Of the normal females, 3 showed increases, 3 a decrease, and 2 remained the same.

The contrast between the effects of prenatal and postnatal treatment is revealed by the results obtained from the animals treated neonatally (Groups 4 and 5) and from those treated during pregnancy (Group 6). During the period after withdrawal of the testosterone propionate, 5 of the 6 normal females which had been injected for 80 days after birth regained the ability to display lordosis, whereas the hermaphrodites did not (Table 4). The effects of the postnatally administered androgen on the mounting behavior displayed by the animals in the two groups were complex and their presentation is being postponed until a further discussion can be given. The females treated with testosterone propionate while pregnant (Group 6) did not, like their "daughters," lose the capacity to display lordosis. Comparison of their behavior in response to estradiol and progesterone in five tests with that of untreated females of the same age (Group 7) (Table 5), revealed that the differences between the groups are not significant for latency, duration of heat, and mounting.

TABLE 5. BEHAVIOR OF NORMAL FEMALES TREATED WITH TESTOSTERONE PROPIONATE FOR 50 DAYS DURING PREGNANCY AND TESTED 10 MONTHS LATER

	Per cent response	Latency of heat in hours	Duration of heat in hours	Mean no. of mounts
Treated females (Group 6)	84	6.7	4.2	17.8
Untreated females (Group 7)	62	7.6	3.2	8.1

Conclusions

1. The suppression of the capacity for displaying the feminine components of the sexual behavior pattern which followed the administration of testosterone propionate prenatally appears to have been permanent.

2. Amounts of testosterone propionate which were effective prenatally had no conspicuous lasting effects when administered postnatally.

Experiment III. *The behavior of gonadectomized hermaphrodites in response to testosterone propionate.*

Subjects

Five hermaphrodites gonadectomized between 86 and 112 days of age.
Five normal females gonadectomized between 80 and 106 days of age.
Eight normal males castrated before 21 days of age.

When the animals were approximately 180 days old all received 2.5 mg. of testosterone propionate daily for 16 consecutive days.

Tests

A sexual behavior test was given the day before the first injection. Additional tests were given on days 1 and 2 of the injection period, and every other day thereafter until each animal had received 9 tests. The ninth test was given the day of the sixteenth injection.

Results

The median value for mounting by the hermaphrodites and females in the single test prior to the injection of testosterone propionate was 0. For the males the median was 5.5.

The remaining data are summarized in Table 6. They demonstrate the masculinizing effect of prenatally administered testosterone propionate on the female. Castrated males and hermaphrodites obtained the highest sexual behavior scores, the control females the lowest. The overall difference in scores was significant (P~.02). The differences between the castrated males and hermaphrodites were not significant, whereas both groups differed significantly from the control females (P = .05). The overall difference in the number of tests to the first display of mounting was significant (P <.01). As with the sexual behavior scores, the difference between males

TABLE 6. MASCULINE BEHAVIOR IN GONADECTOMIZED ADULT ANIMALS
INJECTED WITH TESTOSTERONE PROPIONATE

Group	Mean sexual behavior score	Mean mounts per test	Median number of tests to the first display of mounting	Median mg. of t.p. prior to the display of mounting
Spayed untreated females	2.1	5.8	7.0	30.0
Spayed hermaphrodites	3.6	15.4	3.0	10.0
Males castrated prepuberally	5.0	20.5	1.5	3.8

and hermaphrodites was not significant, but both groups differed significantly from the control females (P = .02). There was a significant overall difference (P < .01) in the amount of testosterone propionate required before the first appearance of mounting. Again, the hermaphrodites resembled the castrated males in that there was no significant difference between these two groups, but both groups displayed mounting with significantly less hormone (P = .02) than the control females.

Conclusions

1. Adult hermaphrodites gonadectomized and injected with testosterone propionate were more responsive to this hormone than gonadectomized normal females.

2. The earlier appearance and greater strength of masculine behavior by the hermaphrodites given testosterone propionate are believed to be effects of the prenatally administered testosterone propionate on the tissues mediating masculine behavior and therefore to be expressions of its organizing action.

Experiment IV. *The behavior of adult male siblings of the hermaphrodites.*

Subjects

Five males from untreated mothers.

Five males born to mothers receiving testosterone propionate during pregnancy. No hormone was administered after birth.

Five males born to mothers receiving testosterone propionate during pregnancy. These animals received 500 μg. of the hormone per 100 gm. body weight daily beginning 1 to 3 days after birth and continuing 80 to 90 days.

Tests

Five tests were given when the animals were 11 months old. In a test the subject was placed with a receptive female of approximately the same size, and the frequency of the display of selected measures of behavior was recorded for a maximum of 10 minutes. These measures included sniffing and nibbling, nuzzling, abortive mounting, mounting, intromissions, and ejaculation. A description of the measures and the method for computing scores are given by Valenstein, Riss and Young (23).

Results

The mean scores are summarized in Table 7. It is clear that any effect of the exogenous testosterone propionate was slight. There was no evidence of suppression of the capacity to display masculine behavior, if anything, the animals receiving the hormone prenatally achieved higher scores than the controls.

TABLE 7. MEAN SEXUAL BEHAVIOR SCORES OBTAINED BY THE
THREE GROUPS OF ADULT MALES

Groups*	Tests				
	I	II	III	IV	V
Untreated	6.9	6.6	9.2	7.2	10.4
Testosterone propionate prenatally	10.4	9.3	9.1	9.3	12.2
Testosterone propionate prenatally and postnatally	10.9	11.2	7.3	11.1	9.4

* Difference among the groups not significant; $F = 1.30$; $df = 2, 12$.

Conclusion

The sexual behavior of adult males which had received testosterone propionate prenatally was not significantly different from that of untreated controls.

DISCUSSION

The data from the four experiments summarized in the preceding sections support the hypothesis that androgenic substances received prenatally have an organizing action on the tissues mediating mating behavior in the sense of altering permanently the responses females normally give as adults. This possibility was suggested by the work of Dantchakoff (13, 14, 15), Raynaud (16), and Wilson, Young and Hamilton (17). Probably, however, because interest in the role of gonadal hormones in the regulation of mating behavior was concentrated so largely on the neonatal individual and adult, the suggestion was never incorporated in our theories of hormonal action. This step may now be taken, but when what has been learned from the present investigation is related to what has long been known with respect to the action of androgens on the genital tracts, a concept much broader than that suggested by the older studies is revealed.

The embryonic and fetal periods, when the genital tracts are exposed to the influence of as yet unidentified morphogenic substances (19, 20, 21, 22, 24), are periods of differentiation. The adult period, when the genital tracts are target organs of the gonadal hormones, is a period of functional response as measured by cyclic growth, secretion, and motility. The response depends on whether Müllerian or Wolffian duct derivatives have developed, and although generally specific for hormones of the corresponding sex, it is not completely specific (25). For the neural tissues mediating mating behavior, corresponding relationships seem to exist. The embryonic and fetal periods are periods of organization or "differentiation" in the direction of masculinization or feminization. Adulthood, when gonadal hormones are being secreted, is a period of activation; neural tissues are the target organs and mating behavior is brought to expression. Like the genital tracts, the neural tissues mediating mating behavior respond to androgens or to estrogens depending on the sex of the individual, but again the specificity is not complete (26, 27).

An extension of this analogy is suggested by the work done on the embryonic differentiation of the genital tracts, particularly that by Burns and Jost and summarized in their reviews (20, 21, 22). It will be recalled from the data reported in the present study that testosterone propionate administered prenatally affected the behavior of the male but slightly, whereas the effects on the female were profound. Not only was there a heightened responsiveness to the male hormone as revealed by the stronger masculine behavior displayed when testosterone propionate was given, but there was a suppression of the capacity to display the feminine components in response to treatment with an estrogen and progesterone. In studies of the genital tracts there were no effects on the male except for a slight acceleration in the development of the prostate and seminal vesicle and an increase in the size of the penis (28). Within the female, the Wolffian duct system was stimulated (13, 14, 15, 28, 29), and locally, when a fetal testis was implanted into a female fetus (20, 21), there was an interruption of the Müllerian duct on that side. What has not been seen when an exogenous androgen was administered, except by Greene and Ivy (30) in some of their rats, is a suppression or inhibition of the Müllerian duct system corresponding to the suppression of the capacity for displaying the feminine component of behavior.

The failure to detect a corresponding suppressing action on the Müllerian duct does not exclude the possibilities 1) that such an effect will be found, and 2) that the suppressing action is in the nature of a reduction in the responsiveness of the genital tract to estrogens rather than in the inhibition of its development. Such an effect was encountered in rats given testosterone propionate prenatally (17) when it was found that uterine as well as behavioral responses to estrogen and progesterone were suppressed.

A final suggestion with respect to the analogy we have postulated arises from a comparison of our results with those reported by Dantchakoff and Raynaud. These investigators stressed the increased responsiveness of their masculinized guinea pigs and mice to exogenous androgens, and seemed to regard the change as the expression of an inherent bisexuality. The possibility that there might have been a suppression of the capacity to respond as females and therefore an inequality of potential does not seem to have been considered. Like Dantchakoff (13, 14, 15), Raynaud (16), and many others (9, 31, 32, 33, 34, 35), the existence of a bisexuality is assumed. We suggest, however, that in the adult this bisexuality is unequal in the neural tissues as it is in the case of the genital tissues. The capacity exists for giving behavioral responses of the opposite sex, but it is variable and, in most mammals that have been studied and in many lower vertebrates as well, it is elicited only with difficulty (27). Structurally, the situation is similar. Vestiges of the genital tracts of the opposite sex persist and are responsive to gonadal hormones (36, 37), but except perhaps in rare instances, equivalence of organs and responses in a single individual is not seen (36, 37, 38, 39).

The concept of a correspondence between the action of gonadal hormones on genital tissues and neural tissues contains much that is new and its full scope is not yet clear. The possibility must be considered that the masculinity or femininity of an animal's behavior beyond that which is purely sexual has developed in response to certain hormonal substances within the embryo and fetus.

Thus far the permanence of the effect achieved when testosterone propionate was received prenatally has not been achieved when the same hormones were administered to adults or to newborn individuals. The dependence of this "permanence" on the action of the hormone during a possible critical period must be ascertained.

The nature of the modifications produced by prenatally administered testosterone propionate on the tissues mediating mating behavior and on the genital tract is challenging. Embryologists interested in the latter have looked for a structural retardation of the Müllerian duct derivatives culminating in their absence, except perhaps for vestigial structures found in any normal male. Neurologists or psychologists interested in the effects of the androgen on neural tissues would hardly think of alterations so drastic. Instead, a more subtle change reflected in function rather than in visible structure would be presumed.

Involved in this suggestion is the view that behavior may be treated as a dependent variable and therefore that we may speak of shaping the behavior by hormone administration just as the psychologist speaks of shaping behavior by manipulating the external environment. An assumption seldom made explicit is that modification of behavior follows an alteration in the structure or function of the neural correlates of the behavior. We are assuming that testosterone or some metabolite acts on those central nervous tissues in which patterns of sexual behavior are organized. We are not prepared to suggest whether the site of action is general or localized.

REFERENCES

1. BEACH, F. A.: *J. Genet. Psychol.* **60**: 121. 1942.
2. ZIMBARDO, P. G.: *J. Comp. & Physiol. Psychol.* **51**: 764. 1958.
3. VALENSTEIN, E. S., W. RISS AND W. C. YOUNG: *J. Comp. & Physiol. Psychol.* **48**: 397. 1955.
4. GOY, R. W. AND W. C. YOUNG: *Psychosom. Med.* **19**: 144. 1957.
5. ROSENBLATT, J. S. AND L. R. ARONSON: *Animal Behav.* **6**: 171. 1958.
6. BALL, J.: *J. Comp. Psychol.* **24**: 135. 1937.
7. BALL, J.: *J. Comp. Psychol.* **28**: 273. 1939.
8. BEACH, F. A.: *Endocrinology* **29**: 409. 1941.

9. BEACH, F. A.: *Anat. Rec.* **92**: 289. 1945.
10. RISS, W., E. S. VALENSTEIN, J. SINKS AND W. C. YOUNG: *Endocrinology* **57**: 139. 1955.
11. KUN, H.: *Endokrinologie* **13**: 311. 1934.
12. WILSON, J. G. AND W. C. YOUNG: *Endocrinology* **29**: 779. 1941.
13. DANTCHAKOFF, V.: *Compt. rend. Acad. sci.* **206**: 945. 1938.
14. DANTCHAKOFF, V.: *Compt. rend. soc. Biol.* **127**: 1255. 1938.
15. DANTCHAKOFF, V.: *Compt. rend. soc. Biol.* **127**: 1259. 1938.
16. RAYNAUD, A.: *Bull. Biol. France et Belgique* **72**: 297. 1938.
17. WILSON, J. G., W. C. YOUNG AND J. B. HAMILTON: *Yale J. Biol. & Med.* **13**: 189. 1940.
18. HAMPSON, J. L. AND J. G. HAMPSON: Allen's Sex and Internal Secretions, ed. by W. C. Young, Baltimore, Williams & Wilkins. In press.
19. JOST, A.: *Arch. Anat. micro. et Morph. exper.* **36**: 271. 1947.
20. JOST, A.: *Rec. Prog. Hormone Res.* **8**: 379. 1953.
21. JOST, A.: Gestation. Transactions of the Third Conference, ed. by C. A. Villee, New York, Josiah Macy, Jr. Foundation, 129. 1957.
22. BURNS, R. K.: Allen's Sex and Internal Secretions, ed. by W. C. Young, Baltimore, Williams & Wilkins. In press.
23. VALENSTEIN, E. S., W. RISS AND W. C. YOUNG: *J. Comp. & Physiol. Psychol.* **47**: 162. 1954.
24. HOLYOKE, E. A. AND B. A. BEBER: *Science* **128**: 1082. 1958.
25. BURROWS, H.: Biological Actions of Sex Hormones, Cambridge, Cambridge University Press. 1949.
26. ANTLIFF, H. R. AND W. C. YOUNG: *Endocrinology* **59**: 74. 1956.
27. YOUNG, W. C.: Allen's Sex and Internal Secretions, ed. by W. C. Young, Baltimore, Williams & Wilkins. In press.
28. GREENE, R. R.: *Biol. Symposia* **9**: 105. 1942.
29. TURNER, C. D.: *J. Morphol.* **65**: 353. 1939.
30. GREENE, R. R. AND A. C. IVY: *Science* **86**: 200. 1937.
31. STEINACH, E.: *Zentrabl. Physiol.* **27**: 717. 1913.
32. STEINACH, E.: *Arch. f. Entwcklngsmechn. d. Organ.* **42**: 307. 1916.
33. LIPSCHÜTZ, A.: The Internal Secretions of the Sex Glands. Baltimore, Williams & Wilkins. 1924.
34. BEACH, F. A.: *J. Comp. Psychol.* **36**: 169. 1942.
35. BEACH, F. A.: *Physiol. Zool.* **18**: 390. 1945.
36. MAHONEY, J. J.: *J. Exper. Zool.* **90**: 413. 1942.
37. PRICE, D.: *Anat. Rec.* **82**: 93. 1942.
38. BURNS, R. K.: *Contr. Embryology*, Carnegie Institution of Washington, **31**: 147. 1945.
39. BURNS, R. K. *Am. J. Anat.*, **98**: 35. 1956.

Reprinted from *Endocrinology*, **68**, 68–79 (1961)

EVIDENCE THAT THE HYPOTHALAMUS IS RESPONSIBLE FOR ANDROGEN-INDUCED STERILITY IN THE FEMALE RAT[1]

36

CHARLES A. BARRACLOUGH AND ROGER A. GORSKI

Department of Anatomy, School of Medicine, University of California at Los Angeles, and Research Division, Veterans Administration Hospital, Sepulveda, California

ABSTRACT

The data presented in the current study suggest that the sterility which ensues from androgen treatment of prepubertal female rats is not the result of malfunction of the adenohypophysis. Rather, the pituitaries of such animals will respond to electrical stimulation of the hypothalamus by discharging sufficient gonadotropin to cause ovulation. It is further suggested from these experiments that there are two regions within the hypothalamus which control adenohypophyseal gonadotropin secretion: (a) a region responsible for activating the cyclic discharge of ovulating hormone: the anterior preoptic (suprachiasmatic) area of the hypothalamus, and (b) a region independently responsible for the tonic discharge of luteinizing hormone in a sufficient quantity to cause estrogen secretion but not ovulation: the arcuate-ventromedial nuclei complex of the median eminence. Evidence is presented which suggests that prepubertal androgen treatment deleteriously alters the function of the "ovulation controlling" preoptic region of the hypothalamus.

INTRODUCTION

PREVIOUS studies have demonstrated a period of steroid sensitivity in the female rat between birth and the tenth day of age during which a single injection of androgen will result in permanent infertility (1). It was originally proposed by Pfeiffer that such sterility was the consequence of a deleterious effect of androgen on the adenohypophysis, resulting in a gland that secreted only follicle stimulating hormone (2). Since the time of these early observations it has become apparent that the pituitary of the androgen-sterilized rat elaborates both follicle stimulating and luteinizing hormone. Not only are vesicular follicles present in the sterile rat ovary but estrogen is secreted as evidenced by the persistence of a cornified vaginal mucosa, cystic enlargement of the uterine endometrial glands and hypertrophy of the ovarian interstitial tissue (1, 3). Seemingly, the particular adenohypophyseal malfunction is a failure to release sufficient gonadotropin to cause ovulation, a phenomenon generally held to be regulated by the hypothalamus. This suggests either that the pituitary of the androgen-sterilized rat is refractory to hypothalamic activation or, more likely, that the malfunction in the ovulatory mechanism is inherent

---

Received June 10, 1960.

[1] Supported by grant RG-5496 from the United States Public Health Service.

68

within the hypothalamus itself. Some support for the latter hypothesis is offered by the observations of Harris and Jacobsohn (4) and Martinez and Bittner (5) that male hypophyses transplanted beneath the median eminence of hypophysectomized female rats would restore normal estrous cycles. Apparently, the sex difference in gonadotropin secretion is not resident within the adenohypophysis as such, but a higher neural level.

The current studies were designed to answer two questions: (a) will the pituitary of the anovulatory female rat respond to stimulation of the hypothalamus by the release of sufficient gonadotropin to cause ovulation and (b) if so, can a specific region of the hypothalamus be implicated as the site of the deleterious androgen action? The results of these experiments have been presented previously in abstract form (6).

MATERIALS AND METHODS

Five-day-old Sprague-Dawley female rats were given a single subcutaneous injection of 1.25 mg. of testosterone propionate (in oil) and then permitted to mature to a weight of 230–250 gms. (4–5 months of age). This experimental regime was chosen since it had been shown previously to induce permanent sterility in the female rat (1).

Following treatment, the rats were maintained in an artificially illuminated, temperature controlled room, on a diet of Purina rat pellets supplemented by a mixture of cod-liver oil and Mazola oil on bread once a week. During the three weeks prior to experimentation, daily vaginal smears were taken to establish the persistence of a cornified vaginal mucosa. This phenomenon can be correlated with the presence of large ovarian vesicular follicles and the absence of corpora lutea and may thus be used as a criterion of sterility; however, laparotomies were performed periodically in randomly selected rats exhibiting persistent vaginal cornification to confirm the absence of corpora lutea.

At the time of experimentation, all rats were given an intraperitoneal injection of 25 mg./kg. of Nembutal. This dosage produced sufficient sedation to permit orientation of the animal in a stereotaxic apparatus, but did not inhibit ovulation in the normal cyclic rat, and it eliminated the problem of hypothalamic refractoriness due to anesthesia. Following procainization of all areas to be subjected to surgery, and orientation of the animal in the stereotaxic apparatus, bipolar concentric electrodes were placed, bilaterally, in various regions of the hypothalamus. The stimulation parameters chosen for this study were 2–3 volts (80–100 microamperes) square wave pulses delivered by a Grass stimulator at a frequency of 100/sec. with a duration of 0.5 msec. for 15 sec. on/off periods over a total of 15 minutes. Such parameters have been used previously by Critchlow (7) and Everett (8) to induce ovulation in the Nembutal-blocked rat. All voltage and current values used for stimulation were monitored by an oscilloscope previously calibrated for such determinations.

Twenty-four hours after stimulation, the rats were sacrificed with ether and the fallopian tubes were examined for the presence of ova and the ovaries for corpora lutea. To confirm the autopsy results, both ovaries of each animal were placed in Bouin's fixative, paraffin-embedded, serially sectioned and stained with hematoxylin and eosin. The presence of newly formed corpora lutea, correlated with the previously identified tubal ova, served as the criterion of an ovulatory discharge of gonadotropin in response to hypothalamic stimulation.

Similarly all brains were perfused and fixed with 10% formalin, frozen-sectioned and stained with thionine for verification of electrode placements. The additional techniques used in this study will be described with the results of these experiments.

RESULTS

To establish that the adenohypophysis of the sterile rat can respond to hypothalamic stimulation, areas were selected for study which had previously induced ovulation when stimulated, namely the arcuate and ventral medial nuclei (7). However, as is shown in Table 1, electrical stimulation of these median eminence structures failed to cause ovulation in any of 8 sterile rats (Fig. 1). Although these results would superficially tend to support the hypothesis that the pituitary of the androgen-sterilized rat

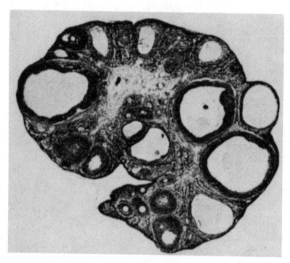

Fig. 1. Ovary of an anovulatory-persistent-estrous rat following electrical stimulation of the median eminence. ×25.

is refractory to hypothalamic activation, a second alternative also deserved consideration. The constant levels of estrogen secreted by the sterile rat ovary might so affect adenohypophyseal physiology as to prevent normal storage of gonadotropin. Thus even though the adenohypophysis was activated by hypothalamic stimulation, gonadotropin was not present in sufficient concentration to cause ovulation. Evidence that the sterile rat pituitary does respond to hypothalamic stimulation by discharging luteinizing hormone is afforded by the increase in uterine weight which occurs 24 hours after such stimulation. Although this is only indirect evidence for the release of LH, the second hypothesis warranted further investigation. Thus a second group of 6 anovulatory rats was given single injections of progesterone in a dosage calculated to interrupt the persistent vaginal cornification (2 mg. s.c. in oil), and presumably the secretion of LH. Such treatment generally induced 2–3 days of diestrus followed by a single day of proestrus. When no further treatment was given, all animals resumed the previous persistent estrous condition. If, however, the ventromedial-arcuate complex was stimulated electrically on the day of vaginal proestrus

TABLE 1

Treatment	No. animals	Body wt. (gms.)	Relative organ wts. mg./100 gm. B. wt.			
			Pituitary wt.	Adrenal wt.	Ovarian wt.	Uterine wt.
Normal Control	10	202.4	5.78	34.7	33.58	168.4
Sterile Control	7	225.2	5.53	35.0	11.75	107.2
Sterile plus median eminence stimulation	8	239.2	5.30	32.4	12.09	113.04
Sterile plus 2 mg. progesterone	8	241.5	5.44	29.35	12.50	107.94
Sterile plus 2 mg. progesterone-hypothalamic stimulation. *No Ovulation*	34	246.5	5.36	30.0	11.20	114.30
Sterile plus 2 mg. progesterone-hypothalamic stimulation. *Ovulation*	11	265.2	5.29	31.7	20.00	135.70

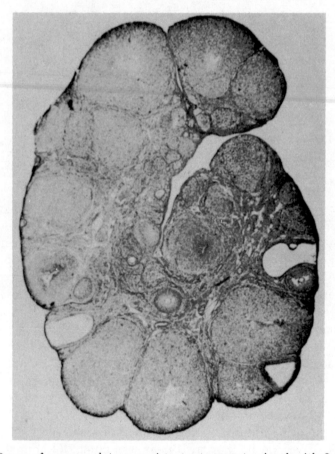

Fig. 2. Ovary of an anovulatory-persistent-estrous rat primed with 2 mg. of progesterone in which electrical stimulation of the median eminence induced ovulation. Note the completeness of luteinization. ×25.

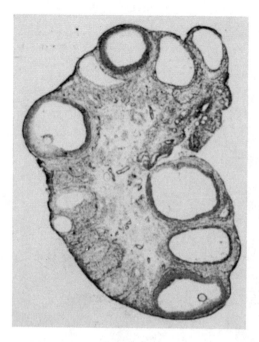

Fig. 3. Ovary of an anovulatory-persistent-estrous rat primed with 2 mg. of progesterone but in which hypothalamic stimulation was not performed. ×25.

all animals ovulated within 24 hours (Table 1, Fig. 2). Progesterone alone failed to cause either luteinization or ovulation of any of the numerous ovarian vesicular follicles (Table 1, Fig. 3).

Once assured that ovulation could be induced in the progesterone-primed sterile rat, a more critical study was undertaken to determine the specific hypothalamic regions in which stimulation would induce ovulation. As is shown in Figure 5, electrical stimulation of the medial or lateral preoptic areas, dorsal to the optic chiasm or in the region of the median forebrain bundle, failed to induce ovulation in 17 of 18 rats. In the one positive response, an incomplete or partial ovulation occurred. Only 2 tubal ova could be identified at autopsy and they were correlated with the presence of 2 newly formed copora lutea. Other ovarian follicles exhibited partial luteinization but contained trapped ova (Fig. 4). In contrast, stimulation of the medial hypothalamic regions immediately rostral or caudal to the arcuate and ventral medial nuclei induced ovulation in 10 of 13 rats (Figs. 5, 6). Stimulation of regions lateral to the ventral medial nuclei or of the medial mamillary nuclei did not induce ovulation in any of the fifteen cases studied (Figs. 5, 6). Thus, in progesterone-primed, anovulatory rats, the hypothalamic areas which would promote an ovulatory discharge of gonadotropin when stimulated could be delineated as extending just rostral and caudal,

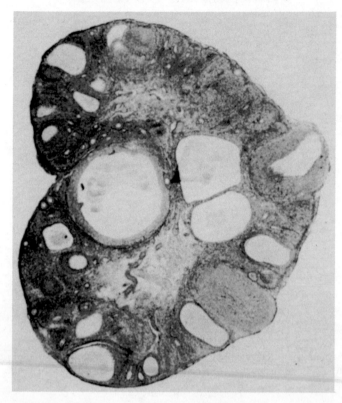

Fig. 4. Ovary of a progesterone-primed, anovulatory-persistent-estrous rat which exhibited only partial ovulation following electrical stimulation of the medial preoptic area. Note the presence of two newly formed corpora lutea and the partial luteinization of the large vesicular follicles. These latter follicles contained trapped ova. ×25.

but not lateral, to the ventral medial nucleus. The antero-posterior extent of this positive region is represented in sagittal section in Figure 7.

DISCUSSION

These results suggest that the adenohypophysis of the androgen-sterilized rat can function normally, provided: (a) proper gonadotropin storage is permitted and (b) an impetus for its release is supplied by the hypothalamus. We have interpreted the failure of the persistent-estrous rat to ovulate on artificial hypothalamic stimulation as due to insufficient pituitary gonadotropin stores. When such animals are primed with progesterone they readily ovulate in response to such hypothalamic stimulation. Presumably progesterone permits sufficient gonadotropin to be accumulated in the adenohypophysis to cause ovulation when released. Segal has reported that the pituitary gonadotropin content of the androgen-sterilized rat is comparable to that of the normal male rat (9). However, his assays are based on glands of 60-day-old rats which have exhibited only

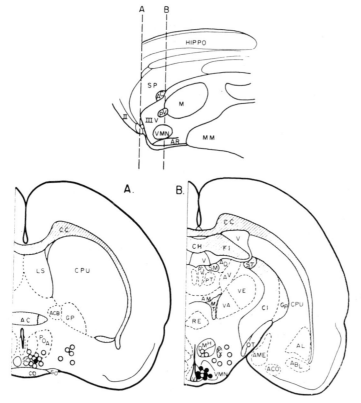

Fig. 5.

2–3 weeks of persistent vaginal cornification, whereas the animals employed in these studies have undergone an additional 60–120 days of persistent estrus (120–180 days of age). It is probable that the pituitary, when subjected to prolonged estrogen stimulation would be depleted of its stored gonadotropin. Preliminary assay data of the pituitary LH content of progesterone-primed persistent estrous rats suggest a twofold increase in content (unpublished observations).

A second mode of progesterone action also deserves consideration. Everett (10) has shown that progesterone will advance ovulation 24 hours in the normal cyclic rat provided this steroid is administered on the last day of diestrus. Furthermore, Kawakami and Sawyer (11) have observed that progesterone, in facilitating ovulation in the rabbit, also lowers various central nervous thresholds to electrical stimulation. There is thus the possibility that treatment of the androgen-sterilized female rat with progesterone may facilitate the electrical stimulus to result in ovulation rather than having a direct effect on pituitary gonadotropin content. Regardless of its site of action, progesterone priming is necessary, prior to hypothalamic stimulation, for ovulation to occur.

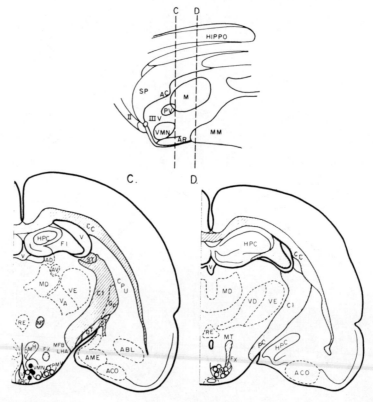

FIG. 6.

FIGS. 5, 6. Midsagittal and coronal sections of rat hypothalamus indicating points of stimulation. Diagrams from DeGroot's "The rat forebrain in stereotaxic coordinates" (15). *Closed circles* indicate hypothalamic sites in which stimulation induced ovulation. *Open circles* represent areas in which stimulation failed to induce ovulation. Abbreviations used in this and subsequent figures: AC, anterior commissure; ACB, area parolfactoria lateralis; AR, arcuate nucleus; CC, corpus callosum; CPU, nucleus caudatus/putamen; CO, optic chiasm; DMH, dorsal medial nucleus; MN, medial mamillary nucleus; MT, mamillothalamic tract; POA, preoptic area; VMN, ventral medial nucleus.

What is the mechanism by which androgen induces sterility? Apparently prepubertal androgen treatment so alters normal hypothalamic function as to render it incapable of activating the cyclic ovulatory discharge of gonadotropin (presumably LH). In contrast, such treatment fails to cause complete cessation of LH secretion. This is evidenced by the syndrome which ensues following prepubertal treatment: persistent vaginal cornification and ovarian interstitial cell, adrenal and pituitary hypertrophy which are either directly or indirectly (through estrogen secretion), the consequence of a tonic discharge of adenohypophyseal LH.

To establish a hypothalamic locus which is deleteriously affected by

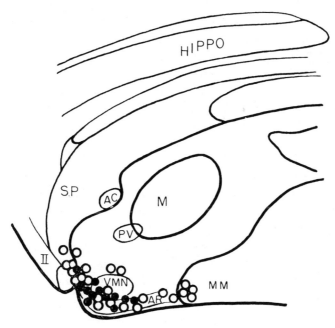

FIG. 7. Midsagittal reconstruction of rat hypothalamus indicating extent of area in which stimulation resulted in ovulation. Abbreviations: hippo, hippocampus; M, massa intermedia; PV, paraventricular nucleus; SP, septum; II, optic nerve.

androgen requires an evaluation of the specific hypothalamic areas proposed to be responsible for the control of ovulation in the normal rat. Critchlow initially demonstrated that electrical stimulation of hypothalamic regions extending from the basal area, from the optic chiasm to the infundibular stalk, consistently induced ovulation in Nembutal-blocked rats (7). The more recent studies of Everett have extended these observations to include the preoptic area rostral to the suprachiasmatic nucleus (8). Furthermore, when small, specific lesions are made in the suprachiasmatic nucleus, reproductive behavior is so affected so as to result in anovulatory-persistent-estrous animals similar in every respect to the androgen-sterilized rat (12).

The hypothalamic locus thus implicated as the site of deleterious androgen action is the midline suprachiasmatic-preoptic area. Lesions of this region result in an imbalance in gonadotropin secretion which imitates that observed in the sterilized rat (12) and ovulation can readily be induced by stimulation of this region in normal but not in androgen-sterilized animals.

The observations of these and previous investigations suggest a dual hypothalamic control of adenohypophyseal gonadotropin secretion in the female rat. The first level of hypothalamic control involves the tonic discharge of gonadotropin in sufficient quantity to maintain estrogen production but cannot independently initiate the ovulatory surge of gonadotropin.

This control is apparently resident in the arcuate-ventromedial nuclei region of the median eminence. Evidence for this primary control and localization is based on the observations that estrogen is secreted in the anovulatory-persistent estrous rat and that electrical stimulation of these structures in the sterile rat will induce LH secretion. Furthermore, destruction of these areas results in the cessation of estrogen production, ovarian atrophy and anestrus (12).

Of fundamental importance is the second and higher control which results in the cyclic discharge of gonadotropin to cause ovulation. The specific region responsible for such control may, most likely, be placed in the preoptic area of the hypothalamus. Furthermore, this region of "ovulation control" is dependent for its activation on exteroceptive (light, etc.) (13), and interoceptive (steroid [14], higher neural control) influences.

Thus it may be that the hypothalamic events which occur during the normal cycle in the female rat are these: the preoptic (suprachiasmatic) area responds under proper environmental and hormonal circumstances,

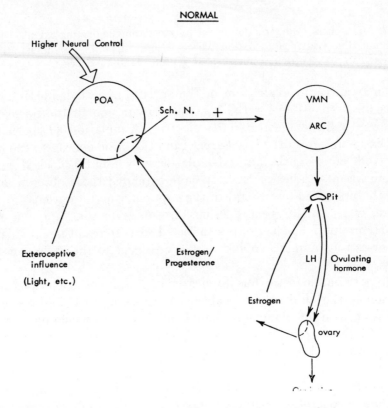

Fig. 8. Diagrammatic representation of hypothalamic events which regulate the discharge of gonadotropin to cause ovulation (ovulating hormone) and/or the tonic release of ovarian estrogen. Abbreviations: POA, preoptic area; Sch. N., suprachiasmatic nucleus; VMN, ventral medial nucleus; ARC, arcuate nucleus.

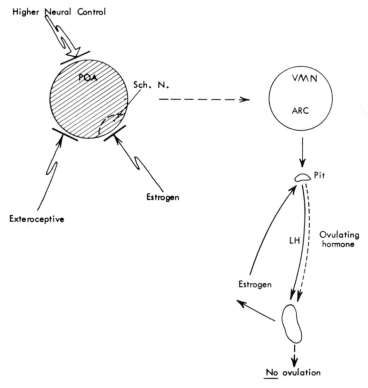

FIG. 9. Hypothalamic events which occur on prepubertal treatment of female rats with androgen or destruction of the anterior preoptic area.

(which are fulfilled on the day of proestrus) by an activation of the more terminal infundibular regions to cause an ovulatory discharge of gonadotropin from the adenohypophysis (Fig. 8). In the absence of this higher control, the terminal structures (arcuate-ventromedial nuclei) still function normally to stimulate LH secretion, but the ovulatory surge of gonadotropin is absent and sterility ensues (Fig. 9).

Seemingly, the anterior preoptic area of the prepubertal female rat is undifferentiated at birth with regard to its subsequent control of gonadotropin secretion. When allowed to differentiate normally it regulates the release of ovulating hormone. However, if differentiated in the presence of androgen, this area becomes refractory to both intrinsic and extrinsic activation and the more tonic type of male gonadotropin secretion is observed.

Acknowledgments

The authors wish to express their appreciation for the technical assistance rendered by Miss Theresa Mangold and Mr. Leroy Brown during this study.

REFERENCES

1. BARRACLOUGH, C. A.: *Endocrinology*. This issue.
2. PFEIFFER, C. A.: *Amer. J. Anat.* **58**: 195. 1936.
3. FEVOLD, H. L.: *Endocrinology* **28**: 33. 1941.
4. HARRIS, G. W. AND D. JACOBSOHN: *J. Physiol.* **113**: 35. 1951.
5. MARTINEZ, C. AND J. J. BITTNER: *Proc. Soc. Exper. Biol. & Med.* **91**: 506. 1956.
6. BARRACLOUGH, C. A.: *Anat. Rec.* **133**: 248. 1959.
7. CRITCHLOW, V.: *Amer. J. Physiol.* **195**: 171. 1958.
8. EVERETT, J.: Harvard Conference on Control of Ovulation, Pergamon Press (in press).
9. SEGAL, S. J. AND D. C. JOHNSON: *Arch. D'Anat. Micro. Morph. Exper.* **48**: 261. 1959.
10. EVERETT, J.: *Endocrinology* **43**: 389. 1948.
11. KAWAKAMI, M. AND C. H. SAWYER: *Endocrinology* **65**: 631. 1959.
12. FLERKÓ, B. AND V. BARDOS: *Acta Neuroveg.* **20**: 248. 1959.
13. CRITCHLOW, V. AND J. DEGROOT: *Anat. Rec.* **136**: 179. 1960.
14. FLERKÓ, B. AND J. SZENTAGOTHAI: *Acta Endoc.* **26**: 121. 1957.
15. DEGROOT, J.: *Trans. Roy. Neth. Acad. Sci.* **52**: 1. 1959.

Author Citation Index

Subject Index